TOUGHENED PLASTICS

TOUGHENED PLASTICS

C. B. BUCKNALL

*Department of Materials, Cranfield Institute of Technology,
Cranfield, Bedford, England*

Distributed Exclusively By

INTERNATIONAL IDEAS INC.
1627 Spruce Street
Philadelphia PA USA 19103

APPLIED SCIENCE PUBLISHERS LTD
LONDON

APPLIED SCIENCE PUBLISHERS LTD
RIPPLE ROAD, BARKING, ESSEX, ENGLAND

ISBN: 0 85334 695 X

WITH 163 ILLUSTRATIONS

Printed in Great Britain by Galliard (Printers) Ltd Great Yarmouth

PREFACE

Rubber-toughened plastics constitute a commercially important class of polymers, which are characterised by a combination of fracture resistance and stiffness. The best known members of the class are toughened polystyrene, or HIPS, and ABS, but there are also toughened grades of polypropylene, PVC, epoxy resin, and a number of other polymers. Each of these materials is a composite polymer, consisting of a rigid matrix and a disperse rubber phase. This book describes their manufacture, analysis, mechanical properties and processing characteristics. The aim is to show how structure can be controlled during manufacture, and how the properties of the product are consequently affected.

The book is divided into three main sections. The first four chapters, which are concerned broadly with the chemistry of toughening, cover phase separation and compatibility, grafting, microscopy, analytical techniques and methods of polymerisation. This section is of interest not only to physical chemists engaged in research and development, but also to all who are involved in the characterisation and quality control of rubber-toughened plastics. The second section, comprising six chapters, deals with mechanical properties, including viscoelastic properties, creep, yield, fatigue, fracture and impact strength. The effects of structure upon properties and the role of crazing and shear deformation are discussed in some detail. This section is of interest to all concerned in the manufacture and applications of rubber-toughened plastics. The third section consists of two chapters on the manufacture of finished components, covering melt rheology, thermoforming, extrusion, injection moulding and electroplating. The emphasis throughout is on the effects of adding rubber on the manufacturing technology and product properties. Topics, such as solvent crazing resistance, which are not considered to be affected significantly by the addition of rubber, have been excluded or treated only briefly.

The field of rubber toughening provides many good examples of the principles of materials science and technology applied to practical ends. For this reason, the subject is of interest not only to specialists involved in the industry, but also to advanced students of materials or polymer science. The treatment followed in this book is designed to introduce the reader to a class of materials which must surely rank amongst the principal inventions of the plastics industry.

I should like to express my thanks to Dr M. J. Folkes, Dr R. P. Kambour, Dr H. Keskkula, Dr T. D. Lewis, Dr A. E. Platt and Professor J. G. Williams, each of whom read a portion of the manuscript and offered valuable comments. I am also grateful to Mrs V. Collier and Mrs S. Skevington for preparing the line diagrams, to Mr I. C. Drinkwater for assistance in preparing the photographs, and to the authors and publishers who granted permission for the use of previously published material. Individual acknowledgements are made in the figure captions.

Department of Materials
Cranfield Institute of Technology
Cranfield   C. B. BUCKNALL

CONTENTS

Chapter 4

MANUFACTURE OF TOUGHENED PLASTICS

Chapter 5

VISCOELASTIC PROPERTIES

Chapter 6
DEFORMATION MECHANISMS IN GLASSY POLYMERS

Chapter 7
MECHANISMS OF RUBBER TOUGHENING

Chapter 8
DEFORMATION AND YIELDING

Chapter 9
FRACTURE MECHANICS

Chapter 10
IMPACT STRENGTH

Chapter 11
MELT RHEOLOGY AND PROCESSING

Chapter 12
ELECTROPLATING

GLOSSARY

a	Cross-sectional area of craze (Chapter 7)
	Crack length (Chapters 9 and 10)
	Major axis of ellipse (Chapter 11)
a_T	Temperature shift factor
a_ϕ	Rubber content shift factor
a_M	Molecular weight shift factor
$\dot{a}$	Shear band velocity (Chapter 6)
	Crack speed (Chapters 9 and 10)
A	Pre-exponential factor (Chapter 6)
	Dummy variable (Chapters 5 and 10)
$A(T)$	Crazing parameter
A^*	Activation area
b	Beam thickness (Chapter 9)
	Minor axis of ellipse (Chapter 11)
B	Crack width (Chapters 9 and 10)
	Die swell ratio (Chapter 11)
$B(T)$	Crazing parameter
C	Specimen compliance (Chapters 9 and 10)
$C(t,T)$	Crazing parameter
C_p	Specific heat at constant pressure
CED	Cohesive energy density
$D(t,T)$	Crazing parameter
e_1, e_2, e_3	Extensions in the 1, 2 and 3 directions
E	Young's modulus
f	Frequency
G	Gibbs free energy (Chapter 2)
ΔG_m	Gibbs free energy of mixing
G, G_0	Shear modulus (Chapter 5)
G_1, G_2, G_i, G_m	Shear moduli of phase 1, phase 2, inclusion, matrix
$\mathscr{G}_B$	Apparent critical strain energy release rate for blunt notch specimen

GLOSSARY

$\mathcal{G}_c$, $\mathcal{G}_{IC}$	Critical strain energy release rate
h	Displacement
H	Enthalpy (Chapters 2, 6 and 8)
	Specimen height (Chapter 9)
ΔH_m	Enthalpy of mixing
ΔH_{vap}	Enthalpy of vaporisation
I	Initiator (Chapter 2)
	Impact strength (Chapter 10)
I_{ke}	Kinetic energy contribution to impact strength
I_0	Intensity of incident light
I_s	Intensity of scattered light
I_{tr}	Intensity of transmitted light
$J(t)$	Compliance at time t
J''	Loss compliance
$\mathcal{J}_{IC}$	Plastic work parameter for ductile fracture
k	Boltzmann's constant
k_d, k_i, k_p, k_{AB}, etc.	Rate coefficients
K	Bulk modulus
$\mathcal{K}$	Stress intensity factor
$\mathcal{K}_C$, $\mathcal{K}_{IC}$	Fracture toughness
L	Length of capillary
m	Integer (Chapter 2)
	Exponent (Chapter 11)
M	Monomer (Chapter 2)
	Beam constant (Chapter 9)
$\overline{M}_w$	Weight average molecular weight
n	Number of cycles (Chapter 9)
	Power law exponent (Chapters 7 and 11)
n_1, n_2	Mole fractions of 1 and 2
N	Fatigue exponent
p	Integer (Chapter 2)
	Probability (Chapter 4)
	Pressure (Chapters 6 and 11)
P	Force
P_c, P_f, P_{gy}	Force at critical stage, at fracture and at general yield
Q	Flow rate
r	Radial distance (Chapters 5, 9 and 11)
	Craze length (Chapter 7)
	Plate radius (Chapter 10)

r^*	Interparticle distance
r_1, r_2	Monomer reactivity ratios
r_0	Root radius of crack or notch
r_w	Radius of capillary or tube
$\dot{r}$	Radial migration rate
R	Gas constant (Chapter 2)
	Particle radius (Chapters 3, 5 and 11)
	Distance between rubber particles (Chapter 7)
R_y	Yield zone length
$\mathscr{R}$	Specific refraction
Re_p	Particle Reynolds number
S	Entropy (Chapter 2)
	Beam span (Chapters 9 and 10)
ΔS^*	Configurational entropy change
ΔS_m	Entropy of mixing
t	Time
T	Temperature
T_e	Effective temperature
T_g	Glass transition temperature
T_0	Reference temperature
$\mathscr{T}$	Interfacial tension
U	Internal energy (Chapters 2, 9 and 10)
ΔU_{vap}	Energy of vaporisation
U_c, U_y	Energy at critical stage and at yield
v	Velocity
	Craze volume (Chapter 7)
v_f, v_p	Velocity of fluid and of particle
V	Volume
V_{liq}	Molar volume of liquid
V_0	Original volume
V^*	Activation volume
ΔV	Volume strain
$\Delta V(0)$	Volume strain at zero time
$\Delta V(t)$	Volume strain at time t
w_1, w_2	Weight fraction of components 1 and 2
w_p	Plastic work
W	Specimen width
x	Integer (Chapter 2)
	Distance (Chapter 3)
X	Rubber content (Chapter 3)

X	Deflection (Chapter 10)
X_c	Critical deflection-to-span ratio
X_p	Deflection subsequent to general yield
y	Integer
Y	Geometrical factor in fracture mechanics
Z	Geometrical factor for impact specimens

GREEK SYMBOLS

β	Volume coefficient of thermal expansion (Chapter 5)
	Angle (Chapter 6)
$\Delta\beta_f$	Temperature coefficient of equilibrium free volume
γ	Shear strain (Chapter 6)
	Stress concentration factor (Chapter 8)
	Fracture surface energy (Chapter 9)
γ_p	Fracture surface energy involving limited plasticity
$\dot{\gamma}$	Shear strain rate
$\dot{\gamma}_w$	Shear strain rate at wall
δ	Solubility parameter (Chapter 2)
	Crack opening displacement (Chapters 9 and 10)
δ_c	Critical crack opening displacement
$\delta_d, \delta_h, \delta_p$	Solubility parameter terms relating to dispersion, hydrogen bonding and dipole interaction contributions
ε	Strain
$\dot{\varepsilon}$	Strain rate
ζ	Craze initiation time
η	Viscosity
η_{21}	Viscosity of suspended phase divided by viscosity of suspending phase
θ	Angle
λ	Wavelength (Chapter 3)
	Natural draw ratio: current length over original length (Chapters 7 and 11)
μ	Refractive index (Chapter 3)
	Plasticity term (Chapter 6)
ν	Poisson's ratio
$\nu_1, \nu_2, \nu_i, \nu_m$	Poisson's ratio of phase 1, phase 2, inclusion, matrix
ν^*	Kinetic chain length
ξ	Frozen free volume

ρ	Density
σ	Stress
$\sigma_{11}, \sigma_{12}, \sigma_{22},$	Components of the stress tensor
$\sigma_{rr}, \sigma_{\theta\theta}, \sigma_{\psi\psi}$	
σ_c	Critical stress
σ_{cy}, σ_{ty}	Yield stress in compression and in tension
σ_y	Yield stress
τ	Turbidity (Chapter 3)
	Shear stress (Chapters 6 and 11)
τ_{oct}	Octahedral shear stress
τ_w	Shear stress at wall
τ_i	Induction period
$\phi_1, \phi_2, \phi_i, \phi_m$	Volume fractions of phase 1, phase 2, inclusion, matrix
χ_{12}	Flory–Huggins interaction parameter
ψ	Angle

CHAPTER 1

THE GROWTH OF A TECHNOLOGY

1.1 THE BALANCE OF PROPERTIES

Toughness is often the deciding factor in materials selection. The continuing growth in the use of plastics for engineering and other applications is due in no small measure to the development, during the past three decades, of new and tougher plastics materials. Plastics are gradually losing their reputation for brittleness as these newer materials, effectively used in well-designed products, become more familiar to the general public.

The problem facing the raw materials manufacturer is not simply to increase toughness. Improvements in fracture resistance must be achieved without undue impairment of other properties; nor must costs be neglected. For many applications, the requirement is for a moderately priced polymer which can be moulded easily, and which exhibits adequate stiffness and toughness over a wide range of temperatures. Most of the major plastics manufacturers have devoted a significant part of their research and development effort to the search for materials with these characteristics.

There are two basic solutions to this problem. One is to produce completely new polymers, based upon novel monomers, as in the case of polycarbonates and polysulphones. The alternative is to modify existing polymers, which already possess many of the desired properties, but are lacking in toughness, stiffness or some other attribute. The latter solution is exemplified by a wide variety of composite materials, including fibre-reinforced plastics, block copolymers, structural foams and rubber-toughened plastics. The two approaches can, of course, be combined: the new, tougher polymers can be toughened further by the addition of rubber or stiffened by means of fibre reinforcement.

One of the most successful methods developed for modifying polymer

1

properties is the rubber-toughening process. In this process, a minor proportion of rubber, typically between 5 and 20%, is incorporated as a disperse phase into a rigid plastics matrix. The resulting composite has a significantly higher fracture resistance than the parent polymer: impact strength, elongation at break, work to break and fracture toughness are all increased severalfold. There is an inevitable reduction in modulus and tensile strength, transparency is usually lost and melt viscosity is increased, but these losses are far outweighed by the gains in fracture resistance. The rubber-toughened polymer has a better balance of properties than the parent polymer, and is therefore a commercially successful product despite its higher price.

TABLE 1.1

PROPERTIES OF A TYPICAL PS AND HIPS (ASTM TEST METHODS D 638-61 T, D 256-56 AND D 648-56)

Property	Units	PS	HIPS
Tensile modulus	GN/m^2	3·5	1·6
Tensile strength at yield	MN/m^2	does not	17·5
Tensile elongation at yield	%	yield	2
Tensile strength at break	MN/m^2	54	21
Tensile elongation at break	%	2·1	40
Notched Izod impact strength	J/cm	1·0	4·5
Deflection temperature	°C	100	96
Light transmission		Clear	Opaque

The effects of the rubber upon mechanical and other properties are discussed in detail in later chapters. Some of these effects are illustrated in Table 1.1, which compares standard data for a typical general purpose polystyrene (PS) with those for a high-impact polystyrene (HIPS). The figures clearly show how the balance between stiffness and fracture resistance is altered by the addition of rubber.

1.2   THE INVENTION OF TOUGHENED POLYSTYRENE

The history of rubber-toughened plastics can be traced back as far as 1927. In that year, Ostromislensky patented a process for making toughened polystyrene by polymerising a solution of rubber in styrene monomer.[1] The process was not developed, because the product was found to be cross-linked, and therefore could not be moulded. Later work has shown that unstirred polymerisations of the type carried out by Ostromislensky produce a cross-linked network of rubber, in which the polystyrene formed during the reaction is embedded as a discrete phase.

The subsequent commercial development of toughened polystyrene has been reviewed by Amos.[2] The availability of cheap styrene monomer led to an expansion of polystyrene production during the 1940s, and revived interest in the rubber-toughening process. With this stimulus, the Dow Chemical Company mounted a research programme which led to the announcement in September 1948 that it was marketing a new, impact-resistant grade of polystyrene.[3]

The new polymer was made by dissolving styrene–butadiene rubber (SBR) in styrene monomer, and polymerising the solution in 10-gal cans. In this respect, the process followed Ostromislensky's basic invention. The key to commercial success lay in the subsequent processing of the cross-linked polymer: Boyer, Rubens and McIntire had developed a method for breaking down the rubber network, by granulating the polymer, and then masticating it on a heated two-roll mill.[2] The cross-linked, intractable polymer of the earlier invention had been converted into a mouldable thermoplastic.

Dow continued the research programme, with the aim of reducing production costs and improving the surface appearance of toughened polystyrene. In 1952, the company introduced a new, continuous process which had been developed by Amos et al.[4] The details are given in Chapter 4. The essential feature of the process is that the solution of rubber in styrene monomer is stirred during the early stages of polymerisation, so that the rubber forms discrete particles rather than a continuous network. In this way, the mastication stage is eliminated, manufacturing costs are reduced and surface gloss is improved. To this day, most HIPS production is based upon continuous polymerisation with stirring during the early stages.

This series of inventions laid the foundations for a substantial new sector of the plastics industry. Rubber-toughened plastics have made a major contribution to the growth of the industry since the 1940s. Not only has the production of toughened polystyrene expanded continuously, but also the principle of rubber toughening has been applied to a number of other plastics, ranging from polypropylene to epoxide resins.

1.3 NEW MATERIALS AND EXPANDING MARKETS

Most polystyrene manufacturers followed Dow's lead in introducing a rubber-toughened grade of polystyrene into their range of products. During the early years, these rival HIPS materials were made by melt blending of polystyrene and rubber, and their properties were generally inferior to those of the solution-grafted polymers made by the two Dow processes. In melt- or latex-blended materials, the rubber particles are only weakly bonded to the surrounding matrix, and are therefore less effective as toughening agents than the well-bonded particles of grafted rubber formed when styrene is polymerised in the presence of SBR. Furthermore, the solution-grafting process increases the effective volume of the rubber particles, with a consequent increase in toughness, whereas blending does not. These factors are discussed more fully in later chapters. The advantages of the solution-grafting process led to its general adoption by the industry, and by the early 1960s very little HIPS was being made by the blending route.[5]

Although there were numerous developments in HIPS production technology after 1952, these were more in the nature of modifications than major alterations of the process. Perhaps the most significant development occurred during the early 1960s, when polybutadiene replaced SBR as the standard toughening rubber. This change, which resulted from advances in rubber polymerisation technology, led to improvements in the fracture resistance of HIPS polymers over a wide range of temperatures. Other developments included the introduction of fire-retardant, high-softening and easy-flow grades, and the widespread use of suspension polymerisation for the finishing stage of the process. Of greater importance than any of these advances, however, was the extension of the principle of rubber toughening to other polymers.

The second rubber-toughened plastic to be manufactured commercially was ABS (acrylonitrile–butadiene–styrene polymer), which was launched onto the market in 1952 by the US Rubber Company. The first ABS was made by melt blending SAN (styrene–acrylonitrile copolymer) with NBR (acrylonitrile–butadiene copolymer rubber).[6] As in the case of HIPS, the melt blending process eventually gave way to graft copolymerisation techniques. The emulsion grafting process, first introduced in 1956, is now the most important route to ABS,[7] with bulk-suspension polymerisation, introduced commercially in 1964,[3, 8] in second place.

Most modern ABS polymers are based upon polybutadiene, although other diene rubbers are used. Unsaturated rubbers are not entirely satisfactory as toughening agents, as they are easily oxidised, especially upon exposure to sunlight, with the result that the ABS becomes brittle. There has therefore been a considerable research effort aimed at replacing the diene polymers with saturated rubbers. This research has resulted in the development of ASA (acrylonitrile–styrene–acrylate)[9] and ACS

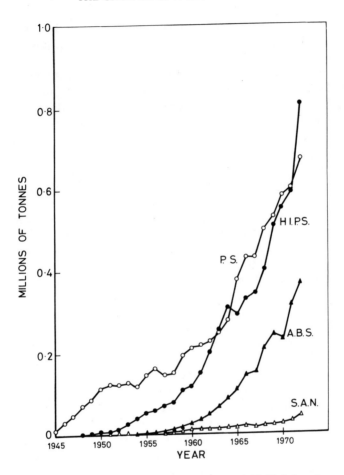

Fig. 1.1   *Annual US production of styrene polymers* 1945–72 (*after Amos, ref.* 2).

(acrylonitrile–chlorinated polyethylene–styrene)[10] polymers, both of which consist of a styrene–acrylonitrile copolymer matrix containing particles of a saturated elastomer.

Both HIPS and ABS are now well established as leading thermoplastics materials. The growth in demand is illustrated by the US annual production figures for styrene polymers recorded in Fig. 1.1. Total world output for each polymer is probably three times greater. Although the properties of HIPS are generally inferior to those of ABS, HIPS maintains its competitive position because it is cheaper. However, the difference in price might become less important if the cost of oil continues to rise, and the ratio between the prices of ABS and HIPS consequently falls.

Unplasticised PVC is a much more ductile polymer than PS or SAN. Nevertheless, it does exhibit brittle fracture, especially under impact loading. For this reason, manufacturers of PVC took an early interest in the rubber-toughening process, and by 1957 were selling toughened grades of PVC. It was a short step from SAN–NBR blends to PVC–NBR blends,[11] which are capable of attaining an excellent balance of properties. However, the next logical step, to a grafting process, presented considerable difficulties which are inherent in the chemistry of vinyl chloride polymerisation. Melt blending has therefore remained an important commercial process for toughening PVC. The demand for a grafted rubber with a well-controlled morphology was met by the development of special ABS and MBS (methyl methacrylate–butadiene–styrene) concentrates, containing a high proportion of grafted rubber dispersed in a glassy matrix that is compatible with PVC. The usual practice is to add about 5% of ABS or MBS concentrate to the PVC by melt blending. The matrix of the blend therefore contains both PVC and a small amount of SAN or acrylic copolymer. The term 'polymeric alloy' is often applied to this type of material.

Polypropylene also suffers from brittle fracture problems, especially at low temperatures. When these problems were recognised, early in the development of the polymer, the obvious solution was to add a rubber as a toughening agent. Rubber-toughened grades of polypropylene became available commercially in 1962, only five years after the homopolymer was first marketed. There are two types: 'impact polypropylene' is a blend of polypropylene with ethylene–propylene rubber (EPR) or other suitable elastomer; whereas 'polypropylene copolymer' is made by polymerising propylene monomer to about 90% conversion, and then adding ethylene monomer to the reactor. Both types of polymer contain a disperse rubber phase, and can therefore be classed with HIPS and ABS. Approximately 30% of all polypropylene produced is of the copolymer type,[12] and a further 10% contains blended rubber.

Poly(methyl methacrylate) (PMMA), like polystyrene, is a brittle, glassy polymer which could obviously benefit from an increase in toughness. However, PMMA is a more expensive polymer than PS, and its main applications are in glazing and signs, which require the transparency and resistance to weathering that it offers. Unfortunately, addition of rubbers tends to make the polymer both opaque and susceptible to weathering. A partial solution to the problem is represented by MBS and MABS (methyl methacrylate–acrylonitrile–butadiene–styrene) polymers, which are transparent because the refractive index of the matrix has been matched with that of the rubber, by suitable combination of monomers.[13] However, these polymers are transparent only over a limited range of temperatures, and are not resistant to weathering. Transparent polymers based upon acrylate rubbers have been developed recently,[14] and appear to offer a solution to the problem. Where transparency is not required, as in dental applications, existing grades of toughened PMMA are satisfactory.

'Modified PPO', introduced in 1966 by the General Electric Company, under the trade name 'Noryl®', is an alloy of HIPS with PPO® resin (poly-2,6-dimethyl-1,4-phenylene oxide).[15-17] The phenylene oxide polymer is compatible with PS, and the alloy is tougher than either of the parent polymers. Owing to its good balance of properties, modified PPO has become established as an important engineering thermoplastic. Another member of this class of materials is Borg-Warner's Cycoloy® polymer, which is an alloy of ABS with polycarbonate.[18]

At the time of writing, one of the most active fields of research and development concerns 'barrier resins'. These are polymers containing 60–75 % of acrylonitrile, and having a low permeability to oxygen, nitrogen and carbon dioxide. The main applications, in bottles for effervescent drinks and in other types of containers, require resistance to impact, and most barrier resins therefore contain between 5 and 20 % of grafted rubber.[19]

Finally, there is an active interest in rubber-toughening of thermosetting resins. Most of these resins are extremely brittle, and at one time it was thought to be impossible to increase their toughness significantly. The work of Sultan and McGarry at Massachusetts Institute of Technology showed that worthwhile improvements could be effected by adding certain liquid rubbers to epoxide resin formulations,[20] and the work was developed further by the B. F. Goodrich Company. In 1972, the company began marketing Hycar® CTBN rubbers as toughening agents for epoxide resins.[21] The developments listed above show how the discovery of rubber toughening in polystyrene has had effects throughout the plastics industry. Doubtless, the principle will be extended further. Many semi-crystalline polymers, including poly(ethylene terephthalate), acetal resin and nylons, suffer from fracture problems which could be overcome if satisfactory methods could be developed for incorporating a rubber-toughening agent. The thermosetting resins have already been mentioned. In view of the level of research activity, there is every reason to expect further important advances in the near future.

REFERENCES

1. I. I. Ostromislensky (to Naugatuck Chem.), US Pat. 1,613,673 (11 Jan. 1927).
2. J. L. Amos, *Poly. Engng Sci.* **14** (1974) 1.
3. H. Keskkula, A. E. Platt and R. F. Boyer, in *Encyclopedia of Chemical Technology*, 2nd edn., Vol. 19, Wiley–Interscience, New York, 1969, p. 85.
4. J. L. Amos, J. L. McCurdy and O. R. McIntire (to Dow), US Pat. 2,694,692 (16 Nov. 1954).
5. US Court of Appeals (9th Circuit), Cases 71-1371 and 71-1372 (1971).
6. L. E. Daly (to US Rubber), US Pat. 2,439,202 (6 Apr. 1948).
7. C. W. Childers and C. F. Fisk (to U.S. Rubber), US Pat. 2,820,773 (21 Jan. 1958).
8. L-H. Lee (to Dow), US Pat. 3,238,275 (1 Mar. 1966).
9. E. Zahn, *Appl. Polymer Symp.* **11** (1969) 209.
10. M. Ogawa and S. Takezoe, *Jap. Plast. Age* **11** (1973) 39.
11. N. E. Davenport, L. W. Hubbard and M. R. Pettit, *Brit. Plast.* **32** (1959) 549.
12. T. G. Heggs, in *Block Copolymers*, D. C. Allport and W. H. Janes (eds.), Applied Science, London, 1973, p. 105.
13. B. D. Gesner, *J. Appl. Polymer Sci.* **11** (1967) 2499.
14. R. G. Bauer, R. M. Pierson, W. C. Mast, N. C. Bletso and L. Shepherd, *ACS Adv. Chem. Ser.* **99** (1971) 251.
15. E. P. Cizek (to General Electric), US Pat. 3,383,435 (14 May 1968).
16. H. E. Bair, *Poly. Engng Sci.* **10** (1970) 247.
17. M. Kramer, *Appl. Polymer Symp.* **15** (1971) 227.
18. T. S. Grabowski (to Borg-Warner), US Pat. 3,130,177 (21 Apr. 1964).
19. E. C. Hughes, J. D. Idol, J. T. Duke and L. M. Wick, *J. Appl. Polymer Sci.* **13** (1969) 2567.
20. J. N. Sultan and F. J. McGarry, *Poly. Engng Sci.* **13** (1973) 29.
21. E. H. Rowe, A. R. Siebert and R. S. Drake, *Mod. Plast.* **49** (Aug. 1970) 110.

CHAPTER 2

COMPATIBILITY AND COPOLYMERISATION

2.1 INTRODUCTION

Rubber-toughened plastics are composite materials, consisting of a rigid matrix with a relatively high T_g and a rubbery disperse phase with a low T_g. The adhesion between these phases should be strong, and the rubber should be broken down into small particles without becoming too finely dispersed to be effective in toughening the rigid polymer. Optimum particle sizes vary, but are usually in the range 0.1 to 2.0μm. The task of the chemist is to control the morphology of the composite, the structure of the separate phases and the interfacial adhesion in such a way as to obtain the best balance of properties.

The early development of HIPS and ABS proceeded along purely empirical lines. More recent work has led to a better understanding of the chemistry of rubber toughening, and defined the limits imposed by chemical considerations upon the choice of rubber and of manufacturing process for each class of toughened polymer. Studies of phase equilibria in polymer/monomer and polymer/polymer systems have formed the basis of this work, which has concentrated especially upon the use of random, graft and block copolymers in controlling phase separation and adhesion. The underlying principles of the subject are set out in this chapter, and their application to the manufacture of rubber-toughened plastics is the subject of Chapter 4.

2.2 COMPATIBILITY

The problem of polymer–polymer compatibility arises in two ways in the field of rubber toughening. On the one hand, it is necessary to produce and

9

maintain phase separation between the rubber and rigid polymer in order to make a product that combines stiffness with toughness. On the other hand, in blending a rubber-toughened thermoplastic with another thermoplastic, it is important to ensure that the two thermoplastics are completely compatible. The point can best be illustrated by reference to the toughening of rigid PVC. One method is to blend the PVC with a rubber, in which case a certain degree of incompatibility is desirable. The alternative is to blend the PVC with ABS, in which case the PVC should be compatible with the SAN matrix of the ABS.

A rather different compatibility problem arises in the manufacture of toughened polystyrene. The process begins with a solution of rubber in styrene monomer, and phase separation occurs as a result of polymerisation. This type of process depends upon the choice of a rubber which is sufficiently compatible with the monomer to form a single homogeneous phase initially, but which forms a separate phase almost as soon as the monomer begins to polymerise.

In a survey of 342 pairs of polymers, Krause listed 33 which were compatible, in the sense that they mixed to form a single stable phase, and 46 which were almost compatible, or ambiguous.[1] Although not unknown, therefore, complete miscibility between polymers must be regarded as somewhat unusual. In this respect, polymers differ from substances of lower molecular weight, a fact which can be explained by statistical thermodynamics.

2.2.1 Flory–Huggins theory

In any closed system at constant pressure, the condition for equilibrium is that the Gibbs free energy be a minimum. The molar Gibbs free energy of isothermal mixing, ΔG_m, is given by

$$\Delta G_m = \Delta H_m - T\Delta S_m \qquad (2.1)$$

where ΔH_m and ΔS_m are the molar enthalpy and entropy of mixing, which together determine whether the components are compatible.

In general, the entropy of mixing is positive, since mixing involves an increase in the disorder of the system. If the mixing is exothermic, so that ΔH_m is negative, the components will be miscible. However, ΔH_m is usually positive, counterbalancing the effect of ΔS_m, and the deciding factor is the relative magnitude of the two terms. Miscibility between pairs of low molecular weight substances is in most cases due to entropy rather than enthalpy effects.

The entropy of mixing of a polymer with another polymer or with a solvent consists of two terms, the configurational entropy and the local pair interaction entropy.[2] The molar configurational entropy of mixing ΔS_m^*, arising from the increased number of ways of arranging the polymer and solvent molecules in the solution, is given by

$$\Delta S_m^* = - R(n_1 \ln \phi_1 + n_2 \ln \phi_2) \tag{2.2}$$

where n_1 and n_2 are mole fractions of solvent and solute, and ϕ_1 and ϕ_2 are volume fractions.

If the molar volumes of the two components are equal, eqn. (2.2) reduces to the equation for an ideal solution:

$$\Delta S_m^* = - R(n_1 \ln n_1 + n_2 \ln n_2) \tag{2.3}$$

which would also be applicable to a pair of low molecular weight liquids.

The local pair interaction entropy is more difficult to quantify. It arises from interactions between neighbouring units which are different from those in the pure component. Segments of the polymer chain may interact either with solvent molecules or with segments of a second polymer. The contribution of this effect to the molar entropy of mixing is proportional to the molar volume of the polymer, and could therefore be appreciable in view of the size of the macromolecule.

The molar enthalpy of mixing, which is also due largely to interaction between neighbouring units, increases in the same way with the molar volume of the polymer. Flory[2] represents the enthalpy and entropy contributions due to local pair interactions by the single dimensionless parameter χ_{12}, and obtains the following expression for ΔG_m:

$$\Delta G_m = RT(n_1 \ln \phi_1 + n_2 \ln \phi_2 + \chi_{12} n_1 \phi_2) \tag{2.4}$$

Equations (2.3) and (2.4) hold the key to the problem of incompatibility between polymers. The molar entropy of mixing ΔS_m^* does not increase with the size of the molecule, whereas the molar enthalpy and interaction entropy terms do increase with chain length, in proportion to the molar volume. Consequently, the interaction effects tend to dominate the configurational entropy effect in the mixing of polymers. Complete compatibility is observed only when the parameter χ_{12} in eqn. (2.4) is either very small or negative. The relative importance of energy and entropy contributions to ΔG_m is thus the reverse of that found in low molecular weight liquids.

2.2.2 Solubility parameter

A rigorous treatment of enthalpy of mixing presents formidable difficulties.

The most widely used approach is due to Hildebrand,[3] following work by Scatchard,[4] and is based upon measurements of cohesive energy density (CED) for liquids. The CED is defined as the energy of vaporisation of unit volume of the liquid, usually expressed in cal/cm^3:

$$CED = \frac{\Delta U_{vap}}{V_{liq}} = \frac{\Delta H_{vap} - RT}{V_{liq}} \qquad (2.5)$$

where ΔU_{vap} and ΔH_{vap} are molar energy and enthalpy of vaporisation, and V_{liq} is the molar volume of the liquid.

The cohesive energy is considered to arise from interactions between neighbours in the liquid, and the contribution of each individual molecule to the cohesive bond is characterised by the solubility parameter δ, equal to $CED^{1/2}$. Demixing of liquids is attributed to the tendency of molecules to attract their own species more strongly than a dissimilar species.

This idea is expressed quantitatively in the equation

$$\Delta H_m = V(\delta_1 - \delta_2)^2 \, \phi_1 \, \phi_2 \qquad (2.6)$$

where V is the molar volume of the liquid mixture. Chemically similar molecules are in most cases found to have similar solubility parameters, and therefore a reduced tendency to demix. Hildebrand's treatment considers only the contribution of dispersion forces to liquid cohesion, and predicts that enthalpies of mixing should always be positive (endothermic) or zero. Specific polar interactions, and hydrogen bonding, both of which might give

TABLE 2.1
SOLUBILITY PARAMETERS OF LIQUIDS

Liquid	δ $(cal/cm^3)^{1/2}$	Liquid	δ $(cal/cm^3)^{1/2}$
Dodecamethyl pentasiloxane	5·35	n-Octanol	10·3
Hexamethyl disiloxane	6·00	Isobutanol	10·5
n-Pentane	7·05	Acrylonitrile	10·5
n-Hexane	7·3	n-Pentanol	10·9
n-Heptane	7·4	n-Butanol	11·4
n-Octane	7.6	Isopropanol	11·5
n-Decane	7·7	n-Propanol	12·0
Methyl isobutyl ketone	8·4	Ethanol	12·7
Carbon tetrachloride	8·6	Propylene glycol	13·7
Methyl n-propyl ketone	8·7	Methanol	14·5
Methyl ethyl ketone	9·3	Glycerol	16·5
Styrene	9·3	Ethylene glycol	17·1
Acetone	9·6	Formamide	19·2
n-Dodecanol	9·8	Water	25·0

rise to exothermic mixing, are not covered by the treatment. Consequently, there are some anomalies. Nevertheless, the approach is useful provided its limitations are recognised. Table 2.1 gives a list of solubility parameters for liquids, showing how δ varies with chemical structure.

The solubility parameter approach was extended to polymers by Gee.[5] Since δ cannot be measured directly, it is estimated indirectly. One method is to cross-link the polymer and measure equilibrium swelling ratios in a range of liquids, equating the solubility parameter of the polymer with those of liquids causing maximum swelling. An alternative method, due to Small,[6] is to sum the contributions of each chemical group in the molecule. Estimates of δ for any given polymer vary as a result of the specific interaction effects mentioned earlier, and some authors have even gone so far as to suggest the use of three different solubility parameters for each material, depending upon the polarity of the liquid with which it is mixed. Typical values of δ for some common polymers are given in Table 2.2.

TABLE 2.2
ESTIMATED SOLUBILITY PARAMETERS OF POLYMERS

Polymer	$\delta \ (cal/cm^3)^{1/2}$
Polybutadiene	8·4
Poly(2,6-dimethyl-1,4-phenylene oxide)	8·6
Poly(butadiene-co-styrene) (75/25)	8·7
Polystyrene	9·1
Poly(vinyl chloride)	9·5
Poly(methyl methacrylate)	9·5
Poly(butadiene-co-acrylonitrile) (75/25)	9·5
Poly(styrene-co-acrylonitrile) (75/25)	9·6
Polycarbonate (Bisphenol A)	9·8
Polysulphone (Bisphenol A)	10·8
Polyacrylonitrile	15·4

In order to quantify specific interaction effects, several authors have advocated the use of a two- or three-dimensional solubility parameter,[7–10] consisting in its most general form of a dispersion term δ_d, a dipole interaction term δ_p and a hydrogen bonding term δ_h, to describe polymer–solvent mixtures. Shaw has shown that a two-dimensional parameter, comprising only dispersion and polar terms, is adequate to represent polymer–polymer interactions.[11] The two terms are compared separately, and the term $(\delta_1 - \delta_2)^2$ in eqn. (2.6) is replaced by

$$(\delta_{d1} - \delta_{d2})^2 + (\delta_{p1} - \delta_{p2})^2$$

In other words, polymers 1 and 2 will mix only when both terms of the solubility parameter are approximately matched. Tables 2.3 and 2.4 give examples of multidimensional solubility parameters for liquids and polymers.

TABLE 2.3
THREE-DIMENSIONAL SOLUBILITY PARAMETERS OF SELECTED LIQUIDS
(after Hansen[8])

Liquid	Solubility parameter $(cal/cm^3)^{1/2}$		
	δ_d	δ_p	δ_h
n-Hexane	7·24	0	0
Benzene	8·95	0·5	1·0
Tetrahydrofuran	8·22	2·8	3·9

Combining eqns. (2.1), (2.2) and (2.6), and neglecting interaction entropy effects, yields the following equation for the free energy of mixing:

$$\Delta G_m = V(\delta_1 - \delta_2)^2 \phi_1 \phi_2 + RT(n_1 \ln \phi_1 + n_2 \ln \phi_2) \qquad (2.7)$$

This equation is the basis of the solubility parameter approach, and illustrates both the strength and the weakness of the treatment. One of its chief merits is simplicity. The term $(\delta_1 - \delta_2)^2$ accounts in a general way for the effects of chemical constitution upon miscibility, and the use of a multidimensional solubility parameter extends the usefulness of the concept. Equation (2.7) predicts an increase in the free energy of mixing with increasing molecular weight (represented by the molar volume V), thus explaining the tendency for polymer–polymer systems to demix as the chain length of either constituent is increased. It does not, however, admit the

TABLE 2.4
TWO-DIMENSIONAL SOLUBILITY PARAMETERS OF SELECTED POLYMERS
(after Shaw[11])

Polymer	Solubility parameter $(cal/cm^3)^{1/2}$	
	δ_d	δ_p
Polystyrene	9·1	0·1
Poly(2,6-dimethyl-1,4-phenylene oxide)	8·9	0·33
Polysulphone (of Bisphenol A)	8·7	5·9
Poly(vinyl chloride)	7·4	6·1
Polyacrylonitrile	6·6	11·6

possibility of exothermic mixing, nor does it explain why some compatible polymer systems exhibit phase separation on heating. The only term in the equation that contains temperature is the configurational entropy of mixing, which increases on heating, and therefore tends to improve miscibility.

One solution to the problem is to discard the solubility parameter concept completely, and to base all analyses on eqn. (2.4), in which enthalpy and interaction entropy terms are represented by the empirical parameter χ_{12}. Most of the observations on partially miscible systems can be accommodated by recognising that χ_{12} varies with temperature and concentration.[12] The weakness of this procedure is that it merely quantifies the observed phenomena without explaining them. A more fundamental solution is outlined in Section 2.2.4. A brief review of partial miscibility is included below as an introduction to the subject.

2.2.3 Partial miscibility

A negative free energy of mixing is a necessary but not sufficient condition for complete miscibility. There is an additional condition, namely that the free energy–composition curve should not pass through a point of inflexion. Systems that satisfy the first requirement but not the second are only partially miscible.

Curve (a) of Fig. 2.1 meets both requirements. The curve is convex downwards over the whole range of n_1, so that at all points

$$\frac{\partial^2 \Delta G_m}{\partial n_1^2} > 0$$

If we choose any arbitrary point A on the curve, we see that phase separation into compositions B and C cannot occur spontaneously, because that would involve an increase in free energy. The components are therefore miscible in all proportions.

Curve (b), on the other hand, passes through points of inflexion defined by $(\partial^2 \Delta G_m/\partial n_1^2) = 0$ at points E and E'. Between these *spinodal* points the system will demix spontaneously into the *binodal* compositions D and D', with a decrease in free energy. Compositions in the ranges F to D and D' to F' form stable single-phase solutions for the reasons given earlier in connection with curve (a). Over the intermediate composition ranges D to E and E' to D' the system is metastable. The lowest free energy state is represented by the binodal compositions D and D', but a small fluctuation in composition produces a rise in free energy because of the increasing slope of the curve. Therefore phase separation does not occur spontaneously:

nucleation is necessary. Spontaneous spinodal decomposition causes phase separation on a finer scale than does nucleation.

In most cases of partial miscibility, the free energy–composition curve varies with temperature. At one temperature, the components are completely miscible, following curve (a). At another temperature, they are only partially miscible, following curve (b). At an intermediate temperature, the upper or lower critical solution temperature (ucst or lcst), there is a

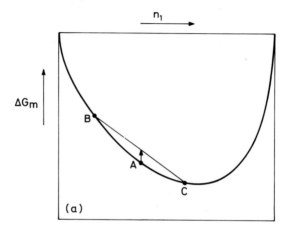

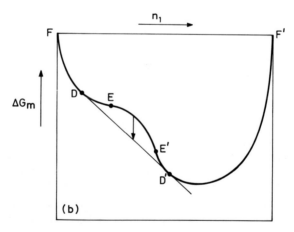

Fig. 2.1   *Free energy of mixing as a function of composition in binary systems, showing (a) complete miscibility and (b) partial miscibility.*

transition from one type of behaviour to the other, so that

$$\frac{\partial^2 \Delta G_m}{\partial n_1^{\,2}} = \frac{\partial^3 \Delta G_m}{\partial n_1^{\,3}} = 0 \qquad (2 \cdot 8)$$

Figure 2.2 is a schematic diagram showing binodal and spinodal curves for both ucst and lcst transitions.

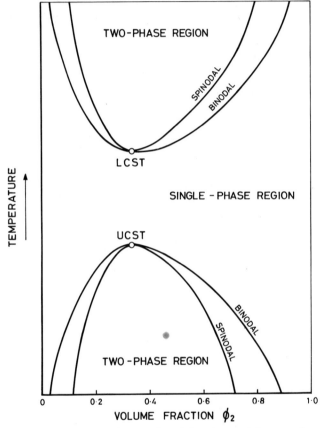

Fig. 2.2 *Schematic diagram showing binodal and spinodal curves for a system exhibiting both LCST and UCST behaviour.*

Low molecular weight liquids usually exhibit ucst behaviour, becoming more compatible with increasing temperature, as the $T\Delta S$ term in eqn. (2.1) becomes larger. There are, however, exceptions. A binary mixture of triethylamine with water has a lcst, and the mixture nicotine–water shows both ucst and lcst behaviour. Polymer–polymer mixtures are unusual in that they more frequently exhibit lcst than ucst behaviour.

2.2.4 Equation of state thermodynamics

Flory's equation of state treatment of polymer solution thermodynamics[13, 14] is based essentially upon a corresponding states model. It has been applied with considerable success by McMaster[15] to the problems of partial miscibility and lcst behaviour in polymer–polymer mixtures.

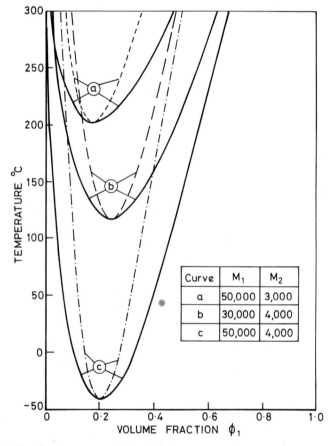

Curve	M_1	M_2
a	50,000	3,000
b	30,000	4,000
c	50,000	4,000

Fig. 2.3   *Binodal (————) and spinodal (– – – – –) curves calculated using the equation of state approach for a binary polymer mixture exhibiting LCST behaviour. Molecular weights M_1 and M_2 of monodisperse components as shown in the table (from L. P. McMaster, ref. 15, reproduced with permission).*

The treatment applies a statistical thermodynamics analysis to volume packing effects in polymer solutions. The partition function is expressed in terms of free volume, volume-dependence of internal energy and thermal pressure coefficient $(\partial P/\partial T)_{V,n_1}$. The partition function is then used to derive an equation of state containing these parameters and the volume coefficient of thermal expansion. In the most general treatment, energy and entropy interaction parameters are also included, as in the Flory–Huggins theory, but the qualitative predictions of the equation of state theory do not depend upon the inclusion of these additional parameters.

Figure 2.3 shows results of calculations made by McMaster using the equation of state approach. The theory predicts lcst behaviour in binary mixtures of monodisperse polymers, in broad agreement with experimental observations on a number of binary pairs. The analysis also shows that the miscibility of the two components decreases with increasing molecular weight of either component, causing a fall in lcst. This is a direct result of a reduced entropy of mixing. The spinodal surface is governed by the weight average molecular weight $\overline{M}_w$ of the polymeric constituents,[16] whereas the compositions of the equilibrated phases are dependent upon number average weight. At a given $\overline{M}_w$, the lcst is relatively unaffected by molecular weight distributions, provided they are similar for both polymers.[15]

One point that emerges clearly from the analysis is the relationship between thermal expansion coefficient and miscibility. A difference in volume expansion coefficient of only 4% is sufficient to cause phase separation in a binary mixture of polymers having a degree of polymerisation of 2000. Differences in thermal pressure coefficient have the opposite effect, enhancing mutual solubility. Exothermic specific interactions between polymers also increase compatibility, allowing larger differences in 'equation of state' properties without phase separation.

2.2.5 Tests for compatibility

It is often difficult to determine whether a polymer–polymer mixture has separated into two phases. The standard methods, which will be discussed in more detail in Chapter 3, are light scattering, microscopy and measurements of glass transition temperatures. These methods are adequate to deal with most binary pairs, in which the polymers are completely incompatible, and are distinguished by differences in refractive index and glass transition temperature T_g; the presence of two phases in such mixtures is usually obvious to the unaided eye. Problems arise, however, when the polymers are partially miscible, especially when they have similar properties. Just above the lcst both phases are of approximately the same composition, and it is not surprising that there has been some disagreement concerning the compatibility of particular polymer pairs.

It is particularly difficult to distinguish a single homogeneous phase from a fine dispersion of one polymer in another. The available analytical techniques differ in their ability to make this distinction, and therefore tend to give contradictory results. Thus calorimetry might indicate a single glass transition for a sample, whilst dynamic mechanical testing detects two separate transitions. Electron microscopy can sometimes resolve the problem, but only if there is sufficient electron contrast between the two components. Differences in chemical reactivity, enabling one constituent to be stained or etched preferentially, are a useful aid to analysis.

The problem is further complicated by kinetic effects. The concept of compatibility is a thermodynamic one, relating to the equilibrium state of the mixture. However, mixing and demixing of polymeric systems are diffusion-controlled processes, which can take a long time to produce equilibrium. A second kinetic effect, which has already been mentioned, is that a partially miscible system between its binodal and spinodal compositions can exist indefinitely as a metastable homogeneous phase in the absence of a nucleation mechanism. Because of these kinetic effects, some caution is necessary in interpreting experimental evidence. Failure to mix is not necessarily an indication of thermodynamic incompatibility, nor is the existence of an homogeneous phase proof of complete thermodynamic compatibility.

2.2.6 Effects of compatibility upon properties

A mixture of two completely compatible polymers has properties intermediate between those of its constituents, and is in many respects similar to a random copolymer of the same composition. The Fox equation,[17] given below, adequately represents the relationship between the weight fraction of the constituents, w_1 and w_2, and the glass transition temperature T_{gm} of a number of mixtures and copolymers:

$$\frac{1}{T_{gm}} = \frac{w_1}{T_{g1}} + \frac{w_2}{T_{g2}} \tag{2.9}$$

A somewhat different relationship is predicted by Gordon and Taylor.[18] Compatibility to this degree between a rubber and a plastic does not produce toughening, but merely serves to plasticise the material.

Complete incompatibility is equally undesirable. A completely incompatible rubber will not form a fine dispersion of the type required for good optical, mechanical and rheological properties, nor will it produce a strong mechanical bond at the rubber–matrix interface. The ideal rubber for the purposes of toughening is neither completely compatible nor completely incompatible.

Graft and block copolymers satisfy this latter requirement. A macromolecule consisting of a polystyrene chain attached chemically to a polybutadiene chain, for example, acts as a polymeric emulsifying agent, allowing the rubber to disperse as small particles in the polystyrene matrix, and also forms a strong mechanical bond between the two phases. These desirable ends are achieved without altering the glass transition temperatures of the constituents significantly.

2.3   COPOLYMERISATION

Random, graft and block copolymers all play an important part in rubber toughening, but in different ways. Random copolymerisation is used to modify the properties of a single phase, which might be either the matrix or the rubbery component of the composite-toughened polymer. Graft and block copolymerisation, on the other hand, are used to modify the properties of the interface between the matrix and the rubber.

The following nomenclature is standard, and is used throughout this book. The term poly(butadiene-*co*-styrene) represents a random copolymer in which butadiene is the major constituent. The term poly(butadiene-*g*-styrene) represents a graft copolymer in which the first-named monomer forms the backbone onto which the second has been grafted. The term poly(butadiene-*b*-styrene) represents a polybutadiene chain to which a polystyrene chain has been added. More specialised nomenclature is introduced where necessary.

2.3.1   Random copolymerisation

A full discussion of random copolymerisation would be inappropriate in this book. Nevertheless, poly(styrene-*co*-acrylonitrile) and other similar random copolymers are of considerable importance in the field of rubber toughening, and a brief outline of the subject is therefore given below.

In the copolymerisation of monomers A and B there are four competing propagation reactions, each with a different rate coefficient k:

$$A\cdot + A \to A\cdot$$
$$A\cdot + B \to B\cdot$$
$$B\cdot + A \to A\cdot$$
$$B\cdot + B \to B\cdot$$

where A· and B· are polymer radicals ending in monomer units A and B respectively.

It is assumed that the rate of attack by a radical upon a monomer molecule is entirely determined by the identity of the last monomer unit added to the propagating chain. If the coefficients k_{AB} and k_{BA} are zero, no copolymer is formed. If, on the other hand, k_{AA} and k_{BB} are zero, an *alternating copolymer* is formed.

The system reaches what is effectively a steady state, in which the rate of removal of A· by reaction with B is exactly balanced by the formation of A· from B·, as follows:

$$k_{AB}[A\cdot][B] = k_{BA}[B\cdot][A] \qquad (2.10)$$

The rates of consumption of monomers A and B are given by eqns. (2.11) and (2.12):

$$-\frac{d[A]}{dt} = k_{AA}[A\cdot][A] + k_{BA}[B\cdot][A] \qquad (2.11)$$

$$-\frac{d[B]}{dt} = k_{AB}[A\cdot][B] + k_{BB}[B\cdot][B] \qquad (2.12)$$

Termination reactions are neglected.

Using eqn. (2.10) to substitute for [A·] in eqns. (2.11) and (2.12), we obtain the following expression for the composition of the copolymer being formed at any given instant:

$$\frac{d[A]}{d[B]} = \frac{[A]}{[B]}\left(\frac{r_A[A] + [B]}{[A] + r_B[B]}\right) \qquad (2.13)$$

where $r_A = k_{AA}/k_{AB}$ and $r_B = k_{BB}/k_{BA}$.

The monomer reactivity ratios r_A and r_B indicate the rate at which a given monomer adds to itself in preference to adding to the comonomer. In the case of styrene (A) and acrylonitrile (B), the ratios are $r_A = 0\cdot37$ and $r_B = 0\cdot05$, which means that both monomer radicals prefer to add to the other monomer rather than to themselves. This preference is especially marked in acrylonitrile.

Figure 2.4 shows the relationship between the composition of the monomer feedstock and the composition of the copolymer being formed from it. At only one point are the two equal, namely at 76 wt. % (62 mole %) styrene. This unique composition will polymerise without producing a change in monomer ratio, and is known as an *azeotropic* mixture, by analogy with constant boiling mixtures. Greek scholars might prefer the

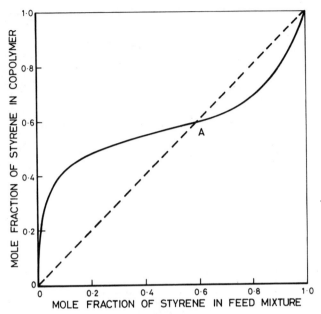

Fig. 2.4 *Copolymer composition in styrene–acrylonitrile copolymerisation, showing atropic composition at A (from A. E. Platt, ref. 19, reproduced with permission).*

term *atropic* mixture, omitting the reference to boiling. All other mixtures of styrene and acrylonitrile change in composition as a result of polymerisation, as shown in the diagram, so that almost pure homopolymer is produced towards the end of the reaction. Figure 2.5 shows how copolymer composition varies with time of formation during the copolymerisation of various mixtures of styrene and acrylonitrile.[19]

Variations in the composition of a styrene–acrylonitrile copolymer are quite significant. A difference in acrylonitrile content of less than 4% is sufficient to cause phase separation.[20] Manufacturers therefore exercise careful control during copolymerisation, either by adjusting the monomer ratio continuously to maintain a constant composition, or by choosing the atropic mixture.

Not all monomer combinations exhibit atropic copolymerisation. On the other hand, few polymer pairs are as incompatible as polystyrene and polyacrylonitrile, so that variations in copolymer composition during formation have a smaller effect upon properties in most cases of interest in the present context.

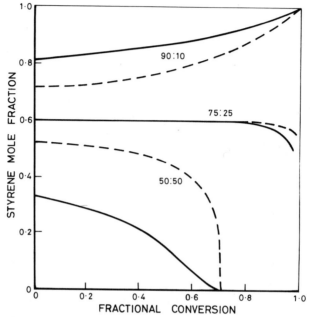

Fig. 2.5 Effect of initial monomer ratio (styrene: acrylonitrile by weight) on change of monomer mixture (————) and instantaneous copolymer composition (– – – – –) with conversion (from A. E. Platt, ref. 19, reproduced with permission).

2.3.2 Graft copolymerisation

Grafting is an important method for obtaining a strong bond between rubber particles and the surrounding resin. For this reason it is an essential feature in the manufacture of HIPS, ABS, MBS and a number of other rubber-toughened plastics. Although almost any rubber can be grafted by the use of suitable techniques, relatively few form satisfactory graft copolymers during free-radical polymerisation of styrene, methyl methacrylate and other monomers. The choice is effectively limited to diene polymers. Several copolymers of butadiene are used commercially, but the preferred rubber is polybutadiene, which has a lower T_g than any copolymer. The following discussion is devoted mainly to the grafting of styrene onto polybutadiene.

Mechanisms of grafting

There are several possible mechanisms of grafting styrene onto polybutadiene. The basic alternatives are (i) addition to the double bond and (ii) abstraction of methylenic hydrogen:

Addition

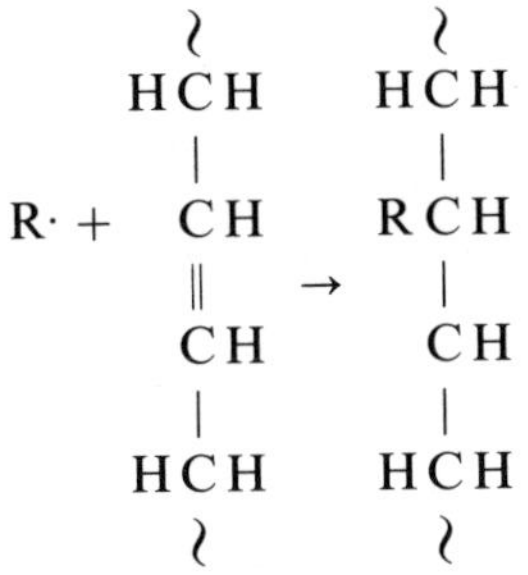

$$
\begin{array}{ccc}
& \wr & \wr \\
& HCH & HCH \\
& | & | \\
R\cdot + & CH & RCH \\
& \parallel \quad \rightarrow & | \\
& CH & CH \\
& | & | \\
& HCH & HCH \\
& \wr & \wr
\end{array}
\tag{i}
$$

Abstraction

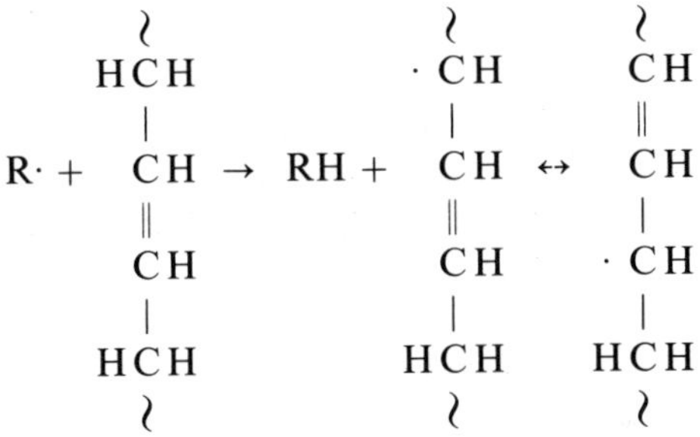

$$
\begin{array}{ccccc}
& \wr & & \wr & \wr \\
& HCH & & \cdot CH & CH \\
& | & & | & \parallel \\
R\cdot + & CH & \rightarrow RH + & CH & \leftrightarrow \quad CH \\
& \parallel & & \parallel & | \\
& CH & & CH & \cdot CH \\
& | & & | & | \\
& HCH & & HCH & HCH \\
& \wr & & \wr & \wr
\end{array}
\tag{ii}
$$

where $R\cdot$ is a polymeric free radical or an initiator fragment. If $R\cdot$ is a polystyryl radical, the addition reaction is a grafting reaction step.

The polybutadiene radicals formed either by (i) or (ii) form graft copolymer by attack on styrene monomer:

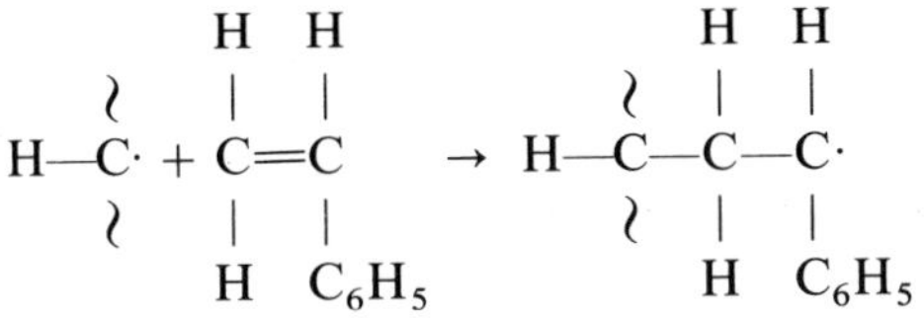

$$
\begin{array}{ccc}
H \quad H & & H \quad H \\
\wr \;\; | \;\; | & & \wr \;\; | \;\; | \\
H-C\cdot + C=C & \rightarrow \; H-C-C-C\cdot \\
\wr \;\; | \;\; | & & \wr \;\; | \;\; | \\
H \quad C_6H_5 & & H \quad C_6H_5
\end{array}
\tag{iii}
$$

Of the two polybutadiene radicals shown, the allylic radical produced in reaction (ii) is the less reactive, since it has a resonance energy of 84 kJ/mol, comparable with that of the styryl radical.[21]

Brydon et al.[22] studied the grafting of styrene onto polybutadiene in benzene solution at 60 °C, in the presence of benzoyl peroxide or azobisisobutyronitrile (AIBN) initiators. They proposed the following kinetic scheme for the polymerisation:

Initiation

Decomposition of initiator $\qquad$ $I_2 \xrightarrow{k_d} 2I\cdot$ $\qquad$ (iv)

Attack by initiator radical
on monomer $\qquad$ $I\cdot + M \xrightarrow{k_{11}} IM\cdot$ $\qquad$ (v)

Attack by initiator radical
on polybutadiene rubber $\qquad$ $I\cdot + R \xrightarrow{k_{12}} IR\cdot$ $\qquad$ (vi)

Attack by rubber radical
on monomer $\qquad$ $R\cdot + M \xrightarrow{k_{13}} RM\cdot$ $\qquad$ (vii)

Propagation

$$RM\cdot + M \xrightarrow{k_p} RM_2\cdot \xrightarrow{k_p} RM_n\cdot \qquad \text{(viii)}$$

$$IM\cdot + M \xrightarrow{k_p} IM_2\cdot \xrightarrow{k_p} IM_n\cdot \qquad \text{(ix)}$$

Termination

$$RM_m\cdot + RM_n\cdot \xrightarrow{k_{t1}} RM_{m+n}R \qquad \text{(x)}$$

$$RM_m\cdot + IM_n\cdot \xrightarrow{k_{t1}} RM_{m+n}I \qquad \text{(xi)}$$

$$IM_m\cdot + IM_n\cdot \xrightarrow{k_{t1}} IM_{m+n}I \qquad \text{(xii)}$$

Reaction (vii) is the basic grafting reaction. The products of the termination reactions (x) and (xi) are graft copolymers, and reaction (x) is also potentially a cross-linking reaction. Following normal practice, all propagation reactions are assumed to be governed by the same rate coefficient. Because of the similarity in the termination reactions, a single rate coefficient is also used for termination.

Application of the steady-state approximation to the four radicals involved in the polymerisation, $I\cdot$, $R\cdot$, $IM\cdot$ and $RM\cdot$, leads to the following expression for the rate of consumption of styrene monomer:

$$\text{Rate} = \frac{d[M]}{dt} = k_p[M]\left(\frac{2k_d[I_2]}{k_{t1}}\right)^{1/2} \qquad (2.14)$$

This equation, which is the normal kinetic relationship for a simple free radical polymerisation, is followed at low or moderate concentrations of polybutadiene. The rate of polymerisation $d[M]/dt$ is linear with monomer concentration and with the square root of initiator concentration, but independent of rubber content. At rubber concentrations above one monomer-mole/litre, however, the rate is lower than is predicted by eqn. (2.14).

In order to explain this effect, it is necessary to introduce two further reactions, in which the polybutadiene radical is directly involved in termination:

$$R\cdot + IM_n\cdot \xrightarrow{k_{t2}} RM_nI \qquad \text{(xiii)}$$

$$R\cdot + RM_n\cdot \xrightarrow{k_{t2}} RM_nR \qquad \text{(xiv)}$$

Both reactions lead to grafting, and reaction (xiv) is also a cross-linking step. These additional termination reactions cause a retardation of the polymerisation as the concentration of rubber, and hence of [R·], increases.

According to the simplified reaction scheme, to which reactions (xiii) and (xiv) do not contribute significantly, the graft fraction GF, defined as the fraction of styrene monomer incorporated into the grafted polymer, is given by

$$GF = \frac{\text{grafted styrene monomer}}{\text{total styrene polymer}}$$

$$GF = 1 - \left(1 + \frac{k_{i2}[R]}{k_{i1}[M]}\right)^{-2} \qquad (2.15)$$

This equation was obeyed at low polybutadiene contents, yielding a value of 0·6 for the ratio k_{i2}/k_{i1}. There was no evidence of a difference in reactivity between *cis*- and *trans*-polybutadiene.[22]

No grafting could be detected when AIBN was used as the initiator at a temperature of 65 °C, indicating that under the conditions of the experiment neither the AIBN initiator fragment nor the polystyryl radical is able to contribute significantly to the grafting reaction. Bevington[23] has shown that AIBN is incapable of abstracting an α-methylenic hydrogen atom. This low reactivity is explained by the resonance stability of the $(CH_3)_2\dot{C}(CN)$ radical formed from AIBN.[24]

The results outlined above are equally applicable to the polybutadiene–methyl methacrylate system. At 65 °C, peroxide initiators abstract hydrogen from the rubber, producing a site for a grafted side-chain.[25] No grafting occurs when AIBN is used as initiator.

Competing reactions begin to appear at higher temperatures. There is clear evidence that styrene forms a graft copolymer with polybutadiene when polymerised at or above 110 °C, even in the absence of an initiator.[26] Figure 2.6 presents results obtained by Fischer for thermal polymerisation of styrene containing 6% polybutadiene. The *grafting efficiency* of the rubber, defined as the weight fraction of PBD that has been grafted at least once, increases with temperature and with 1,2 vinyl content, suggesting either a copolymerisation reaction between the styryl radical and the pendant vinyl group, or abstraction of the tertiary hydrogen atom.

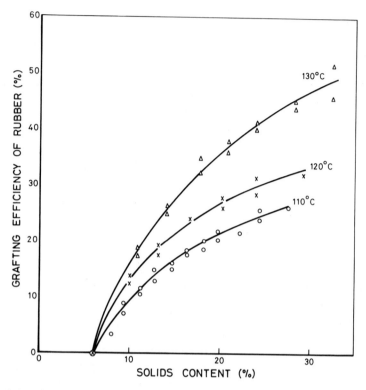

Fig. 2.6 *Thermal grafting of styrene onto* 6% *polybutadiene at various temperatures. Diene composition by NMR:* 9% *vinyl,* 38% *cis,* 53% *trans (from J. P. Fischer, ref.* 26, *reproduced with permission).*

Partitioning

Phase separation begins at an early stage during graft copolymerisation, for the reasons given in Section 2.2. In the styrene–polybutadiene system, for example, polystyrene homopolymer forms a separate disperse phase when its concentration reaches approximately 2%. There is little evidence of mutual solubility. Since it consists of two mutually incompatible macromolecular species, the graft copolymer concentrates at the interface, acting as an emulsifying agent for the resulting polymeric oil-in-oil emulsion.[27]

The partition coefficient for styrene monomer in this system is approximately 1·0. Consequently, as the amount of polystyrene present increases through polymerisation, the concentration of monomer in the polybutadiene phase falls. Rosen has analysed the effects of this removal of available monomer upon the degree of grafting.[28, 29]

Additional effects arise when there are two or more monomers. Locatelli and Riess have made an extensive study of ABS polymerisation, in which styrene and acrylonitrile are grafted onto polybutadiene, and ungrafted styrene-*co*-acrylonitrile is also formed.[30-37] As shown in Table 2.2, the solubility parameter of SAN(75:25) is 9·6, whilst that of PBD is only 8·4. Because of this substantial difference, the concentration of styrene ($\delta = 9·3$) is higher in the PBD phase than in the SAN phase, whilst the concentration of acrylonitrile ($\delta = 10·5$) is greater in the SAN phase than in the PBD phase. As a result, the SAN chains grafted onto the rubber have a higher styrene content than the ungrafted copolymer. The difference can be as high as 7 %, well above the limit of compatibility.[20]

Other reactants are similarly partitioned. In a study of peroxide and hydroperoxide initiators, Locatelli and Riess[35] showed that partition between SAN and PBD phases in benzene solution was dependent upon the solubility parameter of the initiator. Benzoyl peroxide ($\delta = 8·3$) was particularly strongly solvated by the PBD, and gave the highest partition coefficient of 2·20. Since peroxide initiators are principally responsible for grafting, differences in distribution between the two phases affect grafting efficiency.

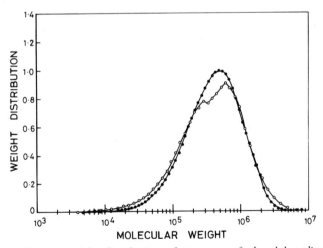

Fig. 2.7   Molecular weight distribution of styrene-grafted polybutadiene: (○) *determined experimentally by GPC;* (●) *calculated assuming identical molecular weight distributions for grafted and ungrafted polystyrene (from L. H. Tung and R. M. Wiley, ref. 38, reproduced with permission).*

Analysis of grafted chains

It is generally agreed that the grafted side chains are similar to the ungrafted polymer chains in molecular weight and molecular weight distribution. Locatelli and Riess found that grafted chains had a higher $\overline{M}_w$ than the 'free' copolymer in ABS.[35] They attributed this difference to a gel effect in the rubber, which is partly offset by an increased termination rate if the initiator is preferentially solvated by PBD.

Grafted chains were also found to have a higher $\overline{M}_w$ than 'free' polystyrene in HIPS,[37] but the difference was less marked than in ABS. Tung and Wiley made a similar study of grafting in the styrene–PBD system, but terminated the polymerisation at 20% conversion, thus avoiding the gel effect.[38] They found that the grafted and ungrafted polystyrene chains had identical molecular weight distributions, as illustrated in Fig. 2.7. Their results were consistent with a random attack upon the polybutadiene backbone.

Cross-linking

Controlled cross-linking of the rubber phase is a desirable feature of rubber-toughened plastics. Uncross-linked particles tend to break down into small fragments under the action of shearing forces during melt

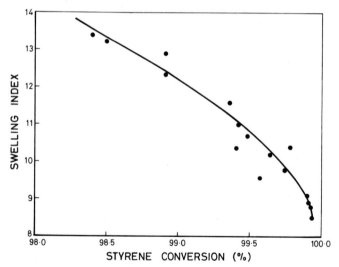

Fig. 2.8   Cross-linking of rubber during the final stages of styrene polymerisation in the HIPS process. Polybutadiene containing 9% vinyl groups (after D. J. Stein et al., ref. 40, reproduced with permission).

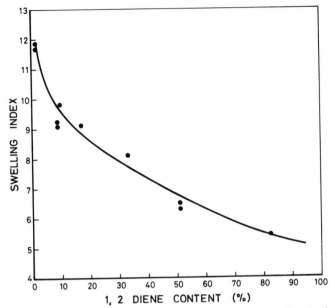

Fig. 2.9   Effect of 1,2-*vinyl content on cross-linking of polybutadiene in the HIPS process at a styrene conversion of* 99·50 % (*after D. J. Stein et al., ref.* 40, *reproduced with permission*).

processing.[39] They also become highly extended during moulding. Both tendencies lead to a reduction in mechanical strength.

Polybutadiene cross-links during the final stages of the grafting reaction in the manufacture of HIPS. As the last 2% of styrene monomer is polymerised, the rate of cross-linking increases, as illustrated in Fig. 2.8.[40] The swelling index of the rubber in toluene gives a useful if imprecise measurement of the progress of the reaction.

At a given styrene conversion, the degree of cross-linking increases with the 1,2 diene content of the rubber. This tendency, which is illustrated in Fig. 2.9, suggests that cross-linking is due mainly to copolymerisation of styrene with pendant vinyl groups to give the network structure shown schematically below:[40]

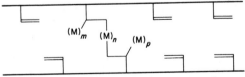

where M = styrene monomer; m, n and p are all > 1.

The reaction can be represented as follows:

$$M_n\cdot + M \overset{k_{MM}}{\rightarrow} M_{n+1}\cdot \qquad \text{(xv)}$$

$$M_n\cdot + R(CH\!=\!\!CH_2) \overset{k_{MP}}{\rightarrow} RCHM_n\!-\!CH_2\cdot \qquad \text{(xvi)}$$

Fischer[26] has shown that the monomer reactivity ratio r, given by k_{MP}/k_{MM}, is $1\cdot5 \times 10^{-3}$ at $110\,^{\circ}C$. The extent of the copolymerisation reaction is therefore low whilst there is an ample supply of styrene monomer with which the polystyryl radical $M_n\cdot$ can react.

It should be noted that these cross-linking reactions are also grafting reactions. The low values of swelling index show that the grafted chains are quite short at this stage of the polymerisation. The styrene homopolymer produced at the same time is also of low molecular weight. A simple kinetic analysis shows that chain length falls as the monomer becomes exhausted because the rate of propagation falls more rapidly than the rate of termination.

2.3.3 Block copolymerisation

In a block copolymer, the monomeric sequences are joined end-to-end along the main chain of the molecule, whereas the molecules of a graft copolymer have a branched structure, one monomer forming the main chain 'backbone' and the other monomer forming the branches. Like graft copolymers, block copolymers concentrate at phase boundaries, acting as polymeric emulsifying agents, and providing a chemical bond between the two phases.

Block copolymers are made in a number of different ways. Perhaps the best known method is sequential anionic polymerisation, which is used, for example, in the manufacture of poly(styrene-b-butadiene). Monomer A is first polymerised anionically, and monomer B is then added. Since there is no intrinsic termination mechanism in anionic polymerisation, a block copolymer AB is formed, by the addition of B to the active anion $A^{\ominus}$. This sequential addition procedure may be used to make three-block (ABA) or multiblock (ABABABA) copolymers.[41]

Block copolymers are also made by polymerising suitable monomers sequentially with the aid of Ziegler–Natta catalysts.[42] This technique is used in the manufacture of polypropylene 'copolymer'. The process begins with the polymerisation of propylene. Ethylene monomer is added at the end of the reaction, and a block copolymer is formed. If all propylene is reacted or removed before introducing the ethylene, the product is poly(propylene-b-ethylene), probably as a mixture with the two homopolymers. If, on the other hand, both monomers are present during the second stage of the reaction, the resulting polymer is poly(propylene-*b*-(ethylene-*co*-propylene)). Since ethylene is considerably more reactive than propylene, and is also less soluble in the hydrocarbons usually employed as solvents, the ethylene content of the copolymer block sequence is initially much higher than that of the monomer feedstock. Consequently, the ratio of propylene to ethylene in the unused monomer supply becomes larger as the reaction proceeds, until finally the ethylene is exhausted, and only propylene remains. The result is a 'tapered' copolymer, poly(propylene-*b*-(ethylene-*t*-propylene)).

Thus different types of propylene–ethylene block copolymer can be produced from a single polymerisation process by appropriate control of the monomer supply during the second stage of the reaction. The manufacturer has the option of feeding propylene monomer continuously with the ethylene in order to adjust the composition of the copolymer to his requirements.

2.3.4 Epoxy resin–CTBN reaction

One of the most interesting developments in rubber-toughened polymers has been the extension of the chemical technology to thermosetting resins, including styrenated polyesters and epoxy resins. Most of the research and development has concerned epoxy resins based on the diglycidyl ether of Bisphenol A (DGEBA), which has the structure

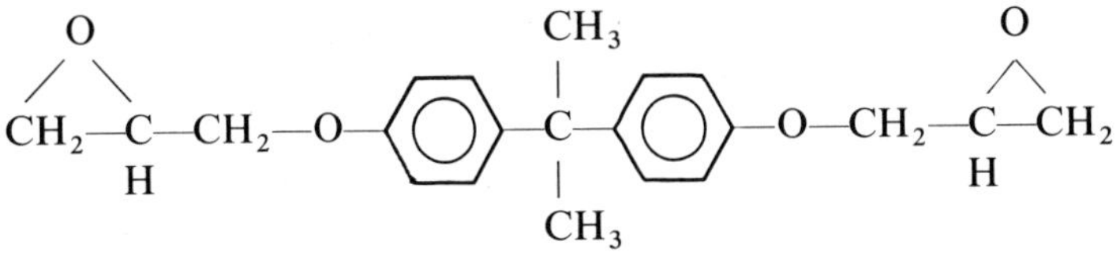

$$CH_2{-}\underset{H}{C}{-}CH_2{-}O{-}\underbrace{\hspace{1em}}_{}{-}\underset{CH_3}{\overset{CH_3}{C}}{-}\underbrace{\hspace{1em}}_{}{-}O{-}CH_2{-}\underset{H}{C}{-}CH_2$$

The chemistry of epoxy resins is described in standard texts on the subject.[43, 44] In the presence of a suitable hardener, the two epoxide rings open, each establishing a branching point in the molecular chain, so that the DGEBA molecule is effectively tetrafunctional. Reactions of the epoxide group with amines and acids are illustrated below:

Primary amine

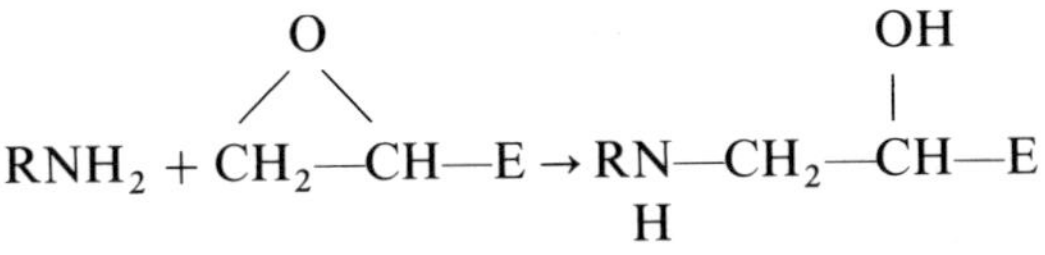

$$RNH_2 + CH_2{-}CH{-}E \rightarrow RN{-}CH_2{-}\underset{H}{\overset{OH}{CH}}{-}E \qquad (xvii)$$

Secondary amine

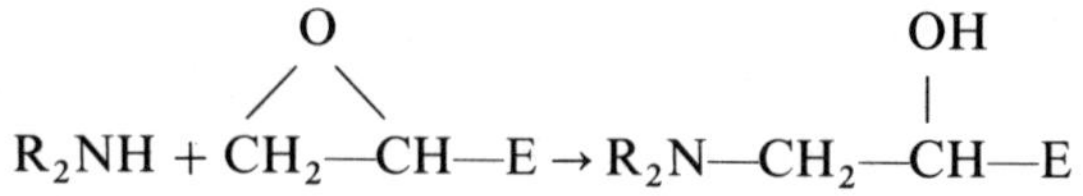

$$R_2NH + \overset{O}{\overset{\diagup\diagdown}{CH_2-CH}}-E \rightarrow R_2N-CH_2-\overset{OH}{\overset{|}{CH}}-E \quad \text{(xviii)}$$

Acid or anhydride

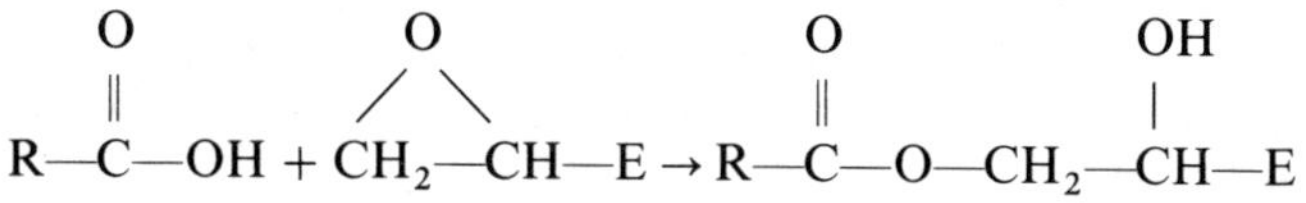

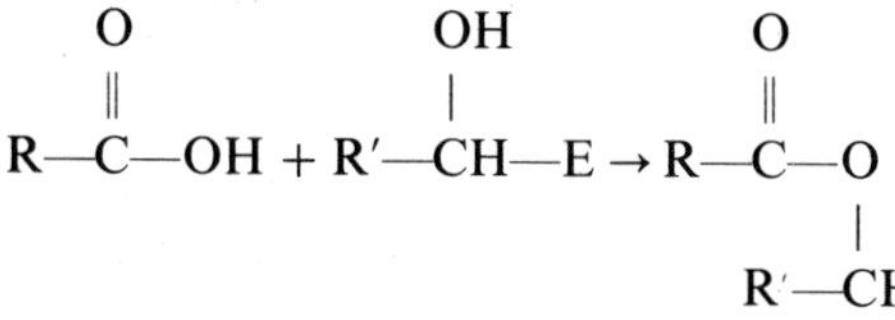

$$R-\overset{O}{\overset{\|}{C}}-OH + \overset{O}{\overset{\diagup\diagdown}{CH_2-CH}}-E \rightarrow R-\overset{O}{\overset{\|}{C}}-O-CH_2-\overset{OH}{\overset{|}{CH}}-E \quad \text{(xix)}$$

$$R-\overset{O}{\overset{\|}{C}}-OH + R'-\overset{OH}{\overset{|}{CH}}-E \rightarrow R-\overset{O}{\overset{\|}{C}}-O \quad \text{(xx)}$$
$$\overset{|}{R'-CH-E}$$

The following cross-linking reaction occurs in the presence of a number of different hardeners:

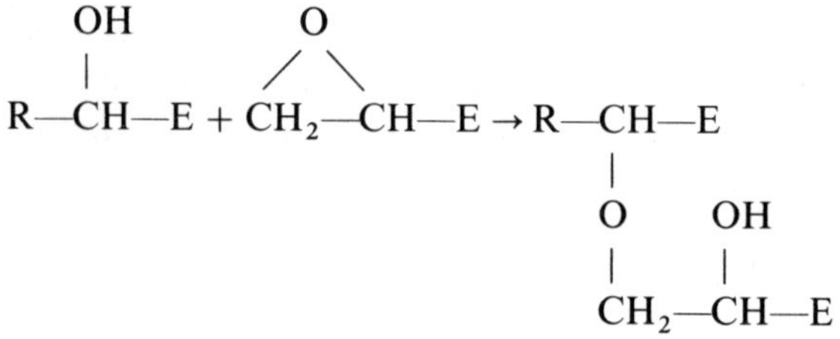

$$R-\overset{OH}{\overset{|}{CH}}-E + \overset{O}{\overset{\diagup\diagdown}{CH_2-CH}}-E \rightarrow R-CH-E \quad \text{(xxi)}$$
$$\overset{|}{O} \quad \overset{OH}{}$$
$$\overset{|}{CH_2}-\overset{|}{CH}-E$$

Reactions with CTBN

Most of the work on rubber-toughened epoxy resins has been based on carboxyl-terminated butadiene–acrylonitrile (CTBN) rubbers. These are liquid compounds of the following structure:

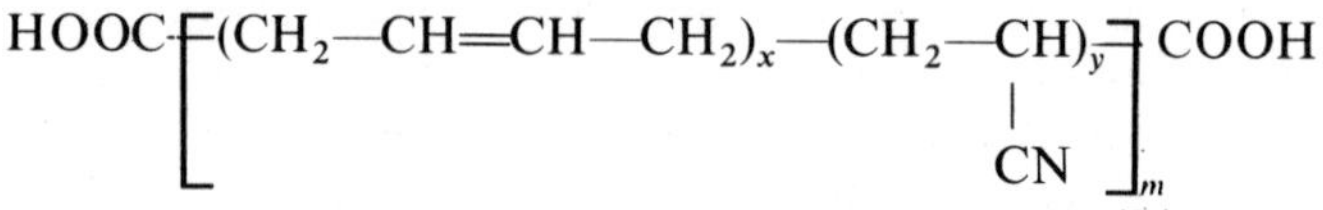

$$HOOC\left[(CH_2-CH=CH-CH_2)_x-(CH_2-CH)_y\right]_m COOH$$
$$\overset{|}{CN}$$

where on average $x = 5$, $y = 1$ and $m = 10$, giving a molecular weight of 3320. Experiments with rubbers of various compositions show that the highest toughness is obtained with acrylonitrile levels between 12 and 18 wt. %.[45]

The choice of rubber is determined by two factors. Firstly, there is a compatibility requirement: the rubber must dissolve and become dispersed in the resin, but precipitate when the resin begins to cure. Secondly, there is a chemical requirement: the rubber must react with the epoxide group. Both requirements are met by CTBN rubbers. The difference in solubility parameter between CTBN and DGEBA is sufficiently small to allow the rubber to dissolve in the resin, but not so small that the polymers will not undergo phase separation on curing. The relatively high molecular weight of the DGEBA molecule is offset by the low molecular weight of the rubber, so that the two are compatible until the reaction begins. The terminal carboxyl groups react with the resin as shown in reactions (xix) and (xx).

Figure 2.10 shows the relationship between solubility parameter and acrylonitrile content in carboxyl-terminated rubbers.[45] An exact match

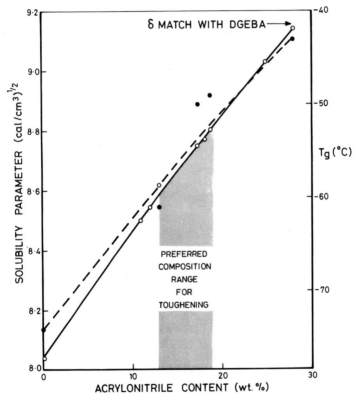

Fig. 2.10 *Relationships between composition and properties of carboxyl-terminated butadiene–acrylonitrile rubbers:* (○) *solubility parameter calculated by Small's method;*[6] (●) *glass transition temperature (after E. H. Rowe et al., ref. 45, reproduced with permission).*

with DGEBA is achieved at 28·5 wt. % of acrylonitrile. The preferred compositions for toughening are 0·4 to 0·6 $(cal/cm^3)^{1/2}$ lower, at only half the acrylonitrile content. Even in this range, the CTBN rubbers have significantly higher glass transition temperatures than polybutadiene, which is one of the disadvantages of using copolymer rubbers: a high T_g impairs toughness.

The following is a typical formulation for a toughened epoxy resin:

100	DGEBA epoxy resin
10	CTBN rubber
5	Piperidine $(CH_2)_5NH$

Because of molecular weight differences there are about 100 epoxy groups and 10 amine groups for every carboxyl group in this formulation. Infra-red studies show that the carboxyl immediately forms a salt with the amine:[46]

$$R'\!-\!\overset{\overset{\textstyle O}{\|}}{C}\!-\!OH + R_3N \rightarrow R'\!-\!\overset{\overset{\textstyle O}{\|}}{C}\!-\!\overset{\ominus}{O}\!-\!\overset{\oplus}{R_3NH} \qquad \text{(xxii)}$$

The carboxylate salt then reacts quite rapidly with the epoxy group:

$$R'\!-\!\overset{\overset{\textstyle O}{\|}}{C}\!-\!\overset{\ominus}{O}\!-\!\overset{\oplus}{R_3NH} + \overset{\overset{\textstyle O}{\diagup\diagdown}}{CH_2\!-\!CH}\!-\!E \rightarrow R'\!-\!\underset{\underset{\textstyle O}{\|}}{\overset{\overset{\textstyle O}{\|}}{C}}\!-\!O\!-\!CH_2\!-\!\overset{\overset{\textstyle O\,(R_3\overset{\oplus}{N}H)\ominus}{|}}{CH}\!-\!E \qquad \text{(xxiii)}$$

The ordinary epoxide curing reactions are of course taking place at the same time. The resulting rubber is probably similar in structure to an ABA block copolymer, where in this case A is composed of DGEBA resin units and hardener, and B is the poly(butadiene-*co*-acrylonitrile) chain of a single CTBN molecule.

Phase separation takes place as the epoxy resin begins to cure, and the molecular weight starts to rise. However, in addition to causing phase separation, the curing reactions lead to an increase in viscosity, and eventually to gelation. The resulting reduction in molecular mobility lowers the rate of attainment of equilibrium. At 120 °C, the epoxy resin–piperidine–CTBN system demixes during the first 60 min of the curing reaction, but shows no evidence of further phase separation during the succeeding 60 min, at the end of which gelation occurs.[45] It therefore appears that phase separation is confined to the early stages of curing, and stops well before the gel time.

REFERENCES

1. S. Krause, *J. Macromol. Sci.* **C7** (1972) 251.
2. P. J. Flory, *Principles of Polymer Chemistry*, Cornell University Press, 1953.
3. J. H. Hildebrand, *J. Am. Chem. Soc.* **51** (1929) 66.
4. G. S. Scatchard, *Chem. Rev.* **8** (1931) 321.
5. G. Gee, *Trans. IRI* **18** (1943) 266.
6. P. J. Small, *J. Appl. Chem.* **3** (1953) 71.
7. J. L. Gardon, *J. Paint Technol.* **38** (1966) 43.
8. C. M. Hansen, *J. Paint Technol.* **39** (1967) 104.
9. J. D. Crowley, G. S. Teague and J. W. Lowe, *J. Paint Technol.* **38** (1966) 269.
10. S. Chen, *J. Appl. Polymer Sci.* **15** (1971) 1247.
11. M. T. Shaw, *J. Appl. Polymer Sci.* **18** (1974) 449.
12. R. Konigsfeld, L. A. Kleintjens and A. R. Schultz, *J. Poly. Sci.* A2, **8** (1970) 1261.
13. P. J. Flory, R. A. Orwoll and A. Vrij, *J. Am. Chem. Soc.* **86** (1964) 3515.
14. P. J. Flory, *J. Am. Chem. Soc.* **87** (1965) 1833.
15. L. P. McMaster, *Macromolecules* **6** (1973) 760.
16. R. Konigsfeld, H. A. G. Chermin and M. Gordon, *Proc. Roy. Soc.* **A319** (1970) 331.
17. T. G. Fox, *Bull. Am. Phys. Soc.* **1** (1956) 123.
18. M. Gordon and J. S. Taylor, *J. Appl. Chem.* **2** (1952) 493.
19. A. E. Platt, in *Encyclopedia of Polymer Science and Technology*, Vol. 13, Wiley, New York, 1970, p. 156.
20. G. E. Molau, *J. Poly. Sci.* **B3** (1965) 1007.
21. A. Brydon, G. M. Burnett and G. G. Cameron, *J. Poly. Sci.* (*Chem.*) **12** (1974) 1011.
22. A. Brydon, G. M. Burnett and G. G. Cameron, *J. Poly. Sci.* (*Chem.*) **11** (1973) 3255.
23. J. C. Bevington, *J. Chem. Soc. Lond.* (1954) 3707.
24. H. A. J. Battaerd and G. W. Tregear, *Graft Copolymers*, Interscience, New York, 1970.
25. R. E. Wetton, J. D. Moore and B. E. Fox, *Makromol. Chem.* **132** (1970) 135.
26. J. P. Fischer, *Angew. Makromol. Chem.* **33** (1973) 35.
27. G. E. Molau, *J. Poly. Sci.* A3 (1965) 4235.
28. S. L. Rosen, *J. Appl. Polymer Sci.* **17** (1973) 1805.
29. W. A. Ludwico and S. L. Rosen, *J. Appl. Polymer Sci.* **19** (1975) 757.
30. J. L. Locatelli and G. Riess, *Angew. Makromol. Chem.* **27** (1972) 201.
31. J. L. Locatelli and G. Riess, *Angew. Makromol. Chem.* **28** (1973) 161.
32. J. L. Locatelli and G. Riess, *Angew. Makromol. Chem.* **32** (1973) 101.
33. J. L. Locatelli and G. Riess, *Angew. Makromol. Chem.* **32** (1973) 117.
34. J. L. Locatelli and G. Riess, *J. Poly. Sci.* **B11** (1973) 257.
35. J. L. Locatelli and G. Riess, *J. Poly. Sci.* (*Chem.*) **11** (1973) 3309.
36. J. L. Locatelli and G. Riess, *Makromol. Chem.* **175** (1974) 3523.
37. J. L. Refregier, J. L. Locatelli and G. Riess, *Eur. Poly. J.* **10** (1974) 139.
38. L. H. Tung and R. M. Wiley, *J. Poly. Sci.* (*Phys.*) **11** (1973) 1413.
39. M. Baer, *J. Appl. Polymer Sci.* **16** (1972) 1109.
40. D. J. Stein, G. Fahrbach and H. Adler, *Angew. Makromol. Chem.* **38** (1974) 67.
41. W. H. Janes, in *Block Copolymers*, D. C. Allport and W. H. Janes (eds.), Applied Science, London, 1973, p. 62.

42. T. G. Heggs, in *Block Copolymers* (see ref. 41), p. 105.
43. H. Lee and K. Neville, *Handbook of Epoxy Resins*, McGraw-Hill, New York, 1967.
44. W. G. Potter, *Epoxy Resins*, Iliffe, London, 1970.
45. E. H. Rowe, A. R. Siebert and R. S. Drake, *Mod. Plast.* **49** (Aug. 1970) 110.
46. C. K. Riew, E. H. Rowe and A. R. Siebert, *ACS Div. Org. Coat. Chem.* **34** (2) (1974) 353.

CHAPTER 3

CHARACTERISATION OF STRUCTURE

Control of the rubber-toughening process depends upon good analytical techniques, capable of characterising the complex structures resulting from phase separation, grafting and cross-linking. The development of suitable techniques has played an important part in the study of structure–property relationships, and hence in improving and extending the range of toughened products.

The information sought from the analysis is of several different kinds. Firstly, we wish to characterise the morphology of the rubber phase by measuring particle sizes and shapes and by examining the internal structure of the particles. Secondly, we wish to determine the type and amount of rubber present, and to measure its glass transition temperature. Thirdly, we wish to know about the chemical microstructure of the composite, including the composition of each phase, the degree of grafting and cross-linking of the rubber, the molecular weight distributions of the grafted and ungrafted polymer and similar details. A wide variety of analytical methods is available for tackling these and related problems; the principal methods are optical and electron microscopy, conventional chemical analysis and dynamic mechanical testing, but some use is also made of light scattering, dielectric loss, NMR and other techniques.

3.1 OPTICAL MICROSCOPY

If the rubber particles are large enough to be resolved by the light microscope, optical microscopy is the simplest and cheapest method for studying morphology. An approximate measurement of particle size in HIPS can be made by placing a speck of the polymer between glass slides, adding a drop of toluene or other solvent, and examining the solution in the phase contrast or interference microscope, which converts small differences in refractive index into differences in light intensity. More viscous and less volatile solvents should be used if photographs are required. In assessing rubber particle size, allowance must of course be made for the swelling of the rubber by the solvent. The technique is most useful for following phase separation during polymerisation. An alternative method for solid polymers is to melt the sample.

The examination of commercial samples by the solvent method is often complicated by the presence of titanium dioxide pigment. A clear distinction between the two types of particle present can be made by staining the rubber red with an azo dye.[1]

Transmission phase contrast or interference techniques are also suitable for the examination of structure in thin sections. The preferred section thickness, between 1 and 5 μm, can be obtained from an ordinary sledge microtome. Sections tend to curl up as a result of distortion introduced during microtoming, and it is usually necessary to relax them by flotation on a heated glycerol bath. The sections are then mounted in glycerol containing dissolved potassium mercury iodide, K_2HgI_4, to raise its refractive index.[2] This solution does not attack HIPS and related polymers, and its concentration can be adjusted to match the refractive index of the matrix polymer (1·59 for PS), so that knife lines and other defects in the section are not seen.

A third method is selectively to etch or stain the sample, and to examine the treated surface in reflection. The specimen can be prepared in a number of ways. In some cases a moulded surface is satisfactory, but it is usually necessary to prepare the surface in some way. The best flat surface is obtained by ultramicrotoming the block.[3] Other satisfactory methods include metallographic polishing on an abrasive wheel, and low temperature fracture.[3]

A typical etch for HIPS is a mixture of chromic and phosphoric acids comprising 100 ml H_2SO_4, 30 ml H_3PO_4, 30 ml H_2O and 5 g CrO_3. This powerful oxidising etch attacks the unsaturated rubber phase selectively, but has little effect upon the polystyrene phase. After 5 min in the etch at 70 °C, the rubber particles in the surface are completely removed. Shorter etching times result in incomplete etching of the rubber, so that the polystyrene sub-inclusions, which are embedded in the rubber particles of most HIPS polymers, can be seen in the microscope, as a resistant residue held in the particle cavity by the remaining rubber. In order to obtain the best results from reflection microscopy of etched HIPS specimens, it is

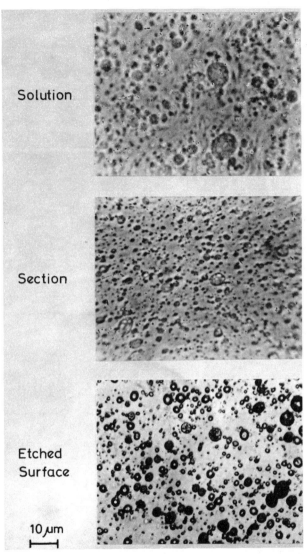

Fig. 3.1 Optical microscopy of HIPS. A single material observed at a fixed magnification by three different techniques: (top) negative phase contrast micrograph of a solution in chloroform; (centre) negative phase contrast micrograph of a thin section; (bottom) reflection micrograph of a microtomed surface etched with chromic acid. Note swelling of rubber particles by solvent.

advisable to apply a thin layer of carbon to the surface by vacuum coating, to eliminate reflection and scattering of light from the subsurface layer.

Results obtained using the techniques described above are compared in Fig. 3.1. Each method has its advantages. The solvent method is the simplest, but loses much of the information to be obtained from solid specimens: the rubber particles relax and swell on addition of the solvent, so that their shapes and spatial distribution cannot be determined. Sectioning overcomes some of these problems, and in some cases allows the internal structure of the particles to be observed, but the shapes of the rubber particles cannot be measured accurately because of the distortion introduced during sectioning. Another effect of distortion is that it is difficult to flatten the section sufficiently to produce good photographs. In many ways, the etch method, which overcomes the problems of shape distortion and of non-planar specimens, is the most satisfactory. The measurement of particle ellipticity is particularly useful in characterising orientation in complex mouldings.

No improvement in specimen preparation techniques, however, can overcome the fundamental limitation of the optical microscope, namely its limited resolution. Most rubber-toughened plastics have rubber particles that are too small to be seen and resolved in the optical microscope. For this reason, most of the important developments have been in the field of electron microscopy.

3.2   TRANSMISSION ELECTRON MICROSCOPY

In order to allow passage of the electron beam, transmission electron microscopy requires much thinner specimens than those used in optical microscopy. The effective upper limit is approximately $0\cdot2\,\mu$m. There are two basic methods of preparing very thin specimens from rubber-toughened plastics, namely replication and ultrasectioning. Solvent-casting is also useful in certain cases.

3.2.1   Replication
Etched surfaces prepared as described above are excellent subjects for one- or two-stage replication. In the one-stage process, a layer of carbon is evaporated onto the surface, the details are highlighted by shadowing and the replica is floated off by immersing in a suitable solvent for the polymer.

In the two-stage process, an aqueous solution of gelatin or poly(vinyl formal) is painted onto the surface and allowed to dry. The replica so formed is stripped off, carbon-coated and shadowed as in the one-stage process. The gelatin is then dissolved away in water, leaving the shadowed carbon replica for examination under the microscope. The advantage of the two-stage method is that the specimen is not destroyed by treatment with solvent.

Several etch methods are available. The chromic–phosphoric acid mixture mentioned earlier is very satisfactory for HIPS, but is less suitable for ABS because it attacks the SAN matrix. Better results are obtained by etching the ABS with 10M CrO_3 for 5 min at 40 °C.[4] These chemical etch methods have the advantage that they record faithfully the shape of the rubber particles, but the internal structure is generally lost.

Ion beam etching has been used to reveal the internal structure of rubber particles in HIPS,[5] but the quality of the results is disappointing. Etching by exposure to isopropanol vapour gives much better results,[6, 7] but suffers from the disadvantage that the solvent vapour relaxes the polymer in the surface, so that particle shapes cannot be determined reliably by this method. In suitable materials, replication of fracture surfaces provides valuable information about particle shape and internal structure,[8] but raises the difficult problem of interpreting fractographs. Abrasive polishing of the type used in metallography yields better results.[9]

Replica methods are useful in studying particle sizes, spatial distributions and shapes. They are less useful in studying the internal structure of rubber particles, although this can be done by appropriate modification of the procedure. Detailed structural studies of this kind are usually made by examination of osmium-stained sections. Figure 3.2 compares results obtained from a single HIPS specimen by replication, sectioning and scanning electron microscopy.

3.2.2 Sectioning

Undoubtedly the most successful method for studying morphology in rubber-toughened plastics is the osmium-staining method developed by Kato.[10–12] The procedure is applicable to polymers containing unsaturated rubbers, and is therefore suitable for most HIPS and ABS polymers, toughened PVC,[13] polypropylene–EPDM blends,[14] toughened epoxy resins,[15] and a number of other rubber-toughened polymers. The basis of the method, which was first used to stain synthetic rubbers by Andrews,[16] is the reaction between osmium tetroxide and the ethylenic unsaturation of the rubber to form a cyclic osmic ester:[17, 18]

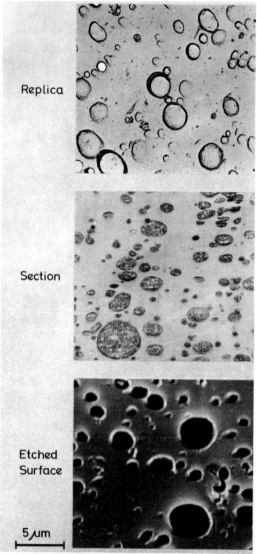

Fig. 3.2   *Electron microscopy of HIPS. The material shown in Fig.* 3.1 *observed at a fixed magnification by three different techniques: (top) two-stage replica of a microtomed and etched surface; (centre) ultrathin section of an osmium-stained block; (bottom) scanning electron micrograph of an etched surface.*

$$\underset{\diagup\quad\diagdown}{\overset{\diagdown\quad\diagup}{C=C}} + OsO_4 \rightarrow \quad (i)$$

This reaction has the dual effect of hardening the rubber, so that it is easier to section, and providing electron contrast with the unstained matrix. Samples are suspended for a few hours in the vapour or for a few days in 1 % aqueous solution at room temperature. As OsO_4 is both toxic and expensive, the solution is preferred. Riew and Smith[19] have shown that faster and deeper staining of thermosetting resins is achieved by dissolving the OsO_4 in an organic solvent, tetrahydrofuran being suitable for use with toughened epoxy resins. Neither the vapour nor the aqueous solution penetrates cross-linked resins sufficiently rapidly to give satisfactory staining. If solid OsO_4 is not available, the tetrahydrofuran can be mixed with an equal volume of 2 % aqueous osmium solution, with good results.

In order to reach the rubber, the OsO_4 must diffuse through the rigid matrix. Polystyrene absorbs substantial quantities of OsO_4, resulting in significant plasticisation.[20] The stain usually penetrates only 10–20 μm into the surface under standard conditions, and it is therefore necessary to prepare the surface to be sectioned quite carefully in order to obtain satisfactory results. The specimen is first trimmed to a pyramidal point, and then carefully ultramicrotomed to produce a truncated pyramid with a perfectly smooth top about 0·2 mm across. Very thin sections cannot be produced from wider specimens. The sample is next immersed in the stain for several days, and then replaced in the ultramicrotome. The knife is carefully aligned with the prepared surface by setting the cutting edge parallel with its reflection, so that acceptable sections are cut immediately, well within the stained region. Best results are obtained with sections below 60 nm thick, which give a silver interference colour. Sections of this thickness usually need a supporting film.

Figures 3.1 and 3.2 present optical and electron micrographs prepared by various techniques from the same HIPS polymer. The ultrasection made by Kato's method reveals the morphology in excellent detail. The dark, stained regions of polybutadiene contrast sharply with the unstained, light areas of polystyrene both outside and inside the rubber particles. The boundaries between the two phases are particularly distinct. The structure shown is typical of HIPS: the rubber particles are relatively large, and contain a high proportion of polystyrene sub-inclusions. The origins of this structure are discussed in Chapter 4.

In blends of ABS with PVC or polycarbonate, electron microscopy sometimes shows differences in electron density within the matrix, suggesting incomplete mixing of the two thermoplastic components. There is clear evidence for this in the work of Stefan and Williams on an ABS–polycarbonate blend:[21] the polycarbonate appears as lighter, rubber-free regions, which are surrounded by rather darker SAN polymer containing osmium-stained rubber particles. The origin of the contrast is not clear. One possibility is a difference in the solubility of OsO_4 in the two thermoplastics. Intrinsic differences in electron density probably make a contribution, especially in PVC blends, but loss of HCl in the electron beam can reduce contrast from this source.

There are numerous variations on Kato's method. Faster and deeper staining is achieved by heating the specimen in a sealed container in the presence of solid OsO_4. A temperature of 40–50 °C is sufficient for this purpose. Another variation is to stain after sectioning. The rubber particles tend to smear and tear unless they are hardened in some way, and it is usually necessary to section at low temperatures, or to subject the specimen to high-energy irradiation, but untreated samples can be ultrasectioned successfullly at room temperature, given a little care.

Saturated rubbers present the greatest difficulties. Although the problems of sectioning can be overcome relatively easily with the aid of the techniques mentioned above, the problem of electron contrast remains. Unstained rubber particles do not show up clearly against the surrounding matrix.[22]

Kanig[23] has developed a specific two-stage stain for butyl acrylate rubber. The polymer is treated with hydrazine or hydroxylamine, followed by osmium tetroxide, in the absence of water. The first step is the formation of a hydrazide:

$$R-C{\overset{\displaystyle O}{\underset{\displaystyle O-C_4H_9}{\big\|}}} + H_2N-NH_2 \rightarrow R-C{\overset{\displaystyle O}{\underset{\displaystyle NH-NH_2}{\big\|}}} + C_4H_9OH \quad \text{(ii)}$$

On addition of OsO_4, metallic osmium is deposited, through the reducing action of the hydrazide.

Another stain due to Kanig produces contrast between crystalline and non-crystalline regions in polyolefines,[24, 25] and is therefore suitable for toughened polypropylene. The polyolefine is treated for 16 h at 60 °C with chlorosulphonic acid, which introduces sulphonic acid groups into polymer chains in the non-crystalline regions of the material, including interlamellar regions. On adding 1 % aqueous uranyl acetate, electron-dense uranyl sulphonate is formed. Other heavy metal salts may also be used.

3.2.3 Analysis of electron micrographs

Most electron micrographs are used in a semi-quantitative way, to obtain approximate information about particle size, volume fraction and other characteristics. More quantitative information can, of course, be obtained by analysing a large number of micrographs, preferably with the aid of an image-analysing computer.

Analyses of thin sections provide direct information about volume fractions of the constituent phases, since a thin section is representative of the bulk polymer. The simplest (but most tedious) method is to draw a set of parallel lines on each micrograph, and to determine manually the proportion of matrix and rubber particles crossed by each line. It is difficult to make reliable measurements of the volume fraction of rubber phase, as the rubber usually contains embedded hard polymer which must be included in the measurement. If the rubber content is already known, the volume

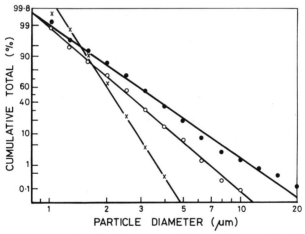

Fig. 3.3 *Particle size distributions measured by microscopy for three HIPS polymers of identical composition but prepared under differing stirring conditions (from H. Willersin, ref. 28, reproduced with permission).*

composition can be calculated. For example, the composition of a typical HIPS might be 70% polystyrene matrix, 6% polybutadiene and 24% polystyrene sub-inclusions. Rubber membranes such as those illustrated in Fig. 3.2 are too thin for accurate direct measurement.

Analyses of thin sections underestimate the diameters of the rubber particles. True diameters are observed only when the section contains the equatorial plane of the particle. A section 50 nm thick, cut at random from a particle 2 μm in diameter is unlikely to show the full diameter of the particle. Holliday[26] recommends multiplying observed mean diameters by $4/\pi$ to compensate for this error. It should be noted, however, that the sections are not infinitely thin, so that the magnitude of the error decreases with decreasing ratio of particle size to section thickness.

There is a strong case for systematic analysis of particle sizes, since a simple visual inspection tends to overestimate particle sizes.[27] A typical particle size distribution in HIPS is illustrated in Fig. 3.3.[28] The log-normal distribution shown in the graph appears to be characteristic of the material.[27]

3.3   SCANNING ELECTRON MICROSCOPY

Scanning electron microscopy has become a standard method for examination of polymer surfaces, replacing replication for many purposes. The surfaces to be examined are prepared in similar ways, by microtoming, abrasive polishing or fracture, but subsequent treatment is simpler. The sample is vacuum-coated with a 20 nm layer of gold or other suitable metal to prevent charging in the electron beam, and inserted in the microscope. This technique avoids some of the difficulties inherent in replication, such as the stripping of polymer films from convoluted surfaces, and removal of residual polymer from the carbon layer.

Figure 3.2 compares a scanning electron micrograph with a replica and a section from the same HIPS polymer. The three-dimensional effect of the scanning picture is evident. Published work on structure in toughened plastics using this method has been based largely upon the examination of fracture surfaces. Sultan et al.[8] observed size distributions and internal structure of rubber particles in epoxy resins by means of fractographs, and a similar approach was adopted by Willmouth and Heggs in their study of toughened polypropylene.[29] The latter authors took the precaution of comparing fractographs from toughened and untoughened polymer, and showing that only the rubber-toughened material had 1 μm hemispherical depressions in the fracture surface. Control experiments are essential in this type of work to eliminate the possibility of artefacts.

3.4 LIGHT SCATTERING

The success of electron microscopy as a method for studying the morphology of rubber-toughened plastics has distracted attention from alternative techniques such as light scattering. Nevertheless, there are materials for which such techniques might be useful, especially those containing rubbers that cannot be stained successfully.

Light scattering from heterogeneous solids was first analysed by Debye and Bueche,[30] whose analysis has been extended by Stein.[31] A simplified treatment shows that the angular dependence of scattering by isotropic spheres of radius R in an isotropic medium is given by

$$I_s = BV^2 \left(\frac{3}{U^3} (\sin U - U \cos U) \right)^2 \tag{3.1}$$

where

$$U = \frac{4\pi R}{\lambda_1} \sin \frac{\theta}{2}$$

I_s = intensity of scattered light, B = constant, V = volume of irradiated material, λ_1 = wavelength of light in the matrix and θ = scattering angle.

This equation provides a basis for the determination of particle size, through the measurement of I_s as a function of θ.

Scattering is caused by variations in polarisability within the system. The incident beam induces dipoles in the material, which then re-radiate the light. In an homogeneous medium, destructive interference eliminates all radiation except that in the forward direction. In an heterogeneous material such as a rubber-toughened polymer, the scattering is determined by the size of the heterogeneities, including not only particle size but also interparticle spacing. A complete analysis of scattering involves correlation and probability functions which take account of both parameters.[32]

Moritani and co-workers[32] found good agreement between light scattering and electron microscopy in measurements of particle size in ABS, over the range 0·1 to 0·3 μm. On the other hand, Visconti and Marchessault[33, 34] obtained consistently higher values from the light scattering method in a study of toughened epoxy resins containing particles in the range 1 to 5 μm. As explained in Section 3.2.3, electron microscopy of thin sections consistently underestimates the sizes of *large* rubber particles. The work demonstrated one of the advantages of light scattering, namely that it is able to follow the process of phase separation during curing of toughened thermosetting resins.

Conaghan and Rosen[35] have made a theoretical analysis of the relationship between particle size and turbidity τ in an ideal monodisperse material containing particles of radius R. The turbidity is defined by Beer's Law:

$$I_{tr} = I_0 \exp(-\tau/x) \tag{3.2}$$

where I_{tr} = intensity of transmitted light, I_0 = intensity of incident light and x = thickness of scattering layer. Their results are shown in Fig. 3.4, in terms of the *specific turbidity* τ/ϕ_2, which takes account of the increase in scattering with the volume fraction ϕ_2 of the particles.

The analysis predicts a maximum in turbidity at particle diameters between 1 and 5 μm, which is the particle size range of many rubber-toughened plastics. Above this diameter, turbidity varies as R^{-1}, and is independent of the refractive index ratio. Below this particle size range, the scattering is predicted by the Rayleigh–Gans equation: turbidity varies as

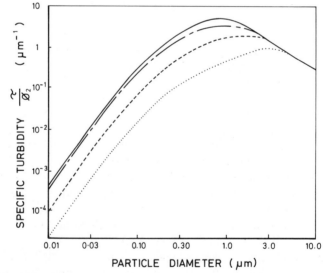

Fig. 3.4 Calculated variation in specific turbidity with particle diameter for dispersions of various refractive index ratios μ_2/μ_1 as follows: (———) 0·80; (− − − −) 0·90; (·······) 1·05; and (− − −) 1·20 (from B. F. Conaghan and S. L. Rosen, ref. 35, reproduced with permission).

R^3, and is highly dependent upon refractive index differences, as follows:

$$\frac{\tau}{\phi_2} = \frac{8}{9}\left(\frac{2\pi}{\lambda_1}\right)^4 \left(\frac{\mu_2}{\mu_1} - 1\right)^2 R^3 \qquad (3.3)$$

where λ_1 = wavelength of light in the matrix and μ_1, μ_2 are the refractive indices of matrix and rubber respectively.

Turbidity is one of the disadvantages of rubber-toughened plastics. It follows from eqn. (3.3) that in order to produce a transparent two-phase polymer it is necessary either to reduce particle size or to match refractive indices. Unfortunately, both methods present problems. Very small particles are ineffective in toughening brittle polymers; and refractive index matching cannot be achieved over a wide range of temperatures.[36] The best solution is a combination of the two principles, using the smallest particles that will confer toughness, and matching refractive indices as far as possible.

Refractive index matching in styrene-based polymers is achieved by adding methyl methacrylate to the formulation. The refractive index of polybutadiene is 1·52, lower than that of polystyrene (1·59), but higher than that of poly(methyl methacrylate) (1·50). Suitable combinations of styrene and methyl methacrylate produce a random copolymer having the same refractive index as the rubber. The composite is known as MBS, or AMBS if acrylonitrile is also added.

The refractive index of a material is given by the Lorentz–Lorenz equation:

$$\left(\frac{\mu^2 - 1}{\mu^2 + 2}\right)\frac{1}{\rho} = \mathscr{R} \qquad (3.4)$$

where ρ is density and $\mathscr{R}$ is a temperature-independent quantity called the specific refraction, which depends only upon the polarisability of the molecules. Variations of μ with temperature are the result of thermal expansion. Since rubbers have higher coefficients of thermal expansion than plastics, ρ varies more rapidly with temperature in the particles than in the matrix. Consequently, a refractive index match is obtained at only one temperature for a given system. Manufacturers usually aim for a match at $20\,^\circ$C. Above and below this temperature, turbidity varies as predicted by eqn. (3.3).

3.5 COULTER COUNTER

The Coulter Counter is an electronic instrument which measures particle sizes in the range 0·5 to 400 μm. The particles are suspended in an electrolyte and passed through a small aperture between two electrodes. The particles displace electrolyte, and as each particle enters the aperture the resistance between the electrodes increases by an amount proportional to the volume of the particle. The changes in resistance are registered as voltage pulses which are counted electronically.

The method is well suited to measuring the relatively large particles in HIPS. The suspending liquid must be an electrolyte that is capable of dissolving the polystyrene matrix without swelling the rubber appreciably. A 1 % solution of tetrabutylammonium perchlorate in dimethyl formamide is satisfactory for the purpose.[27] This solvent causes a slight swelling of the rubber particles, which must be measured in a separate experiment if absolute values of diameter are required.

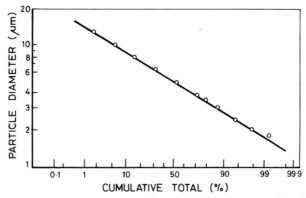

Fig. 3.5   Particle size distribution in a typical HIPS as determined by the Coulter Counter (from D. E. James, ref. 27, reproduced with permission).

Figure 3.5 shows results obtained from Coulter Counter measurements on HIPS. The logarithm of rubber particle diameter is plotted against cumulative weight percentage of rubber on a Gaussian probability scale. The recommended method for presenting the data is to quote the weight average median diameter, and the difference between the upper and lower quartile diameters, so that both particle size and dispersity of the distribution are indicated.[27]

3.6   GEL SEPARATION

Gel separation is a standard technique for the analysis of styrene-based rubber-toughened plastics. A weighed quantity of the polymer is dissolved in a suitable solvent and centrifuged at 20,000–30,000 rpm. Cross-linked rubber gel becomes separated by sedimentation, leaving the uncross-linked polymer and low molecular weight additives in solution. The soluble fraction is decanted, and the gel is washed several times with fresh solvent. Standard solvents include toluene for HIPS and methyl ethyl ketone for ABS.[37] Additional solvent treatments may be used to separate the soluble components of the material, as illustrated in Fig. 3.6.[38, 39]

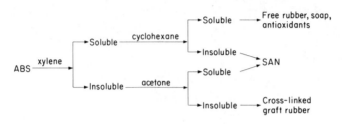

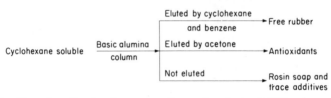

Fig. 3.6   Flow sheet for the phase separation and analysis of ABS polymers (from B. D. Gesner, ref. 38, reproduced with permission).

Figure 3.7, which shows the structure of gel from a typical HIPS polymer, emphasises the main problem in gel separation analysis. There is a substantial amount of polystyrene phase associated with the rubber.[40] In this case, most of it is probably grafted chemically to the polybutadiene, but it is possible that some polystyrene homopolymer remains. Extraction of occluded polystyrene from within the particles is hindered by the low permeability of the rubber, especially if the chosen solvent does not swell the rubber membranes sufficiently. Indeed, by suitable choice of solvent, the rubber particles can be precipitated with almost all of the occluded polystyrene still inside. Careful selection of solvents and repeated washing are necessary in order to obtain reproducible results.

The *gel content* of the toughened polymer is expressed as a weight percentage. The gel content of a typical HIPS polymer is approximately

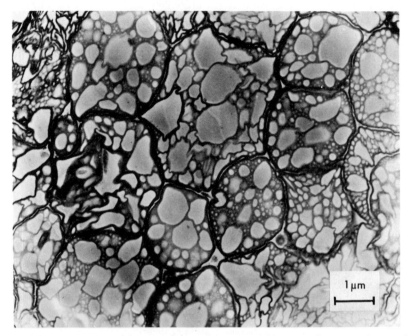

Fig. 3.7   Osmium-stained section from gel precipitated by centrifuging a solution of HIPS in 50/50 MEK/acetone mixture (from H. Keskkula and S. G. Turley, ref. 40, reproduced with permission).

twice the rubber content, which means that the precipitated graft copolymer contains roughly equal amounts of polystyrene and rubber, since very little of the diene elastomer remains soluble at the end of the polymerisation reaction. Comparison with electron micrographs shows that graft copolymer accounts for rather less than half of the volume of rubber particles, and it is therefore clear that substantial amounts of styrene homopolymer are contained in the sub-inclusions within the particles, as illustrated in Fig. 3.2. The amount of soluble polymer occluded depends upon the method of manufacture, and is much smaller in most ABS polymers, for reasons that are discussed in Chapter 4.

The *swelling index*, usually defined as the weight of swollen gel divided by the weight of dry gel, gives an approximate indication of the degree of cross-linking of the rubber. Standard theory for the swelling of rubbers by solvents cannot be applied to graft rubbers. The swelling behaviour is radically altered by the presence of the grafted chains and of ungrafted sub-inclusions within the rubber particles. Stein and co-workers[41] found that the swelling index of uncross-linked polybutadiene in methyl ethyl ketone was 1·9, whereas the figure for cross-linked styrene-grafted polybutadiene was between 4 and 8, reflecting the greater solubility of polystyrene in MEK. It is obvious from this example that swelling index data must be treated with some caution.

Further analysis of the soluble fraction presents few problems. Standard techniques such as gel permeation chromatography, gradient elution fractionation and turbidimetric titration are suitable for separating any soluble rubber or graft copolymer. Where necessary, infra-red spectroscopy is used to identify the soluble fractions. These techniques are especially useful for analysing prepolymer from the early stages of the polymerisation: the finished polymer rarely contains much soluble rubber.[42] Fischer has shown that the two-phase structure of graft copolymer prevents accurate quantitative analysis by infra-red spectro-scopy, and recommends high-resolution nuclear magnetic resonance as an alternative: styrene-grafted polybutadiene is dissolved in hexachloro-butadiene, and the signals from the phenyl proton of the styrene are compared with those from the olefin proton of the rubber.[43]

Analysis of the insoluble gel fraction is less straightforward. Perhaps the most informative technique is selective degradation of the unsaturated rubber, which renders the grafted side chains soluble, so that they can be examined by standard methods. Several different oxidising agents have been used for this purpose, including perbenzoic acid followed by periodic acid,[44] and osmium tetroxide followed by tertiary butyl hydroperoxide with benzaldehyde.[45] These reagents produce chain scission in the rubber without affecting the molecular weight of the polystyrene side chains:

$$
\begin{array}{c}
\text{C--O} \quad \text{O} \\
\diagdown \quad \diagup\!\!\diagup \\
\text{Os} \xrightarrow{\text{(CH}_3)_3\text{COOH}} \text{OsO}_4 + \underset{\overset{|}{\text{H}}}{\text{C}}{=}\text{O} + \underset{\overset{|}{\text{R}}}{\text{C}}{=}\text{O} \\
\diagup \quad \diagdown\!\!\diagdown \\
\text{C--O} \quad \text{O}
\end{array} \qquad \text{(iii)}
$$

An alternative but less informative method of analysis is thermogravimetry combined with mass spectrometry.[46]

3.7   THE SECONDARY GLASS TRANSITION

Rubber-toughened plastics normally exhibit two glass transitions, a primary transition above room temperature, due to the matrix component of the composite, and a secondary transition below room temperature, due to the disperse rubber phase. The temperature of this secondary transition is not only an important property in itself, but is also a valuable aid in identifying the rubber, and especially in differentiating between copolymers. For example, infra-red spectroscopy of HIPS usually shows that only styrene and butadiene monomer units are present: a secondary transition at $-90\,^\circ C$ clearly indicates that the rubber is polybutadiene, whereas a transition at $-50\,^\circ C$ is characteristic of a 75:25 butadiene–styrene copolymer rubber.

The principal methods for measuring the secondary glass transition are dynamic mechanical testing and differential scanning calorimetry. Dielectric methods and nuclear magnetic resonance also find some application. There are advantages in using more than one method, as they do not yield identical information.

3.7.1 Dynamic mechanical testing

There are several methods for measuring the dynamic mechanical properties of polymers, including torsion pendulum, vibrating reed and cyclic tension tests. Each of these methods consists in measuring modulus and loss tangent over a range of temperatures in small-amplitude oscillations. Some instruments provide for the oscillation frequency to be selected and maintained throughout a test, whilst others are free-oscillation machines, in which the frequency is determined by the modulus of the sample, and therefore varies with temperature.

Figure 3.8 illustrates the use of the method in characterising structure. The toughened polypropylene shows a prominent secondary loss peak between $-50\,^\circ C$ and $-10\,^\circ C$ which is not present in the homopolymer.[29] This additional peak is clear evidence that ethylene–propylene rubber (EPR) is formed when ethylene is introduced to the reactor towards the end of the polymerisation (*see* Section 2.3.3), and that the EPR separates to form a discrete phase.

The area of the secondary loss peak increases with rubber content, and is therefore of use in quantitative analysis. However, it is important to note that the area is dependent upon morphology as well as upon composition, a theme that will be developed in Chapter 5. For the present, it is sufficient to observe that a given quantity of rubber generates a small loss peak if it forms homogeneous particles, but a much larger peak if it occludes rigid polymer. The sub-inclusions illustrated in Fig. 3.2 increase the volume of the rubber particles, and this increase is reflected in the dynamic mechanical properties. The area of the secondary loss peak is a measure of rubber *content* only if all of the materials tested have been made in a similar way. In the more general case, it provides information about rubber *particle volume fraction*.

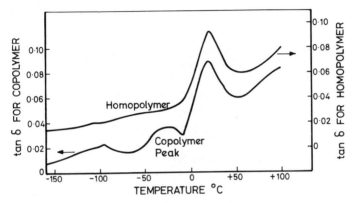

Fig. 3.8   Dynamic loss data for two propylene polymers showing a secondary loss peak in the copolymer at approximately $-25°C$ *due to the presence of a separate poly(ethylene-co-propylene) rubber phase (from T. G. Heggs, ref. 29, reproduced with permission).*

Cross-linking and grafting broaden the loss peak by altering the relaxation behaviour of sections of the rubber chain. Excessive cross-linking also causes the peak to shift to higher temperatures. Large upward shifts are avoided in the manufacture of toughened polymers, but are observed during the subsequent service life of these polymers, as a result of degradation (*see* Section 3.8).

Partial mixing of the rubber and resin phases causes the secondary loss peak to move upwards and the primary loss peak to move downwards in temperature, and as the two polymers approach complete miscibility their loss peaks merge into one. The same effect is also observed when the matrix consists of two partially or completely miscible thermoplastics.

3.7.2 Dielectric relaxation measurements
Dielectric loss measurements are analogous to mechanical loss studies, but are of less general application, since they depend upon the presence of polar

groups in the polymer molecule. In the field of rubber-toughened plastics, their use has been restricted largely to blends of PVC with polar rubbers.

The relaxation behaviour of two-phase polymers is studied by measuring relative permittivity and dielectric loss tangent as functions of frequency over a range of temperatures. The presence of a polar polymer is indicated by a prominent loss peak at the glass transition.

An alternative technique, employed by Hedvig and Foldes,[47] is to characterise the dielectric depolarisation spectrum. The sample is heated above its principal glass transition, and then cooled in a static electric field, thus producing an *electret*. The field is removed, and the sample is heated in a programmed manner through the transition regions of interest. Relaxation of a polar rubber phase at its glass transition causes a partial depolarisation of the electret, which generates a signal that is detected by the instrument. Other relaxations of polar molecules are detected in the same way. Tests on PVC blends showed good agreement between dielectric depolarisation, dielectric loss and dynamic mechanical measurements.[47]

3.7.3 Differential scanning calorimetry

Differential scanning calorimetry (DSC) measures the change in enthalpy of a sample with temperature. The method is widely used to measure glass transition temperatures in polymers, since the specific heat C_p increases sharply over the glass transition region, so that T_g can be defined fairly accurately.

Bair has demonstrated the use of DSC in analysing rubber-toughened plastics.[48-51] These materials show at least two steps in the specific heat curve, at the glass transitions of the two components. Typical traces are presented in Fig. 3.9. The magnitude of the specific heat rise ΔC_p at the secondary glass transition is proportional to the concentration of rubber present, independent of morphology. In this respect, the information provided by DSC is complementary to that obtained from dynamic mechanical tests: calorimetry measures rubber content, whilst mechanical tests measure effective rubber volume.

For example, the rubber content X of a standard ABS polymer is determined by comparing the specific heat change at low temperature with that of pure polybutadiene:

$$X = \frac{\Delta C_p(\text{ABS})}{\Delta C_p(\text{PBD})} \qquad (3.5)$$

The same principle applies to the glass transition of the SAN matrix. Since the temperature coefficients of C_p above and below T_g are usually different,

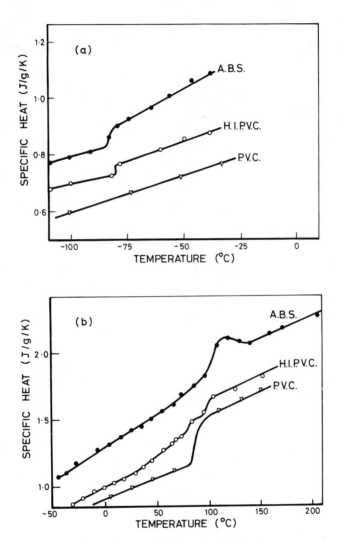

Fig. 3.9 *Differential scanning calorimetry traces for PVC, toughened PVC and ABS showing* (a) *the rise in specific heat at low temperatures due to the glass transition of the rubber phase and* (b) *the rise in specific heat at higher temperatures due to the glass transition of the matrix (from H. E. Bair, ref.* 50, *reproduced with permission).*

ΔC_p must be measured at the same temperature in each case. The accuracy of the method has not been fully evaluated in the literature, but Bair[50] has obtained a value of 15 % for the rubber content of an ABS polymer by DSC, compared with 18 % by chemical analysis. The discrepancy is probably due to errors in both methods.

The thermograms often show endothermic peaks in addition to those due to glass transitions in the polymeric constituents. One possible cause is the melting of crystalline additives such as mould lubricants or fire retardants. Another is the packing effect associated with prolonged annealing of polymeric glasses: additional enthalpy is required at T_g to heat an annealed polymer through the glass transition.[52, 53]

Thermograms obtained by DSC at much higher temperatures reflect the degradation behaviour of the polymeric constituents. Characteristic degradation patterns can be used to identify these constituents.[54]

3.7.4 Nuclear magnetic resonance

Broad-line nuclear magnetic resonance has so far been used relatively little for analysis of rubber-toughened plastics, but initial results suggest that the technique deserves further investigation. The basis of the method is the observation that the magnetic resonance peaks of polymers are broad below T_g but sharp and narrow above T_g. Line broadening occurs in glassy polymers, as in other solids, because of magnetic interactions between neighbouring nuclei. The applied magnetic field is homogeneous, but the local fields experienced by the nuclei are inhomogeneous, owing to the influence of the surrounding 'lattice'. Each proton in the sample precesses at its 'Larmor frequency' in the strong magnetic field, and in a solid the period of precession is short compared with the period of molecular vibration or rotation, so that inhomogeneities in the field are detected as a broadening of the resonance peak. In the liquid or rubbery states, on the other hand, the period of molecular motion is short compared with the nuclear precession period, so that the Larmor frequency of the nucleus is no longer affected by local inhomogeneities in magnetic field due to the lattice. The resulting change in the resonance peak is termed 'motional narrowing'.

The addition of a rubber to a glassy polymer produces a narrow NMR resonance peak, the magnitude of which is directly related to the number of protons in the rubbery state. This principle was employed by Elmquist and Svanson to study phase separation in PVC blends containing ethylene–vinyl acetate copolymer rubber.[55] The system exhibited partial miscibility, becoming further demixed on heating.

An alternative technique is to apply a radio-frequency pulse to the sample and measure the decay of magnetisation. Signal decay is characterised by a spin-lattice relaxation time, which passes through a minimum when molecular motions generate magnetic fields at the Larmor frequency. Minima are observed in the relaxation time curve at T_g and at other transitions in the polymer.

Stefan and Williams used the pulsed NMR method to study a blend containing equal amounts of ABS and polycarbonate.[21] Electron microscopy showed that the polycarbonate formed separate domains within the ABS matrix, so that the blend consisted of three phases: polybutadiene, SAN and polycarbonate. In the NMR experiments, the material exhibited a non-exponential decay of magnetisation, which was resolved into two exponential decays of differing strengths. These were assigned to the constituent polymers, and the concentrations of the polymers were estimated from the signal strengths. Over a wide range of temperatures, the method considerably overestimated the amounts of polycarbonate and polybutadiene present, suggesting that these dispersed polymers modify the relaxation behaviour of the surrounding SAN matrix.

3.7.5 Comparison of techniques

In general, there is good agreement between the methods outlined above.[47, 56, 57] Dynamic mechanical tests and DSC measurements detect the same transitions, at essentially the same temperatures. Dielectric loss and NMR techniques provide supporting evidence relating to those transitions for which they are active. However, conflicting results have been reported for some materials.

Where problems arise, they are usually associated with compatible or nearly compatible systems. The blend of PVC with poly(butadiene-co-acrylonitrile) provides a good example. Dynamic mechanical and other tests indicate that PVC is completely compatible with diene rubber containing 40 wt. % of acrylonitrile,[13, 57] but electron microscopy reveals inhomogeneity on a very fine scale, of the order of 10 nm.[13] The merging of loss peaks suggests a considerable amount of intermixing between the two polymers, but it is doubtful whether the system should be regarded as completely compatible.

Figure 3.10 illustrates a slightly different problem, this time in blends of two thermoplastics. Calorimetry indicates a single T_g for each of a series of blends of polystyrene with poly(phenylene oxide).[58] Dielectric loss tests also show a single T_g for each mixture, but not at the same temperature as the DSC peak. Dielectric loss peaks are broader in the blends than in the pure polymers. Dynamic mechanical tests also show broad loss peaks,

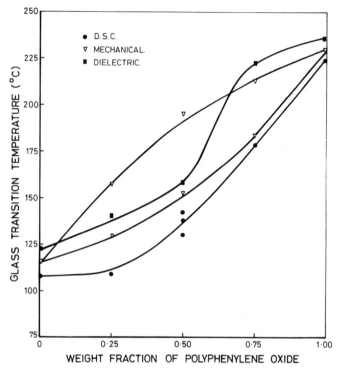

Fig. 3.10   *Glass transition temperatures of binary mixtures of polystyrene with poly(2,6-dimethyl-1,4-phenylene oxide) measured by various techniques (from W. J.ʹ MacKnight et al., ref. 58, reproduced with permission).*

which can be resolved into two components, in contrast to the other measurements. These observations suggest that there is some compositional heterogeneity in the blends on a relatively fine scale.

3.8   CHARACTERISATION OF STRUCTURE IN DEGRADED POLYMERS

Most rubber-toughened plastics are based on unsaturated rubbers, which are very susceptible to oxidation. Antioxidants are added to protect the polymer against thermal oxidation during processing, but this protection is less effective against photo-oxidation during the subsequent service life of the polymer. Exposure to ultraviolet (u.v.) light, either from the sun or from terrestrial sources, usually causes rapid embrittlement, an effect known as *ageing*. The techniques described in this chapter have proved valuable aids in the study of the ageing process.

Priebe and Stabenow[59] used the osmium-staining procedure to follow oxidation in the surface layers of ABS polymers exposed to u.v. radiation. They showed that oxidation rendered the rubber particles resistant to staining, and used this resistance as a means of measuring the thickness of the degraded layer. Their results are presented in Fig. 3.11, which shows that the thickness of the oxidised zone increases linearly with log (exposure time).

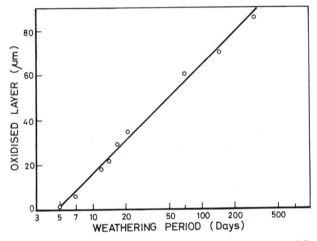

Fig. 3.11   *Thickness of oxidised surface layer as a function of time in ABS exposed to natural sunlight. Depth of oxidation measured by osmium staining (from E. Priebe and J. Stabenow, ref. 59, reproduced with permission).*

Ghaffar and co-workers[60] exposed cast HIPS films to u.v. radiation, and measured a number of properties including dynamic mechanical loss. Their results are illustrated in Fig. 3.12. The secondary loss maximum at $-83\,°C$, due to the glass transition of polybutadiene, disappears over an exposure period of 14 h, as the peak broadens and shifts to higher temperatures. There is a corresponding increase in tan δ at $+20\,°C$. Infra-red spectroscopy shows that these changes are accompanied by the disappearance of *trans*-1,4-polybutadiene groups, and the formation of carbonyl and hydroxyl groups. The relationships between these effects and fracture behaviour are discussed in Chapter 10.

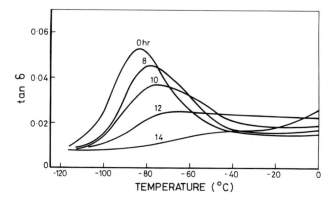

Fig. 3.12 *Ageing of HIPS. Effects upon the secondary loss peak of exposure to ultraviolet light from a sunlamp for periods of up to 14 h (from A. Ghaffar et al., ref. 60, reproduced with permission).*

REFERENCES

1. K. V. C. Rao, *Angew. Makromol. Chem.* **12** (1970) 131.
2. P. A. Traylor, *Anal. Chem.* **33** (1961) 1629.
3. C. B. Bucknall, I. C. Drinkwater and W. E. Keast, *Polymer* **13** (1972) 115.
4. C. B. Bucknall and I. C. Drinkwater, *Polymer* **15** (1974) 254.
5. B. J. Spit, *Polymer* **4** (1963) 109.
6. H. Keskkula and P. A. Traylor, *J. Appl. Polymer Sci.* **11** (1967) 2361.
7. R. J. Williams and R. W. A. Hudson, *Polymer* **8** (1967) 643.
8. J. N. Sultan, R. C. Liable and F. J. McGarry, *Appl. Polymer Symp.* **16** (1971) 127.
9. H. D. Moskowitz and D. T. Turner, *J. Appl. Polymer Sci.* **18** (1974) 143.
10. K. Kato, *J. Electron Micros.* **14** (1965) 220.
11. K. Kato, *Poly. Engng Sci.* **7** (1967) 38.
12. K. Kato, *Koll. Z. Z. Polym.* **220** (1967) 24.
13. M. Matsuo, C. Nozaki and Y. Jyo, *Poly. Engng Sci.* **9** (1969) 197.
14. W. M. Speri and G. R. Patrick, *Poly. Engng Sci.* **15** (1975) 668.
15. A. C. Soldatos and A. S. Burhans, *ACS Adv. Chem. Ser.* **99** (1971) 531.
16. E. H. Andrews, *Proc. Roy. Soc.* **A227** (1964) 562.
17. L. F. Fieser and M. Fieser, *Advanced Organic Chemistry*, Reinhold, New York, 1962.
18. H. B. Henblest, W. R. Jackson and B. C. G. Robb, *J. Chem. Soc. Lond.* B (1966) 803.
19. C. K. Riew and R. W. Smith, *J. Poly. Sci.* A1, **9** (1971) 2739.
20. A. A. Donatelli, D. A. Thomas and L. H. Sperling, *ACS Poly. Prepr.* **14**(2) (1973) 1080.
21. D. Stefan and H. L. Williams, *J. Appl. Polymer Sci.* **18** (1974) 1451.
22. M. Niinomi, T. Katsuta and T. Kotani, *J. Appl. Polymer Sci.* **19** (1975) 2919.
23. G. Kanig, *Prog. Coll. Polym. Sci.*, **57** (1975) 176.
24. G. Kanig, *Koll. Z. Z. Polym.* **251** (1973) 782.
25. G. Kanig, *Kunststoffe* **64** (1974) 470.
26. L. Holliday, *Composite Materials*, Elsevier, Amsterdam, 1966.
27. D. E. James, *Poly. Engng Sci.* **8** (1968) 241.
28. H. Willersin, *Makromol. Chem.* **101** (1967) 296.

29. T. G. Heggs, in *Block Copolymers*, D. C. Allport and W. H. Janes (eds.) Applied Science, London, 1973, p. 493.
30. P. Debye and A. M. Bueche, *J. Appl. Phys.* **20** (1949) 518.
31. R. S. Stein, in *Newer Methods of Polymer Characterisation*, B. Ke (ed.), Interscience, New York, 1964, p. 155.
32. M. Moritani, T. Inoue, M. Motegi, H. Kawai and K. Kato, in *Colloidal and Morphological Behaviour of Block and Graft Copolymers*, G. E. Molau (ed.), Plenum, New York, 1971, p. 33.
33. S. Visconti and R. H. Marchessault, *Macromolecules* **7** (1974) 913.
34. S. Visconti and R. H. Marchessault, *ACS Poly. Prepr.* **15**(2) (1974) 66.
35. B. F. Conaghan and S. L. Rosen, *Poly. Engng Sci.* **12** (1972) 134.
36. C. F. Ryan, *Appl. Polymer Symp.* **15** (1970) 165.
37. L. D. Moore, W. W. Moyer and W. J. Frazer, *Appl. Polymer Symp.* **7** (1968) 67.
38. B. D. Gesner, *J. Poly. Sci.* A3 (1965) 3825.
39. B. D. Gesner, *J. Appl. Polymer Sci.* **11** (1967) 2499.
40. H. Keskkula and S. G. Turley, *Polymer Letters* **7** (1969) 697.
41. D. J. Stein, G. Fahrbach and H. Adler, *Angew. Makromol. Chem.* **38** (1974) 67.
42. B. Chauvel and J. C. Daniel, *ACS Poly. Prepr.* **15**(1) (1974) 329.
43. J. P. Fischer, *Angew. Makromol. Chem.* **33** (1973) 35.
44. J. A. Blanchette and L. E. Nielsen, *J. Poly. Sci.* **20** (1956) 317.
45. P. Hubin-Eschger, *Angew. Makromol. Chem.* **26** (1972) 107.
46. G. J. Mol, *Thermochimica Acta* **10** (1974) 259.
47. P. Hedvig and E. Foldes, *Angew. Makromol. Chem.* **35** (1974) 147.
48. H. E. Bair, *Poly. Engng Sci.* **10** (1970) 247.
49. H. E. Bair, *Poly. Engng Sci.* **14** (1974) 202.
50. H. E. Bair, in *Analytical Chemistry*, Vol. 2, R. S. Porter and J. F. Johnson (eds.), Plenum, New York, 1970, p. 51.
51. H. E. Bair, in *Analytical Chemistry*, Vol. 3, R. S. Porter and J. F. Johnson (eds.), Plenum, New York, 1974, p. 797.
52. S. E. B. Petrie, *J. Poly. Sci.* A2, **10** (1972) 1255.
53. S. E. B. Petrie, *Bull. Am. Phys. Soc.* **17** (1972) 373.
54. A. K. Sircar and T. G. Lamond, *Thermochimica Acta* **7** (1973) 287.
55. C. Elmquist and S. E. Svanson, *Koll. Z. Z. Polym.* **253** (1975) 327.
56. L. Bohn, *Angew. Makromol. Chem.* **29/30** (1973) 25.
57. G. A. Zakrzewski, *Polymer* **14** (1973) 347.
58. W. J. MacKnight, J. Stoelting and F. E. Karasz, *ACS Adv. Chem. Ser.* **99** (1971) 29.
59. E. Priebe and J. Stabenow, *Kunststoffe* **64** (1974) 497.
60. A. Ghaffar, A. Scott and G. Scott, *Eur. Poly. J.* **11** (1975) 271.

CHAPTER 4

MANUFACTURE OF TOUGHENED PLASTICS

There are several different routes to toughened plastics. The most important processes are bulk or bulk-suspension polymerisation, emulsion polymerisation and the Ziegler–Natta process. Significant amounts of toughened plastics are also made by melt blending. Each process has its own technical and economic advantages, which are often specific to the type of material being produced. The bulk and bulk-suspension processes are especially suitable for HIPS, whilst emulsion polymerisation is preferred for ABS, MBS and related products. Ziegler–Natta catalysis has been adapted for the manufacture of toughened polypropylene, and melt blending is used extensively to make toughened PVC. Each of these processes is analysed in this chapter, with particular reference to the relationship between method of manufacture and structure of product.

4.1 BULK AND BULK-SUSPENSION POLYMERISATION OF HIPS

Bulk polymerisation is the standard process for the manufacture of high-impact polystyrene, and is also used to some extent for ABS production. The HIPS process comprises three basic steps:

(i) *Feedstock preparation*—uncross-linked rubber is dissolved in styrene monomer;

(ii) *Prepolymerisation*—the monomer is partially polymerised, with stirring;

(iii) *Finishing cycle*—the polymerisation is completed, with or without stirring.

In the bulk-suspension process, the prepolymer from step (ii) is dispersed as droplets in water, and the finishing cycle consists of a suspension polymerisation. Attempts to make rubber-toughened plastics by direct suspension polymerisation, omitting the prepolymerisation stage, have proved unsuccessful: unless the solution is stirred during the early stages of the reaction, the product is a cross-linked gel. Another distinction of commercial importance is between continuous and batch processes. Most experimental studies have been made on batch polymerisations, which are easier to analyse, but it must be recognised that most industrial manufacturers choose continuous processes for economic reasons.

Rubber-toughened plastics made by the bulk or bulk-suspension processes have a characteristic structure, which is easily recognisable in the electron microscope. The details of this structure, which determine the physical properties of the product, are controlled by chemical changes such as grafting and cross-linking, and physical changes such as phase separation, that take place during the polymerisation. Recent investigations have emphasised the complex way in which these chemical and physical factors interact during the reaction. Rubber, monomers, additives and reactor conditions must all be chosen with care in order to optimise the process.

4.1.1 Outline of process

Ingredients
The following is a typical recipe for toughened polystyrene:

- 92·0 styrene monomer
- 8·0 polybutadiene
- 0·05 benzoyl peroxide (initiator)
- 0·05 dicumyl peroxide (initiator)
- 0·20 tertiary dodecyl mercaptan (chain transfer agent)

Polybutadiene is the preferred rubber for a number of reasons. It is soluble in styrene monomer, participates in grafting and cross-linking reactions, and has a low glass transition temperature. Earlier products were based mainly on poly(butadiene-*co*-styrene), but this rubber is no longer used by most manufacturers because of its higher T_g.

The formulation includes two peroxides. Benzoyl peroxide is the initiator during prepolymerisation, which takes place at a relatively low temperature, whilst the more stable dicumyl peroxide is active during the finishing cycle, which takes place at a higher temperature.

Phase separation

The solution of rubber in styrene is heated to 90 °C in a stirred reactor, and polymerisation begins. Both polystyrene and poly(butadiene-*g*-styrene) are formed, generating 70 kJ of heat for every mole of styrene converted.[1] Stirring is essential to aid the removal of heat and thus control the exotherm.

The first 1–2% of styrene polymer remains in solution with the polybutadiene, but at a conversion of approximately 2% the phases begin to separate, and the solution becomes turbid.[2] The relevant phase diagram is shown in Fig. 4.1. As explained in Section 2.2.2, the miscibility of two polymers is determined not only by their molecular constitution, but also by their molecular weight characteristics, so that a family of curves would be required for a full description of the phase relationships.

As the minor constituent, polystyrene forms the disperse phase, separating as monomer-laden droplets within the rubber solution.[3] The

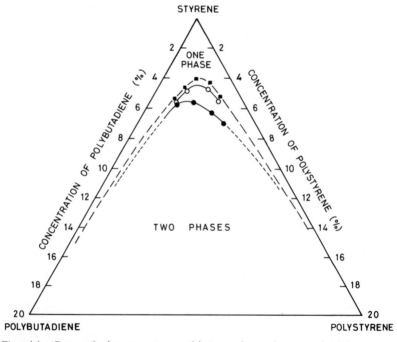

Fig. 4.1 *Part of the ternary equilibrium phase diagram for the system polystyrene–styrene–polybutadiene at 60 °C. Rubber polymerised by:* (●) *Ziegler process, high* cis-*content;* (○) *emulsion process;* (■) *alkyl-lithium process (from J. L. White and R. D. Patel, ref. 2, reproduced with permission).*

graft copolymer concentrates at the phase boundary, where it acts as a surfactant. Molau[4] has given the name 'polymeric oil-in-oil' (POO) emulsions to dispersions of this type, in which two incompatible polymers are emulsified with the aid of a graft or block copolymer. Styrene monomer polymerising in the rubber phase forms new droplets rather than adding to existing droplets, especially at the beginning of the reaction, when the phase ratio is high, and most of the monomer is in the rubber phase.

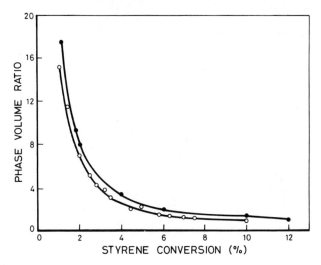

Fig. 4.2   *Variation of phase volume ratio during the polymerisation of styrene containing dissolved polybutadiene:* (○) 6% *rubber solution at* 50°C; (●) 8% *rubber solution at* 70°C *(from G. F. Freeguard and M. Karmarkar, ref.* 5, *reproduced with permission).*

Styrene monomer is distributed between the two phases as required by the partition coefficient: the concentration of styrene in the PBD phase is 1·05 times higher than in the PS phase at a total solids content of 10%.[3] As polystyrene is formed, monomer migrates from the polybutadiene, and the phase volume ratio falls, as shown in Fig. 4.2. However, despite the rapid change in this ratio during the first 10% conversion of monomer, the *concentration* of both polymer solutions changes by only 10%, as a result of styrene depletion. Consequently, a polystyrene droplet formed during the early stages of polymerisation shows little change in size or volume due to monomer migration.

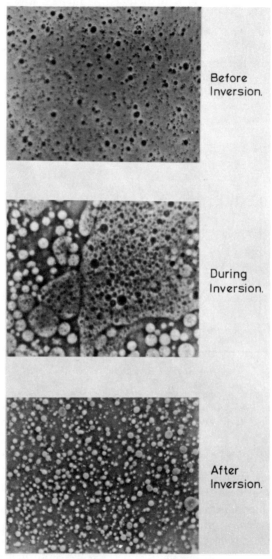

Fig. 4.3 *Stages in the prepolymerisation of HIPS. Positive phase contrast micrographs showing: (top) dispersion of polystyrene droplets in polybutadiene solution; (centre) phase inversion; (bottom) dispersion of polybutadiene droplets in polystyrene solution (from G. E. Molau and H. Keskkula,* J. Poly. Sci. *A1, **4** (1966) 1595, reproduced with permission).*

Phase inversion

Continued polymerisation accompanied by stirring produces a *phase inversion*. Polystyrene becomes the continuous phase, and polybutadiene the disperse phase. The sequence of events is illustrated in Fig. 4.3. As often happens near a phase inversion point, a multiple emulsion[6] is formed: the newly formed polybutadiene particles themselves contain sub-inclusions of polystyrene. In other words, some of the structure produced during the early stages of reaction is retained after phase inversion. In a sense, the phase inversion is incomplete. Occasionally, more complex multiple emulsions are observed, with polystyrene sub-inclusions containing tiny rubber particles that in turn contain polystyrene sub-inclusions.

Phase inversion takes place when the volume of the disperse phase is comparable with that of the continuous phase. Usually, the volume fraction of the disperse phase exceeds 0·5 before inversion. In the case of the polystyrene–polybutadiene–styrene system, this means that the conversion of polystyrene in the prepolymer must exceed the concentration of the rubber. Published data show phase inversions at between 9 and 12 % styrene conversion in solutions containing 8 % rubber.[7]

Within the molecular weight range used in commercial practice, the polybutadiene solution has a higher viscosity than the polystyrene solution with which it is in equilibrium. Consequently, there is a drop in viscosity when the emulsion goes through a phase inversion, as illustrated in Fig. 4.4.

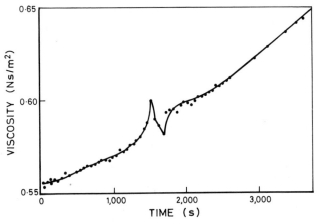

Fig. 4.4   Viscosity changes in HIPS prepolymer during prepolymerisation at 70 °C in a cone-and-plate viscometer showing discontinuity at phase inversion. Shear rate 130 s⁻¹ (from G. F. Freeguard and M. Karmarkar, ref. 5, reproduced with permission).

Viscosity measurements show that phase inversion takes place over a relatively narrow range of conversions. Molau and co-workers[8] simulated phase inversion due to polymerisation by mixing styrene solutions of polystyrene and polybutadiene at appropriate concentrations, using a small amount of styrene–butadiene block copolymer as an emulsifying agent. The relationship between viscosity and polystyrene concentration in the mixture was similar to that shown in Fig. 4.4.

Phase inversion is reversible. Re-inversion occurs if rubber solution is added to a system that has already undergone phase inversion, indicating that the phenomenon is primarily concentration-dependent.

Post-inversion stage
Polymerisation with stirring is continued until the conversion of styrene reaches approximately 30 %. At this stage, the viscosity is sufficiently high to prevent further changes in the basic structure of the complex POO emulsion, and the prepolymerisation can be considered complete. The standard practice is then to transfer the prepolymer to a second reactor in order to complete the polymerisation. The choice is essentially between unstirred bulk polymerisation and a suspension polymerisation. Details of these finishing stages are given in Section 4.1.3.

4.1.2 Factors controlling structure
The structure of a toughened polymer made by the bulk or bulk-suspension processes is determined by the conditions of manufacture and by the choice of ingredients. The size, shape and internal structure of the rubber particles, the molecular weight of the matrix and the type of grafting can all be controlled within wide limits by the manufacturer. The principal factors affecting structure are discussed in this section.

Stirrer speed during prepolymerisation
Stirring conditions are of primary importance during prepolymerisation, since phase inversion will not take place if the shear rate is too low. Freeguard and Karmarkar[5] studied phase inversion in the system styrene–polystyrene–polybutadiene by conducting the polymerisation in a cone-and-plate viscometer, which has the advantage that shear rates are uniform throughout the sample. They found that phase inversion occurred only if the shear rate was above a minimum level, which varied with rubber content as shown in Fig. 4.5. With increasing rubber content, the critical shear rate fell, but inversion took place at higher conversions because of the

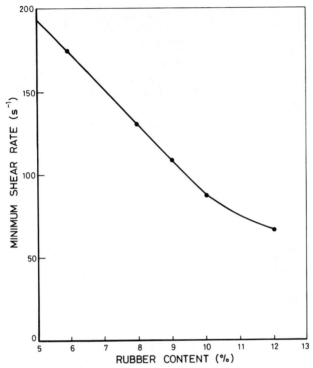

Fig. 4.5   Minimum shear rate for phase inversion in HIPS prepolymer as a function of polybutadiene content (from G. F. Freeguard and M. Karmarkar, ref. 5, reproduced with permission).

requirement that the volume fraction of polystyrene phase be greater than 0·5. As a result, the critical shear *stress* rose with rubber content.

Resistance to phase inversion comes partly from the stabilising influence of the graft copolymer, but is due mainly to the high viscosity of the solutions, especially at high rubber contents. If the shear rate is too low, the system becomes stabilised in the uninverted state, and the finished polymer is a dispersion of polystyrene in a matrix of rubber rather than a dispersion of rubber in a matrix of polystyrene. The structure produced in an unstirred polymerisation is illustrated in Fig. 4.6.

In any practical prepolymerisation reactor, shear rates vary from one point to another, and it is essential to design both reactor and stirrer with care, in order to avoid 'dead spots' in which the shear rates are below the critical value for phase inversion. Early HIPS polymers contained lumps of

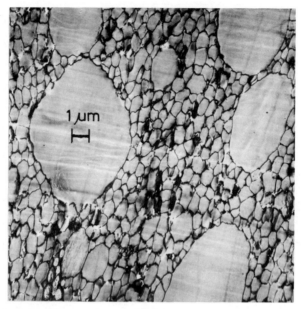

Fig. 4.6   *Osmium-stained section showing rubber network structure produced by polymerising a solution of polybutadiene in styrene without stirring (from G. F. Freeguard,* Polymer **13** (1972) 366, *reproduced with permission).*

uninverted polymer produced in such 'dead spots'. The resulting problems of poor flow characteristics and irregularities on moulded or extruded surfaces have now been eliminated almost completely by improvements in polymerisation technology.

The final size and composition of the rubber particles is determined by the shear rate during the period immediately following phase inversion, before the viscosity is too high to allow further change. Wagner and Robeson[9] showed that increased agitation produced smaller rubber particles containing less occluded polystyrene, and concluded that the shearing action of the stirrer allowed the trapped polystyrene to join the continuous phase, by rupturing the rubber membranes. This process is essentially an extension of the phase inversion, which thereby becomes more complete. Freeguard and Karmarkar[10] found that high average shear rates, well in excess of those required for phase inversion, resulted in small rubber particles of a very distinctive type, consisting of a single small droplet of polystyrene enclosed within a skin of rubber. Larger, more complex rubber particles were formed at lower shear rates, as illustrated in Fig. 3.2. They suggested that the larger particles were formed by the aggregation of the smaller simple particles.

Shear during the finishing cycle
Shearing agitation is usually confined to the first part of the polymerisation reaction, up to a conversion of 30–40%. Ordinary paddle stirrers are ineffective at higher conversions. However, it is possible to continue stirring during the finishing cycle by employing sufficiently powerful agitators. This procedure produces rubber particles of distinctive structure, in which the polystyrene sub-inclusions have distorted shapes, and the rubber particles themselves have irregular shapes, showing evidence of having been torn and broken at a late stage of the polymerisation reaction.

Grafting
The extent of grafting during prepolymerisation is an important factor affecting the structure of the finished polymer. Since the graft copolymer acts as a polymeric emulsifying agent, particle sizes are directly related to graft concentration. Before phase inversion, the size of the polystyrene droplets decreases with increased grafting;[3] after phase inversion, the size of the rubber particles decreases with increased grafting.

The chemical factors affecting grafting were discussed in Chapter 2: the type of rubber, type of initiator and temperature of polymerisation all influence the course of the reaction. Physical control of graft concentration is also possible. For example, polystyrene homopolymer can be added to the feedstock, so that phase inversion takes place at a lower graft level, and larger rubber particles with larger sub-inclusions are formed.[11] The opposite effect is achieved by adding a styrene–butadiene block copolymer to the formulation in place of some or all of the polybutadiene.

Grafting affects not only the morphology but also the chemical structure of the rubber, and hence its relaxation characteristics. Long grafted side-chains appear to have relatively little effect, but the very short chains formed towards the end of the polymerisation are miscible with the rubber, and may raise T_g significantly.

Molecular weight of rubber
The molecular weight of the rubber controls the viscosity of the rubber phase throughout the prepolymerisation. One result is to delay phase inversion if the viscosity is particularly high, but in most cases there is no net effect upon the structure of the finished polymer. In exceptional cases, the

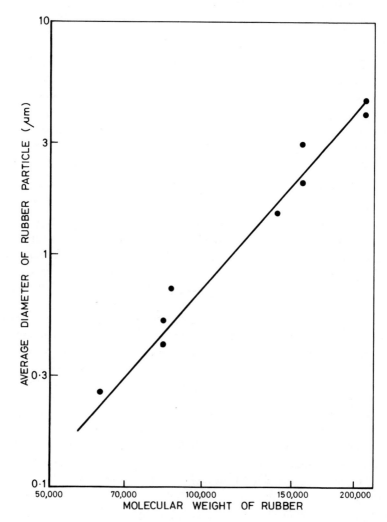

Fig. 4.7 Relationship between number average molecular weight $\overline{M}_n$ of polybutadiene and rubber particle size in HIPS (from M. Baer, ref. 11, reproduced with permission).

critical shear rate may be raised sufficiently to prevent phase inversion at normal stirrer speeds.

A more significant effect of molecular weight is illustrated in Fig. 4.7: under given conditions of prepolymerisation, there is a linear relationship between log (average rubber particle size) and log ($\overline{M}_n$ of rubber).[7, 11] As the solution viscosity of the rubber increases, it becomes more difficult to break the rubber phase down into small droplets.

The molecular weight of the polystyrene is equally important, since it is the ratio of the viscosities of the two solutions that determines the rubber particle size. However, the molecular weight of the matrix polymer cannot be regarded as a control variable; it is chosen on the basis of the melt-flow and fracture properties required of the finished product, and is best regarded as a fixed quantity, against which the molecular weight of the rubber is varied.

Rubber content
The viscosity of the rubber solution also increases with rubber concentration, and with similar results to those described above. The critical shear stress rises, and at rubber concentrations above about 15% the prepolymer is usually so viscous at the expected point of phase inversion that it will not invert. For this reason, few HIPS or ABS polymers made by the bulk process contain more than 12% of polybutadiene. This is a major limitation of the process.

Prepolymers containing less than 12% of polybutadiene will invert under normal stirring conditions, the point of inversion being determined by the rubber content. Higher rubber contents mean higher conversions of styrene at phase inversion, and therefore higher viscosities in the critical post-inversion stage of polymerisation. Because of the increased viscosity, rubber particle sizes tend to increase with rubber content.[12] In the limit, the system completely fails to invert, as already discussed.

Temperature programme
The polymerisation temperature, together with the concentrations of initiator and chain transfer agent, determines the molecular weight of the polystyrene being produced and the course of the grafting reaction. These factors have a considerable influence over the morphology and properties of the finished polymer.

During prepolymerisation, the most important effect of raising temperature is to increase the degree of grafting. This causes a reduction in particle size.[12] The degree of grafting is increased by peroxide initiators, and

decreased by chain transfer agents.[13] Another effect of raising temperature
is to decrease the polystyrene chain length, which is directly related to the
kinetic chain length v^*, defined as the mean number of monomer molecules
polymerised as a result of each initiation step, and given by

$$v^* = \frac{k_p[M]}{(2k_t k_d [I_2])^{1/2}} \tag{4.1}$$

where the terms have the meanings noted in Section 2.3.2.

The activation energy for decomposition of the initiator is greater than
those for propagation and termination, so that the principal result of raising
the temperature is to increase k_d. Shorter chains are produced as the
temperature and rate of initiation increase. The resulting reduction in the
viscosity of the polystyrene solution tends to produce an increase in rubber
particle size.

Very short polystyrene chains, including dimers, trimers and other
oligomers, are produced during the final stages of polymerisation, as the
monomer becomes depleted. In an initiated polymerisation, this tendency is
partly compensated for by depletion of the initiator. The kinetics are
complicated by chain transfer reactions and other effects, and are not easily
analysed, especially under the non-isothermal conditions prevailing in most
commercial bulk polymerisations, which often reach temperatures well over
200 °C. Nevertheless, the trend is clear: very low molecular weight products,
including polystyrene oligomers and short grafted chains, are formed at the
high temperatures necessary for completion of a bulk polymerisation.

Monomer and solvent stripping
The formation of low molecular weight polymer during the final stages of
bulk polymerisation is avoided in some commercial processes by removing
monomer under reduced pressure. An inert solvent such as ethylbenzene is
often added to the monomer feedstock at a concentration of 5–25 %, and
removed with the monomer at the end of the reaction.[14] The solvent lowers
the viscosity of the polymerising mass throughout the process, and thus
reduces the problems of heat transfer and mass transport. Both monomer
and solvent are recycled.

Removal of the last 5 % of monomer in this way significantly reduces the
extent of cross-linking in the rubber, and also avoids the formation of short
grafted chains that would otherwise be produced. Excessive cross-linking
and grafting by short polystyrene side-chains are to be avoided, since they
raise the T_g of the rubber.

By removing much larger amounts of monomer, it is possible to produce a toughened polystyrene with a very high polybutadiene content. This technique circumvents the problem of phase inversion at high rubber concentrations. A normal feedstock is used, and the monomer is evacuated after phase inversion has taken place. The polymerisation must then be allowed to continue until the rubber is sufficiently cross-linked to prevent disintegration during moulding. Some commercial manufacturing processes for polystyrene appear to be based on this principle of concentrating the rubber by monomer stripping.

4.1.3 Commercial processes

Both batch and continuous processes are used in the commercial production of HIPS and ABS polymers, but the preference is for continuous operation because of lower running costs. It is therefore necessary to add some comments on continuous operation on a large scale to the foregoing description of the bulk and bulk-suspension processes.

Prepolymerisation

Figure 4.8 shows a prepolymerisation kettle used in the manufacture of HIPS by the continuous process. The styrene conversion is maintained constant at approximately 30 % in the kettle, which supplies a continuous

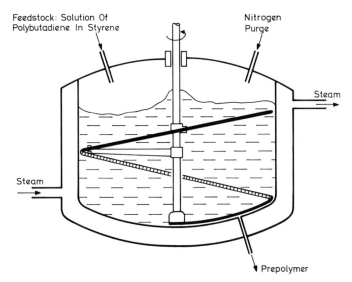

Fig. 4.8 *Kettle reactor equipped with helical stirrer, suitable for prepolymerisation of styrene polymers to* 30–40 % *conversion.*

stream of prepolymer to the second-stage reactor. The level of reactants in the kettle is maintained by a feedstock consisting of styrene monomer with dissolved rubber, initiators, chain transfer agents, solvents and other necessary ingredients.

In bringing a kettle into operation, it is usual to begin with a solution of rubber in monomer, and to go through the phase inversion and post-inversion steps as for a batch polymerisation. Continuous operation can then start. A slow stream of rubber solution is fed to the agitated kettle, where it undergoes immediate phase inversion. The structure of the finished polymer shows that multiple emulsions are formed at phase inversion, as in batch polymerisation; in both cases, the rubber particles contain a high volume fraction of polystyrene sub-inclusions.

Operating conditions vary. Prepolymerisation temperatures are usually between 80 °C and 120 °C, depending upon the choice of initiator. Purely thermal polymerisation, in the absence of initiators requires relatively high temperatures. Average residence times are of the order of 4 h, but some material spends considerably longer in the kettle. If the volume of the kettle is V, and the flow rate Q, the average residence time is V/Q. Assuming good mixing, the probability p that any given particle remains in the kettle for a period t is given by[1]

$$p = \frac{Qt}{V}\exp\left(-\frac{Qt}{V}\right) \qquad (4.2)$$

The resulting spread in residence times is a principal disadvantage of the continuous process, since it causes a broadening of distributions of molecular weight, particle size, grafting and other characteristics of the prepolymer.

Finishing cycle
Both bulk and suspension finishing processes can be operated continuously. Figure 4.9 shows a continuous bulk polymerisation process based upon an unstirred 'tower' reactor, in which the temperature rises from 100 °C to 200 °C as the reacting mixture of monomer and polymer passes downwards under gravity. The increasing temperature compensates for the depletion of monomer, so that a satisfactory rate of conversion is achieved. Average residence times are of the order of 10 h. Problems sometimes arise from material that is delayed for much longer periods in the tower, either in contact with the walls, or trapped on heat-exchanger coils. The resulting cross-linked polymer produces unsightly 'hard spots' on mouldings or extrusions. Regular cleaning reduces the problem.

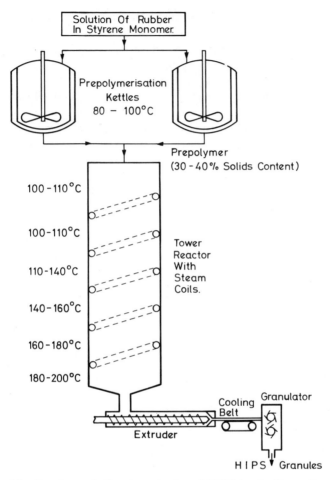

Fig. 4.9   *Continuous bulk polymerisation of HIPS by the Tower Process.*

Heat exchangers help to control the highly exothermic polymerisation reaction, and heat transfer is further improved by the presence of an inert solvent and by agitation. These modifications also reduce channelling within the reactor.

Various methods are employed to remove monomer and solvent at the end of the reaction. In the stationary devolatiliser, the reaction mixture is extruded vertically downwards from a multi-orifice die into an evacuated chamber, and the polymeric melt is then extruded from the bottom of the devolatiliser. Another process uses a two-roll mill with threaded rolls, which introduce considerable shear deformation into the melt, as well as aiding the vacuum extraction of monomer and solvent. A third method is to employ a vented extruder to produce the sheet or rod that is chopped into pellets ready for shipment at the end of the production process.

The suspension process avoids the problems of channelling and heat transfer. The prepolymer is dispersed in water, with stirring, in the form of droplets between 0·1 and 1·0 mm in diameter. Water to monomer ratios are usually between 1:1 and 3:1. The suspension is stabilised by the addition of a water-soluble organic polymer such as polyvinyl alcohol, or a water-insoluble inorganic compound such as magnesium silicate, to prevent agglomeration of the suspended droplets. Each droplet is effectively an unstirred bulk polymerisation vessel. Heat exchange with the surrounding water provides excellent temperature control. During the finishing cycle, temperatures in the autoclave are usually raised to about 140 °C, and the polymerisation is completed with the aid of high-temperature initiators such as tertiary butyl hydroperoxide or dicumyl peroxide. Conditions during the finishing cycle are particularly important in determining the degree of cross-linking of the rubber, which begins at a conversion of about 80 %. Suspension polymerisation is often operated as a batch process, but can be made continuous by cascading the suspension down a flooded tower. At the end of the reaction, water is separated from the polymeric beads by means of a settling tank, continuous centrifuge and, finally, a rotary dryer.

4.2   MANUFACTURE OF TOUGHENED EPOXY RESINS

The manufacture of toughened epoxy resins, which is of necessity a bulk polymerisation process, is in many ways similar to the bulk process for producing HIPS. Both processes begin with a solution of uncross-linked rubber in a low molecular weight liquid, which is then polymerised. During this reaction, the rubber is precipitated as a disperse phase, and chemical bonds are formed between rubber and matrix. There are, however, several important differences. In particular, the chemistry of a thermosetting resin system is under the control of the moulder or component manufacturer rather than the raw materials supplier. It is therefore necessary to define curing procedures precisely, in order to obtain optimum properties. There is great scope for variation both in composition and in curing schedules, and the factors affecting structure and properties are not as well understood as in the case of HIPS. As the technology of toughening thermosets is relatively new, considerable improvements are to be expected as a result of further investigation.

A typical formulation is as follows:

  100 DGEBA epoxy resin
  10 CTBN rubber
  5 piperidine

The rubber is first dissolved in the resin at 50–80 °C, and the solution is subjected to vacuum to remove trapped air bubbles. The hardener is then dissolved with stirring, and the resin–rubber–hardener mixture is cast. A second application of vacuum removes air bubbles entrapped at this stage, but care is necessary with volatile hardeners such as piperidine. After precuring at 100 °C for 2 h, the resin is cured at 120 °C for 16 h.

The formulation and procedure can be varied in a number of ways. The type and molecular weight of resin, the acrylonitrile content and molecular weight of the rubber, and the type and concentration of hardener are all open to choice, as is the precise curing schedule. Published studies show that most of these variables affect mechanical properties, especially impact strength.[15–17]

Initially, the rubber is completely soluble in the resin. Phase separation takes place as a result of molecular weight increases in both the resin and the rubber. Molecular weight distributions are rather different from those obtaining during the early stages of styrene polymerisation: condensation polymerisation produces a broad distribution of low molecular weight species, whereas the free-radical reaction produces polystyrene of high molecular weight, which is sharply differentiated from styrene monomer. Consequently, the phase separation follows a different path: the rubber precipitates directly from the epoxy resin as a discrete phase, rather than forming the continuous phase first and then going through a phase inversion. There is never a stage at which epoxy resin of high molecular weight constitutes a minor component of the mixture and precipitates as a disperse phase in the rubber. The reactants are best considered as comprising only two species, rubber and epoxy resin, the shorter chains of both species being partitioned between the phases.

The rubber particles formed during the cure of epoxy resins are usually perfectly spherical, as illustrated in Fig. 4.10, indicating that phase separation occurred whilst the system was still fairly fluid. Larger particles sometimes contain resin sub-inclusions, but many particles, especially the smaller ones, appear to be homogeneous. This does not mean that the rubber contains no epoxy resin, only that the resin is very well dispersed,

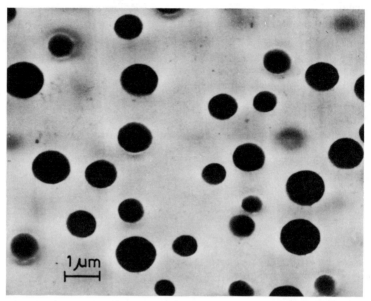

Fig. 4.10 *Osmium-stained section of toughened epoxy resin containing* 8·7 % *CTBN rubber (from T. Yoshii, ref. 18, reproduced with permission).*

perhaps on the molecular level, as a result of reactions with the carboxyl-terminated butadiene–acrylonitrile copolymer.

Phase separation is the key to success in making a rubber-toughened epoxy resin. The separation predicted on thermodynamic grounds is often incomplete, and might not occur at all in some systems. The most important factor is the difference in solubility parameter between the rubber and the resin. When the difference is too small, phase separation tends to be incomplete; on the other hand, when the difference is too large, the rubber will not disperse in the resin to form an homogeneous solution as a starting point for the reaction, and the results are equally poor. The preferred difference in δ is about 0·5 $(cal/cm^3)^{1\,2}$

Similar considerations apply to the molecular weights of the rubber and the resin used for the process. If they are too high, it is difficult to dissolve the rubber in the resin. If, on the other hand, they are too low, phase separation will not take place, and the result will be a plasticised epoxy resin. A rubber of intermediate molecular weight, preferably over 3800, gives adequate phase separation and consequently good fracture resistance.[17] The molecular weight of the resin is less critical, but nevertheless has some influence over the phase separation.

The choice of hardener is also important. It is difficult to achieve phase equilibrium in a highly reactive system of rapidly increasing viscosity, especially in the absence of stirring. The less reactive hardeners such as piperidine therefore give better phase separation than the more reactive hardeners such as triethylene tetramine.[16,18] However, when phase separation does occur, the more reactive systems show a greater tendency to form sub-inclusions within the rubber, thus increasing the volume fraction of rubber particles.

Riew *et al.* showed that the addition of Bisphenol A to the resin formulation increased the toughening effect of the rubber.[19] Yoshii[18] has shown that this is also a phase separation effect, which can be explained by a condensation reaction between the Bisphenol and epoxy resin. In the presence of a base, oligomers of DGEBA are produced in the following manner:[20]

$$Ar-OH + CH_2\overset{O}{\overset{/\backslash}{-}}CH \sim \; \rightarrow \; Ar-O-CH_2-\overset{OH}{\overset{|}{C}}H \sim \tag{i}$$

This increase in molecular weight in the resin promotes separation of the rubber phase. The preferred concentration of Bisphenol A is 24 pph; quantities in excess of this tend to remain unreacted and therefore to impair properties.

Slowly curing resin systems tend to form large rubber particles between 1 and 10 μm in diameter. In some cases, smaller particles less than 0·5 μm in diameter are also formed, giving a bimodal size distribution. The smaller particles are probably produced by phase separation at a later stage of the reaction, when molecular mobility is reduced. Bimodal distributions of particle size appear to be associated with a wide distribution of resin molecular weight during the early stages of curing, and have been observed particularly in formulations containing Bisphenol A[19] and in materials made from mixtures of high and low molecular weight epoxy resins.[18] Rapidly curing resins tend to form small rubber particles if they do undergo phase separation. Again the small size can be related to reduced molecular mobility at the time of separation.

Owing to the problems of phase separation, the volume fraction of rubber particles produced from a given amount of added rubber varies very considerably. For example, 10 wt. % of CTBN rubber forms between 0 and 25 vol. % of particles, depending upon conditions and formulation. In general, those factors that favour phase separation give rise to large rubber particles, often in a bimodal distribution, whereas the factors that inhibit phase separation tend at the same time to produce small rubber particles. In discussing structure–property relationships, it is important to distinguish the effects of particle size and size distribution from effects due simply to the volume fraction of rubber.

A disperse rubber phase is obtained only when the rubber content is below 20 wt. % (25 vol. %).[15, 21, 22] Above this concentration, the rubber forms the matrix, and the resin separates as spherical particles. As the rubber content approaches the critical level, the proportion of epoxy resin sub-inclusions within the particles rises, so that the volume fraction of rubber particles tends towards 50 %.

There is some evidence that the rubber particles remain uncross-linked throughout a standard curing reaction. Yoshii has shown with the aid of scanning electron microscopy that CTBN rubber particles are etched by, and therefore soluble in, methanol.[18] The reason for this behaviour is not known. One possibility is that partition of the curing agent leaves the rubber deficient in piperidine, so that the cross-linking reaction of the epoxy resin is hindered.

Cycloaliphatic epoxy resins can be toughened by adding rubber in a similar way to that described above.[15, 23] The formulation can also be modified by replacing the CTBN polymer with another rubber. Amongst the butadiene–acrylonitrile copolymers, rubbers containing terminal carboxyl groups (—COOH) give the highest fracture resistance,[19] followed in decreasing order of toughening effect by phenol- (—C_6H_5OH), epoxy-

$$\overset{O}{\underset{\diagup\diagdown}{}}$$

(—CH—CH$_2$) and hydroxyl- (−CH$_2$OH) terminated rubbers. The toughening effect of mercaptan-terminated rubbers is very small. This list places the rubbers in order of increasing reactivity and decreasing selectivity in reacting with the epoxy resin,[19] a result that again emphasises the importance of the phase separation. The most reactive rubber, containing terminal mercaptan groups (—CH$_2$SH), gives a transparent polymer with little sign of phase separation between resin and rubber. The same considerations apply to other rubbers: carboxyl-terminated polycaprolactone and polypropylene oxide rubbers also toughen epoxy resins, provided the molecular weight of the rubber is high enough to ensure phase separation.[23] An alternative procedure, which avoids the problem of phase separation, is to add an ABS polymer to the resin.[16] Satisfactory toughening can be achieved by using an ABS masterbatch polymer containing a high concentration of polybutadiene. The rubber is already present as a discrete phase of controlled particle size, and has the advantage of a lower T_g than CTBN rubbers.

4.3 TOUGHENED POLYPROPYLENE

The principal method of manufacturing toughened polypropylene is block copolymerisation using Ziegler–Natta catalysts. The method consists of modifying the standard process for the manufacture of polypropylene homopolymer by adding a second monomer, usually ethylene, during the final stages of polymerisation, with the aim of producing rubbery chains that are attached chemically to the existing polypropylene chains. The reaction is discussed in Section 2.3.3. The relevant literature, especially the patent literature, is reviewed by Heggs.[24]

The Ziegler–Natta process is based on a family of catalysts consisting of salts of metals in Groups IV–VIII of the Periodic Table, in combination with alkyls of metals in Groups I–III, which catalyse the stereoregular polymerisation of α-olefines, including propylene. The preferred catalyst combination for the manufacture of propylene block copolymers is γ-TiCl$_3$ with Al(C$_2$H$_5$)$_2$Cl, but other combinations are used.

Typically, the reaction takes place in a liquid suspension medium such as heptane. The pure, dry liquid is first purged of oxygen, and a heptane solution of Al(C$_2$H$_5$)$_2$Cl is added, followed by a slurry of TiCl$_3$ in heptane. The reactor is then heated to 50 °C, and propylene is fed into it for a period of 2–3 h. In the presence of the catalyst, the monomer polymerises to form a stereoregular polymer which is insoluble in the suspension medium. Stirring brings monomer and catalyst into contact, and also aids the removal of heat produced in the highly exothermic reaction.

During the final stages of the process, ethylene is introduced into the feedstock, either alone or in combination with propylene, to form a block copolymer. This process for block copolymerisation by sequential addition of monomers depends upon the lifetimes of the chains formed during the first stage of polymerisation.[25] There is evidence for chain lifetimes of 70 min,[26] so that at least some block copolymer is to be expected. In some processes, however, the lifetime of the polypropylene chain is reduced by a factor of 2–4, owing to the addition of a chain transfer agent, which in almost all cases is hydrogen. Writing R for the polymer and M for the monomer, the reaction can be represented as follows:

$$\text{Ti—R} + \text{H}_2 \rightarrow \text{Ti—H} + \text{RH} \qquad \text{(ii)}$$

$$\text{Ti—H} + n\text{M} \rightarrow \text{Ti—(M)}_n\text{H} \qquad \text{(iii)}$$

Chain transfer agents are necessary ingredients, since they control molecular weight, which would otherwise be too high for satisfactory injection moulding.

The structure of the block copolymer is determined principally by the composition of the available monomer supply during the second stage of the polymerisation. A pure ethylene block is produced by removing all propylene monomer from the reactor at the end of the first stage, and then adding pure ethylene monomer. Propylene may be removed by purging,

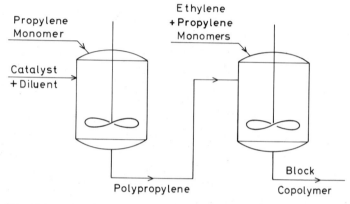

Fig. 4.11 *Schematic diagram of plant for the manufacture of toughened polypropylene 'copolymer' (from T. G. Heggs, ref.* 24, *reproduced with permission).*

evacuation or polymerisation, but complete removal is difficult to achieve because propylene is very soluble in suspension media such as heptane. Where removal is complete, the product is poly(propylene-*b*-ethylene), probably with an admixture of polypropylene and polyethylene homopolymers. The use of hydrogen as a chain transfer agent increases the tendency to form homopolymers.

In a batch process, the addition of ethylene to a reactor containing residual propylene monomer leads to the formation of a tapered polymer, poly(propylene-*b*-(ethylene-*t*-propylene)), as explained in Section 2.3.3, because the ethylene polymerises rapidly and becomes progressively depleted, so that the propylene:ethylene ratio increases with time. The most effective method for controlling this monomer ratio during the second stage of the reaction is to employ a continuous polymerisation technique.

Figure 4.11 is a schematic diagram of the continuous polymerisation process. Polypropylene is formed in the first-stage reactor, and transferred to the second-stage reactor, which is fed by a mixture of ethylene and propylene monomers. The monomer ratio in the second-stage reactor is fixed by the composition of the feedstock, so that a random copolymer of constant composition is formed. The copolymer adds to the active polypropylene chains generated in the first-stage reactor, thus producing poly(propylene-*b*-(ethylene-*co*-propylene)). The polypropylene sections of the chains have a broad distribution of lengths owing to the wide range of

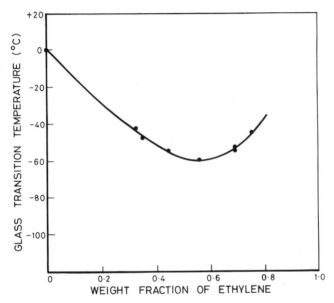

Fig. 4.12 Glass transition temperature of ethylene–propylene copolymers: relationship between composition and T_g, as indicated by maximum in the β loss peak (from C. A. F. Tuijnman, ref. 28, reproduced with permission).

residence times of the suspended polypropylene chains in the first-stage reactor. Batch polymerisation produces a narrower distribution of chain lengths.

The most satisfactory balance of properties is obtained when the second-stage polymerisation is programmed to produce an elastomeric block with a low T_g. Pure polyethylene blocks form crystalline domains, which are much less effective than rubber particles in toughening polypropylene.[27] In a random copolymer of ethylene with propylene, crystallisation is inhibited, and the material is rubbery. Figure 4.12 shows the relationship between T_g and copolymer composition. There is a minimum of $-61\,°C$ in T_g at 57 wt. % (66 mole %) of ethylene,[28] which is the preferred composition for the second segment of the block copolymer. Experiments on toughened polypropylene show that the highest impact strengths are obtained when the second block segment of the chain contains approximately 60 wt. % of ethylene.[27]

There is little published information concerning the structure and morphology of toughened polypropylene. Since the block copolymer is uncross-linked, its particle size and shape are not fixed during polymerisation, as they are in the HIPS and ABS processes. It must therefore be assumed that particle sizes change during processing, as a result of aggregation or splitting processes taking place under shear. These processes are discussed further in Section 4.5, which deals with blends. On cooling, the polypropylene crystallises around the rubber particles, which can be seen under the optical microscope as small spheres embedded in the spherulites. Scanning electron microscope studies of fracture surfaces also indicate that the rubber particles are perfectly spherical in slowly cooled, isotropic toughened polypropylene.

4.4 EMULSION POLYMERISATION—ABS MANUFACTURE

Emulsion polymerisation is the standard method of manufacturing ABS polymers and a number of related materials, including AMBS (acrylonitrile–methacrylate–butadiene–styrene), MBS (meth-acrylate–butadiene–styrene) and toughened PMMA. The process comprises two basic stages, as follows:

(i) a rubber latex is prepared by emulsion polymerisation;
(ii) monomers (*e.g.* styrene and acrylonitrile) are charged to the reactor and polymerised in the presence of the latex.

The patent literature concerning ABS manufacture, especially by the emulsion process, is comprehensively reviewed by Placek.[29] Since routes to other toughened polymers differ only in detail from those used to make ABS polymers, the following discussion will be restricted to the ABS process, in the interests of clarity.

4.4.1 Outline of process

Preparation of rubber latex
A typical recipe for the production of the rubber latex is as follows:[29]

    200   water
    100   butadiene monomer

2   divinyl benzene
5   sodium stearate (soap)
0·4 tert-dodecyl mercaptan (chain transfer agent)
0·3 potassium persulphate (initiator)

The water and soap are first charged to a pressure vessel fitted with an anchor stirrer, boiled to remove dissolved oxygen and cooled under nitrogen. The remaining ingredients are added, the reactor is sealed and the temperature is raised to 50 °C. Polymerisation takes place with stirring over a period of 48 h, at the end of which the unreacted butadiene monomer, amounting to approximately 5% of the original monomer charge, is removed as vapour. The polybutadiene latex is then ready for the grafting stage.

Grafting stage
In the second stage of the ABS process, styrene and acrylonitrile monomers are polymerised in the presence of the polybutadiene latex, to form a grafted polymer. This stage of the process requires additional water, soap, initiator and chain transfer agent. The following is a typical recipe:[29]

300   water (including water from rubber latex)
20   polybutadiene (in latex form)
62   styrene monomer
18   acrylonitrile monomer
0·1 tert-dodecyl mercaptan
0·5 sodium stearate
0·5 potassium persulphate

The polymerisation is carried out to 100% conversion at 50 °C under a nitrogen atmosphere. At the end of the reaction, 1% of a phenolic anti-oxidant is added. The latex is then coagulated by adding aqueous calcium chloride solution, washed with water and finally spray-dried. One function of the anti-oxidant is to prevent oxidative degradation at the drying stage

Both stages of the ABS emulsion process outlined above are free-radical polymerisations initiated by the water-soluble persulphate ion. The mercaptan controls the molecular weight of the product by acting as a chain transfer agent, and the water provides a means for removing the heat produced in the highly exothermic reactions. These features are common to both the emulsion and suspension processes for the manufacture of ABS polymer. However, the differences between the two techniques are equally significant. Firstly, in the emulsion process the monomer droplets and polymer particles are all less than 1 μm in diameter. Secondly, the rubber latex is cross-linked as a result of adding the difunctional monomer divinyl benzene. Thirdly, the rubber particle size is predetermined during the preparation of the rubber latex, in contrast to the phase inversion mechanism described in Section 4.1. Fourthly, emulsions and latices are stabilised by electrical repulsions between charged particles. Fifthly, the reaction mechanism is significantly different in the case of emulsion polymerisation.

4.4.2 Mechanism of emulsion polymerisation

The kinetics and mechanisms of emulsion polymerisation are significantly different from those of homogeneous bulk polymerisation. In addition to the initiation, propagation and termination steps of the free-radical reaction, it is necessary to consider a fourth process, the nucleation of latex particles, in which the surfactant plays a major role. A quantitative treatment of the emulsion process was first developed by Smith and Ewart,[30] on the basis of a qualitative model proposed by Harkins.[31] The Smith–Ewart theory was later comprehensively reviewed and extended by Gardon,[32–37] and other modifications have been proposed more recently. The mechanism of emulsion polymerisation is complex, and there is still some controversy concerning important features of the reaction. The following account covers the main points of the Smith–Ewart theory.

The surfactant, represented by sodium stearate in the foregoing example, is an essential ingredient in the reaction. At low concentrations, it forms a monomolecular layer at the surface of the aqueous solution, lowering the surface tension of the water. At higher concentrations, the surface becomes saturated, and the hydrophobic stearate anions begin to aggregate into molecular clusters called *micelles*, consisting of about 100 molecules and having diameters of approximately 5 nm. The point at which aggregation begins is called the *critical micelle concentration* (cmc).

When a water-insoluble monomer is added to the aqueous soap solution, a small amount of the monomer enters the micelles, roughly doubling their size. The remainder is dispersed as small droplets approximately 1 μm in diameter, which are also stabilised by adsorbed stearate ions. A small amount of monomer dissolves in the water: the solubilities of styrene and acrylonitrile are 0·04% and 5% respectively. The emulsion formed in this way is stable even when unstirred, since aggregation of the monomer droplets is prevented by their negative charge.

Polymerisation begins when the persulphate initiator is added and the reactor is heated to a temperature at which the persulphate ion decomposes into two sulphate radical-ions:

$$S_2O_8^{2-} \rightarrow 2SO_4^{-} \qquad \text{(iv)}$$

The radical enters a micelle and initiates polymerisation, which proceeds rapidly, supported by monomer diffusing through the water from the dispersed monomer droplets in response to the gradient in chemical potential. The growing particle adsorbs surfactant from the aqueous phase, causing some of the micelles in which there is no growth to disintegrate. Termination occurs when a second radical enters the latex particle. It is a feature of the Smith–Ewart theory that this termination step takes place immediately, because of the small size of the emulsion particles. The result is a 'stop–go' polymerisation reaction, starting and stopping in each particle at intervals of the order of 10 s, as radicals arrive from the aqueous phase. Stage I of the process, during which the rate of reaction increases with the number of monomer-swollen latex particles, ends at a conversion of 10–20%, when the supply of stearate ions from the micelles becomes exhausted. At this point the surface tension suddenly rises.

Once the surfactant is exhausted, no more latex particles can form. The monomer droplets feed a constant number of latex particles, in which polymerisation takes place. During Stage II of the process, the rate of reaction is constant because the number of particles is constant. At any given time, half of the latex particles contain propagating free radicals, and the remainder are inactive. Under these conditions, the rate of polymerisation is independent of monomer concentration.

Stage III begins at a conversion of 60–70%, when the monomer droplets become exhausted, so that polymerisation must be maintained by monomer already present within the particles. Because of the decreasing availability of monomer, the rate of reaction decreases with time.

Some authors have suggested that polymerisation is initiated in the water phase,[38] and that the polymer molecule precipitates as a particle when it reaches a certain molecular weight. After absorbing monomer and adsorbing surfactant, a particle formed in this way would be indistinguishable from one originating from a micelle. The point of this modification of the Smith–Ewart theory is that negatively charged micelles and emulsion particles would repel anions such as the sulphate radical-ion. On the other hand, hydroxyl radicals formed by reaction with water would not suffer this repulsion:

$$SO_4^{-} + H_2O \rightarrow HSO_4^{-} + OH^{\cdot} \qquad \text{(v)}$$

Gardon developed the Smith–Ewart theory to take account of slow termination reactions during Stage II of the process. This extended treatment explains the increasing rate of reaction observed during some emulsion polymerisations,[35] by relating the time required for termination to the size of the latex particle.[34]

On the basis of kinetic evidence supported by electron microscopy, Williams and co-workers concluded that the monomer is not distributed uniformly within a swollen latex particle, but is concentrated near the surface, with the polymer molecules tending towards the centre of the particle.[39–41] They argued that a 'core-shell' structure of this kind is to be expected on thermodynamic grounds, because the configurational entropy of the polymer molecules is reduced in the surface zone of the particle. The case for this model is not proven. The kinetic data can equally well be explained by the modified Smith–Ewart theory,[42, 43] and the core-shell structures that are observed in latex particles, including ABS particles, can be related to the use of water-soluble initiators.

4.4.3 Factors controlling structure

Emulsion polymerisation offers considerable scope for the control of structure and morphology in ABS polymers. The particle size and size distribution of the rubber phase are determined during the preparation of the rubber latex. The degree of grafting and cross-linking of the rubber, and the internal structure of the particles, are determined during the grafting stage.

Rubber particle size distribution

The distribution of particle sizes in the finished ABS polymer is determined largely by the conditions under which the rubber latex is made. The subsequent grafting reaction increases particle sizes because of the formation of internal inclusions, but this is a relatively minor effect. Both particle size and particle size distribution are controlled essentially by adjusting the surfactant concentration during the polymerisation of the butadiene. Increasing the amount of sodium stearate decreases the particle size. In general there is a wide distribution of particle sizes in a rubber latex because the particles are formed at different times during the reaction. This range of 'ages' of particles is especially large when an excess of surfactant is used. However, by suitable control of conditions it is possible to make monodisperse latices.

The first stage in the manufacture of a monodisperse latex is to prepare a 'seed latex' of very small particles. Additional monomer, surfactant and initiator are then added, the surfactant concentration being kept below the cmc. In this way, the formation of new particles is avoided, so that polymerisation results simply in an increase in particle size. Adding surfactant during the second stage of the reaction, in order to exceed the cmc again and form new particles, gives a bimodal distribution of particle sizes.

Thus by adjustment of monomer and surfactant concentrations, the manufacturer can vary average sizes and size distributions within wide limits. The chief restriction arises at the upper end of the particle size range. Particles over about 1 μm in diameter tend to coagulate because electrical repulsions due to the surfactant are insufficient to stabilise the system.

Cross-linking of latex

Since butadiene is a difunctional monomer, polymerisation to a conversion of 70 % or more produces a cross-linked rubber. However, the reaction is poorly controlled, and manufacturers prefer to add about 2 % of a difunctional comonomer to produce cross-linking. The usual choices are divinyl benzene $C_6H_4(CH:CH_2)_2$ and ethylene glycol dimethacrylate $(CH_2:CCH_3COO)_2C_2H_4$. Excessive, uncontrolled cross-linking of the rubber through the reactions of the diene unit is prevented by adding an inhibitor such as hydroquinone to stop the polymerisation at a conversion of 75–95 %, and removing the remaining butadiene monomer as vapour.

The extent of cross-linking in polybutadiene latices used for ABS manufacture varies considerably. In some processes, uncross-linked rubber is used. The difunctional comonomer is omitted from the recipe, and the polymerisation is stopped before cross-links have begun to form.

Core-shell structure of ABS latex

In the second stage of the ABS emulsion process, styrene and acrylonitrile monomers are added to the polybutadiene latex, together with additional water, surfactant, initiators, chain transfer agents and other ingredients. Some of the monomer is absorbed by the rubber particles, causing them to swell, and the remainder forms separate emulsion droplets. As in the first stage of the process, polymerisation is initiated in the aqueous phase, following the decomposition of the water-soluble initiator molecules. The growing polymer radical either forms the nucleus of a new SAN particle or deposits on the surface of a polybutadiene particle. In the latter case, there is a strong possibility that grafting will take place.

The grafting reaction produces an ABS latex particle with a 'core-shell' morphology. A polybutadiene core is surrounded by a shell of

styrene–acrylonitrile copolymer. Some SAN copolymer also forms within
the rubber core. Kato has developed a technique for observing structure
within ABS latex particles. The particles are embedded in a mixture of
agar–agar and polybutadiene, and the mounted specimen is treated with
osmium tetroxide.[44] Figure 4.13 shows a latex mounted in this way. Both
the polybutadiene cores of the particles and the rubbery embedding

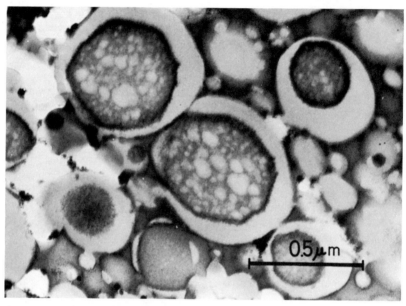

Fig. 4.13 *Osmium-stained section of ABS latex particles, showing core-shell
structure. Particles embedded in stained mixture of agar–agar and uncross-linked
polybutadiene (from K. Kato, ref. 44, reproduced with permission).*

medium are stained and hardened by the osmium, giving a sample that can
be sectioned in an ultramicrotome. The section illustrated reveals a number
of interesting features. The ABS particles consist of polybutadiene cores
about 0·5 μm in diameter and SAN shells approximately 0·1 μm thick. SAN
sub-inclusions up to 0·1 μm in diameter can be seen within the cores of some
particles. The micrograph also shows homogeneous SAN particles without
rubbery cores.

Homogeneous SAN particles of this type form only when the surfactant
concentration is above the cmc. At lower surfactant concentrations, all

SAN copolymer is formed in, or precipitated onto, a polybutadiene particle. Some of this copolymer forms sub-inclusions within the rubber, but most of it is to be found in the shell zones of the latex. When the surfactant concentration is above the cmc, there is competition for monomer between the rubber particles and the new homogeneous SAN particles. In many cases, this competition is of little consequence, but when the

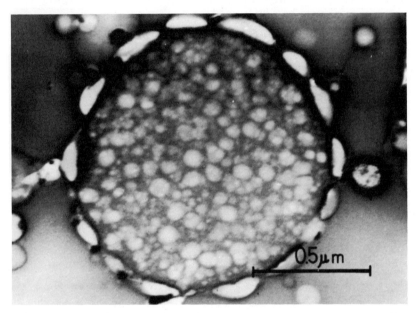

Fig. 4.14 *Osmium-stained section of ABS latex particle showing 'warty' surface formed by maintaining a low monomer: rubber ratio during polymerisation. Particle embedded as in Fig.* 4.13 *(from K. Kato, ref.* 44, *reproduced with permission).*

monomer: rubber ratio is low there is a possibility that the SAN shells around the rubber particles will become depleted, leaving part of the polybutadiene surface ungrafted.

Figure 4.14 illustrates the type of latex particle structure produced by reducing the monomer: rubber ratio. The surface layer of SAN copolymer is discontinuous, leaving areas of ungrafted polybutadiene. The critical monomer: rubber ratio necessary for the formation of a complete shell depends upon two basic factors. The first has already been discussed: competition for monomer due to the formation of new SAN particles arises when the cmc is exceeded. The second is the particle size of the rubber. As the particle size decreases, the surface area to be covered rises rapidly.

Huguet and Paxton showed that a shell of SAN approximately 10 nm thick is necessary to prevent the rubber particles from aggregating in ABS polymer.[45] This figure was obtained from experiments on two polybutadiene latices, one with a particle diameter of 88 nm, and the other with a diameter of 280 nm. The critical monomer:rubber ratios were 0·85 and 0·27 wt. % respectively, which correspond to a layer 10 nm thick on all particles, assuming that all of the SAN was in the shell layer.

A low monomer:rubber ratio during graft copolymerisation does not necessarily mean a high rubber content in the resulting ABS. It is common practice to prepare ABS masterbatch materials, having high rubber contents, for subsequent dilution with SAN copolymer. Thus the degree of grafting, especially at the surface of the rubber particle, can be varied independently of rubber content in the final product. Some commercial ABS polymers appear to contain very small rubber particles which have become aggregated, presumably as a result of incomplete shell formation.

The choice of initiator also affects the core-shell structure of the latex particles. Water-soluble initiators such as the persulphate ion tend to produce shell structures, but oil-soluble initiators such as organic peroxides and hydroperoxides have a greater tendency to generate sub-inclusions within the core regions of the particles.[46] Organic initiators are able to diffuse into monomer-swollen rubber latex particles, and initiate polymerisation well within the interior. The resulting competition for monomer leads to a depletion of the SAN shell. This tendency to initiate polymerisation within the core is modified by combining an inorganic reducing salt with the organic initiator, as in the redox system cumyl hydroperoxide–ferrous pyrophosphate. A mixed organic–inorganic initiator system of this type does not form new SAN particles, even when the surfactant concentration is above the cmc, because the peroxide is insoluble in water. Copolymerisation of styrene and acrylonitrile is therefore confined to the rubber latex particles,[47] increasing the thickness of the shell.

The observed differences in core-shell structure between ABS polymers made with organic initiators and those made with purely inorganic initiators suggests that the core-shell morphology is determined by the site of initiation, rather than by the monomer distribution, as proposed by Williams.[39−41] There is little evidence from studies on ABS polymerisation to support the view that added monomers are concentrated at the surface of the rubber latex particles before polymerisation begins, except in the case of tightly cross-linked rubber particles which are unable to swell sufficiently to absorb all of the available monomer.

Latex aggregation
The aggregation of rubber particles mentioned above is beneficial if the particles are very small, since they would otherwise be ineffective in toughening the polymer. However, the poor grafting that causes the aggregation is undesirable. It is preferable to graft the rubber fully, establishing a good bond between the rubber and the matrix, but to produce the aggregation in some other way. The result is a polymer with the toughness of a material containing large particles, but the transparency of a material containing small particles.[48] Matsuo and Sagaye obtained a polymer with these properties by adding a controlled amount of a water-soluble salt to an ABS emulsion polymer during the course of the reaction.[48] The rubber particles, which were initially 50 nm in diameter, formed clusters between 100 and 300 nm across, containing both ABS latex and SAN latex particles. Addition of an electrolyte is a standard procedure for coagulating an emulsion.

Styrene: acrylonitrile ratio
A distinctive feature of emulsion polymerisation is the ease with which comonomer ratios can be adjusted during the reaction. This facility is especially useful in the manufacture of ABS polymers. As explained in Section 2.3.1, there is only one 'atropic' monomer ratio that polymerises to a copolymer of the same composition. In general, the monomer ratio changes with time as the reaction proceeds, as illustrated in Fig. 2.5. However, in emulsion polymerisation, the drift in composition can be checked by adding styrene or acrylonitrile as required, and it is therefore possible to prepare ABS polymers with high acrylonitrile contents without producing an inhomogeneous matrix. The problem of differential monomer partitioning remains: acrylonitrile concentrations are presumably higher in the shell than in the core of an ABS latex particle, for the reasons discussed in Section 2.3.2.

4.5 BLENDING

The advances in bulk, bulk-suspension and emulsion polymerisation techniques have to some extent displaced blending as a method for manufacturing toughened plastics, but blending is still used in commercial processes. The method is used both for adding rubbers to thermoplastics and for diluting rubber-toughened polymers made by other processes.

Several blending procedures are available. The simplest and most commonly employed is melt blending in an extruder, Banbury mixer or heated two-roll mill. The homogeneity of the product/depends upon the efficiency of the mixing equipment. A twin-screw extruder offers the advantages of a good shearing action combined with continuous operation, but a single-screw extruder is equally effective if used in line with a static mixer.[49] More intimate mixing can be achieved by blending in solution, but the subsequent removal of solvent is inconvenient and expensive. A third possibility is to mix the polymers in latex form, and to coagulate the blended latices; this method is of course limited to those polymers that can be made by the emulsion process.

4.5.1 Blending directly with rubber

The principal application of direct blending of a rubber with a thermoplastic polymer is in the manufacture of toughened PVC. The bulk and emulsion polymerisation processes used for incorporating rubbers into styrene-based polymers are not easily adapted for incorporating rubbers into vinyl chloride-based polymers. The standard commercial practice, therefore, is to prepare the PVC separately, and to add the rubber later.

The rubbers usually chosen for toughening PVC are poly(butadiene-*co*-acrylonitrile) (nitrile rubber) and chlorinated polyethylene.[50] Several other elastomers, including grafted ethylene–vinyl acetate (EVA) copolymer[51] and grafted ethylene–propylene rubber (EPR),[52] are also suitable. The rubber should preferably have a solubility parameter between 0.2 and $0.4 \, (\text{cal}/\text{cm}^3)^{1/2}$ above or below that of PVC, to ensure that there is a good mechanical bond between the two phases, and that the rubber forms a fine dispersion without being completely compatible with the matrix.[53] This requirement limits the choice of rubber, often at the expense of mechanical properties.

The other major application of direct blending with rubber is in the manufacture of toughened polypropylene. This process will probably become obsolete as a result of developments in the Ziegler–Natta process described in Section 4.3. At the time of writing, a minor but substantial amount of toughened polypropylene is made by mechanical blending, usually with EPR or EPDM (a terpolymer of ethylene with propylene and a diene monomer).

Rubber particles formed in mechanical blending operations are often irregular in shape. In contrast to grafted rubber particles produced in bulk and emulsion polymerisation processes, they are usually free from sub-inclusions, and consist simply of the homogeneous rubber phase. At high concentrations, the rubber sometimes forms a continuous network: this type of structure was observed by Matsuo *et al.* in PVC containing 15 % of a nitrile rubber.[54] This is an extreme example of the difficulty of dispersing a moderately compatible uncross-linked rubber by mechanical mixing with a glassy thermoplastic. In contrast, EPR does form spherical particles in blends with polypropylene, which is a crystalline polymer having a sharp melting point and a relatively low viscosity in the melt state. Because of the low viscosity of the continuous polypropylene phase, particle shape is determined principally by surface tension. Spherical rubber particles are also observed in materials made by latex blending,[55] which furthermore offers the opportunity of controlling particle size and degree of cross-linking before making the blend.

The structure of a mechanical blend is determined partly by the chemical behaviour of the rubber. Rubbers must initially be uncross-linked, but some, including polybutadiene, become cross-linked during blending as a result of thermal and mechanochemical oxidation reactions. Rubbers that are unable to cross-link, either because of their structure or because of the presence of anti-oxidants and other free-radical scavengers, tend to form large sheets rather than discrete particles.[56] The resulting laminar structure has an adverse effect upon fracture resistance.

The selection of a rubber for mechanical blending often requires a compromise between compatibility considerations and the mechanical properties of the rubber. In particular, it is necessary in many cases to accept an increase in T_g in return for good interfacial adhesion. The use of poly(butadiene-*co*-acrylonitrile) in place of polybutadiene as a toughening agent for PVC is a typical example. However, developments in block copolymerisation technology have provided an alternative solution. A rubber can be selected for its intrinsic merits, and the required level of compatibility with a given thermoplastic polymer can then be added by making a block copolymer.

Very high values of fracture resistance have been obtained by blending thermoplastics with block copolymers. For example, the notched Izod impact strength of polystyrene was raised by a factor of thirty on adding 20 wt. % of polybutadiene in the form of a block copolymer with styrene.[57] The main problem in making this type of blend is to prevent the rubber from becoming too finely dispersed: rubber particles less than about 1 μm in diameter are ineffective in toughening polystyrene. Particle size is increased either by raising the molecular weight of the block copolymer or by adding butadiene homopolymer to the blend.[57] The block copolymer concentrates at the interface between the rubber and the polystyrene, so that the particle size is determined by the composition of the blend. The same principle has been applied to other plastics, and is equally effective when a graft copolymer is used in place of the block copolymer.

4.5.2 Blending with grafted rubbers

In view of the effectiveness of grafted rubbers as toughening agents, it is a logical step to prepare these rubbers in the most convenient form for adding to polymers such as PVC. This is done on a commercial scale by making masterbatches consisting of rubber and grafted thermoplastic in roughly equal amounts. The best-known examples are ABS and MBS masterbatches for addition to PVC. In these materials the polybutadiene rubber particles are dispersed in a styrene–acrylonitrile or styrene–methyl methacrylate matrix which is compatible with PVC. A typical toughened PVC contains 5–10% of an ABS or MBS masterbatch.

The principal advantage of using a masterbatch is that particle size, size distribution, degree of cross-linking and grafting, and other important structural features are all controlled separately during the preparation of the masterbatch by emulsion polymerisation. The blending operation is then simply a matter of ensuring that the polymers are thoroughly mixed, so that the rubber particles are homogeneously distributed throughout the matrix.

The masterbatch principle is used in making a number of other toughened plastics. Polycarbonate is blended with ABS masterbatch rubber concentrates,[58] and polyphenylene oxide is blended with HIPS.[59] These blends are not only tougher than the parent polymers, but also offer the advantage of improved moulding performance. Indeed, the lower viscosity appears to be the main consideration in some cases.

4.6   COMPARISON OF PROCESSES

The choice of a process for making a toughened polymer is governed by a variety of technical and economic factors. For toughened thermosetting resins, there is obviously no alternative to bulk polymerisation. On the other hand, ABS is made by bulk, emulsion and blending processes, and a meaningful comparison can be made between the products.

The structures produced by these three processes are illustrated in Fig. 4.15. The most obvious differences between the materials are in the shapes

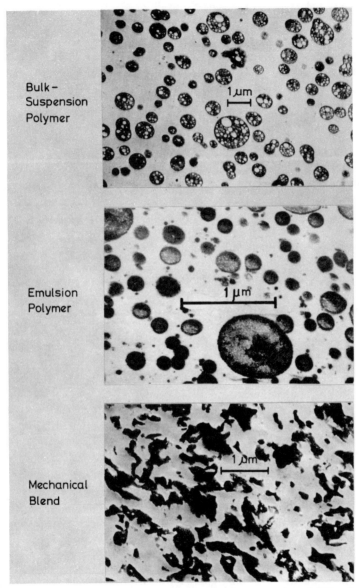

Fig. 4.15   Osmium-stained sections showing contrasting structures of ABS polymers made by three different processes. Note differences in magnification.

and internal structures of the rubber particles. Bulk and emulsion polymerisations generally produce spherical particles, whereas blending tends to disperse the rubber as irregularly shaped particles. There are, however, notable exceptions to these generalisations: as already noted in Sections 4.1.2 and 4.5.1, shear during the finishing cycle of a bulk polymerisation gives non-spherical particles, whilst EPR forms spheres on blending with polypropylene.

In a typical HIPS or ABS made by the bulk process, the rubber particles contain 80–90 vol. % of hard sub-inclusions. This structure is characteristic of a material that has gone through a phase inversion, and is not found to the same extent in toughened epoxy resins, which do not undergo phase inversion in the same way. The presence of the sub-inclusions means that the effective rubber content is much higher than the actual concentration of added rubber. However, it cannot be assumed that a composite sphere is equivalent to an homogeneous rubber particle in its effect upon fracture resistance. This problem will be discussed in later chapters.

In a typical ABS emulsion polymer, the rubber particles probably contain approximately 50 vol. % of hard sub-inclusions. In contrast to ABS bulk polymer, the sub-inclusions in an emulsion polymer are usually less than 100 nm in diameter, and the composition of the particle can be varied within wide limits. The structure is determined by the choice of an initiator, degree of cross-linking of the rubber and other factors discussed in Section 4.4.3.

The rubber content of a HIPS bulk polymer is limited to about 14 % by the requirement that the prepolymer should undergo phase inversion, but this limitation is largely compensated for by the filling of the rubber particles with polystyrene sub-inclusions. Monomer removal after phase inversion permits the use of higher rubber contents. There is no restriction on the rubber content of an emulsion polymer or a blend.

A major advantage of the bulk and emulsion processes is the control that they offer over grafting and cross-linking. In mechanical or latex blending, there is no grafting and little control over cross-linking, with the result that the product is inferior in mechanical properties to materials made by the other processes.

In bulk polymerisation, it is possible to vary rubber particle size within wide limits. Mean particle diameters between 0·3 and 30 μm are obtained by control of stirring, grafting and other factors. On the other hand, the manufacturer has little control over particle size *distribution*. In emulsion polymerisation, the particle size is limited to a maximum of about 1 μm by the low stability of latices containing larger particles. For this reason, emulsion polymerisation is unsuitable for the manufacture of HIPS, which requires rubber particles over 2 μm for maximum toughness. The critical particle size is smaller in ABS and PVC. In contrast to the bulk process, emulsion polymerisation offers considerable control over particle size distribution.

As discussed in Section 4.4.3, the emulsion process also offers a ready means of controlling comonomer ratios, so that ABS manufacturers are not restricted to the atropic styrene: acrylonitrile composition. This facility is not generally available in other processes, although coil reactors capable of producing non-atropic compositions by bulk polymerisation have been developed.[60]

One of the disadvantages of the emulsion process is that it involves the use of iron and other transition metals as components of redox systems. Unless these impurities are removed by thorough washing, they cause degradation problems during drying and subsequent processing. Bulk polymers contain the smallest amounts of impurities, and suspension polymers occupy the middle place.

Economic factors are a major consideration in choosing a process for manufacturing toughened plastics. Continuous bulk polymerisation is the cheapest process to run, followed by bulk-suspension polymerisation. Emulsion polymerisation is relatively expensive because it involves washing and drying operations.

REFERENCES

1. N. Platzer, *Ind. Eng. Chem.* **62** (1970) 6.
2. J. L. White and R. D. Patel, *J. Appl. Polymer Sci.* **19** (1975) 1775.
3. R. L. Kruse, *ACS Poly. Prepr.* **15**(1) (1974) 271.
4. G. E. Molau, *J. Poly. Sci.* A3 (1965) 4235.
5. G. F. Freeguard and M. Karmarkar, *J. Appl. Polymer Sci.* **15** (1971) 1657.
6. P. Becher, *Emulsions: Theory and Practice* (2nd edn.), ACS Monograph No. 162, Reinhold, New York, 1965.
7. B. W. Bender, *J. Appl. Polymer Sci.* **9** (1965) 2887.
8. G. E. Molau, W. M. Wittbrodt and V. E. Meyer, *J. Appl. Polymer Sci.* **13** (1969) 2735.
9. E. R. Wagner and L. M. Robeson, *Rubber Chem. Technol.* **43** (1970) 1129.
10. G. F. Freeguard and M. Karmarkar, *J. Appl. Polymer Sci.* **16** (1972) 69.
11. M. Baer, *J. Appl. Polymer Sci* **16** (1972) 1109.
12. J. D. Moore, *Polymer* **12** (1971) 478.
13. B. Chauvel and J. C. Daniel, *ACS Poly. Prepr.* **15**(1) (1974) 329.
14. A. E. Platt, in *Encyclopedia of Polymer Science and Technology* Vol. 13, Wiley, New York, 1970, p. 156.
15. A. C. Soldatos and A. S. Burhans, *ACS Adv. Chem. Ser.* **99** (1971) 531.
16. A. C. Meeks, *Polymer* **15** (1974) 675.

17. E. H. Rowe, A. R. Siebert and R. S. Drake, *Mod. Plast.* **49** (Aug. 1970) 110.
18. T. Yoshii, Ph.D. Thesis, Cranfield, England, 1975.
19. C. K. Riew, E. H. Rowe and A. R. Siebert, *ACS Div. Org. Coat. Plast. Prepr.* **34**(2) (1974) 353.
20. H. Batzer and S. A. Zahir, *J. Appl. Polymer Sci.* **19** (1975) 585.
21. S. Visconti and R. H. Marchessault, *Macromolecules* **7** (1974) 913.
22. W. D. Bascom, R. L. Cottington, R. L. Jones and P. Peyser, *ACS Div. Org. Coat. Plast. Prepr.* **34**(2) (1974) 300.
23. A. Noshay and L. M. Robeson, *ACS Poly. Prepr.* **15**(1) (1974) 613.
24. T. G. Heggs, in *Block Copolymers*, D. C. Allport and W. H. Janes (eds.), Applied Science, London, 1973, Chapter 4.
25. G. Bier and G. Lehmann, in *Copolymerization*, G. E. Ham (ed.), Wiley-Interscience (1964) 149.
26. A. D. Caunt, *J. Poly. Sci.* C4 (1963) 49.
27. T. G. Heggs, in *Block Copolymers* (*see* ref. 24), Chapter 8.
28. C. A. F. Tuijnman, *J. Poly. Sci.* C16 (1967) 2379.
29. C. Placek, *Chemical Process Review No.* 46, Noyes Data Corp., Park Ridge, New Jersey, 1970.
30. W. V. Smith and R. H. Ewart, *J. Chem. Phys.* **16** (1948) 592.
31. W. D. Harkins, *J. Am. Chem. Soc.* **69** (1947) 1428.
32. J. L. Gardon, *J. Poly. Sci.* A1, **6** (1968) 623.
33. J. L. Gardon, *J. Poly. Sci.* A1, **6** (1968) 643.
34. J. L. Gardon, *J. Poly Sci.* A1, **6** (1968) 665.
35. J. L. Gardon, *J. Poly. Sci.* A1, **6** (1968) 687.
36. J. L. Gardon, *J. Poly. Sci.* A1, **6** (1968) 2853.
37. J. L. Gardon, *J. Poly. Sci.* A1, **6** (1968) 2859.
38. P. I. Lee, *Plast. Polym.* **39** (1971) 111.
39. M. R. Grancio and D. J. Williams, *J. Poly. Sci.* A1, **8** (1970) 2617.
40. D. J. Williams, *J. Poly. Sci.* (*Chem.*) **12** (1974) 2123.
41. P. Keusch, R. A. Graff and D. J. Williams, *Macromolecules* **7** (1974) 304.
42. N. Friis and A. E. Hamielec, *J. Poly. Sci.* (*Chem.*) **11** (1973) 3321.
43. J. L. Gardon, *J. Poly. Sci.* (*Chem.*) **12** (1974) 2133.
44. K. Kato, *Koll. Z. Z. Polym.* **220** (1967) 24.
45. M. G. Huguet and T. R. Paxton, in *Colloidal and Morphological Behaviour of Block and Graft Copolymers*, G. E. Molau (ed.), Plenum, New York, 1971, p. 183.
46. K. Kato, paper presented to Royal Microscopical Society, Oxford, September 1968.
47. Z. Kromolicki and J. G. Robinson, *SCI Monograph No.* 26 (1967) 16.
48. M. Matsuo and S. Sagaye, in *Colloidal and Morphological Behaviour of Block and Graft Copolymers*, G. E. Molau (ed.), Plenum, New York, 1971, p. 1.
49. C. D. Han, Y. W. Kim and S. J. Chen, *J. Appl. Polymer Sci.* **19** (1975) 2831.
50. H. H. Frey, *Kunststoffe* **49** (1959) 50.
51. W. Göbel, *Kaut Gummi* **22** (1969) 116.
52. F. Severini, E. Mariani and A. Pagliari, *ACS Adv. Chem. Ser.* **99** (1971) 260.
53. J. E. Bramfitt and J. M. Heaps, in *Advances in PVC Compounding and Processing*, M. Kaufmann (ed.), Maclaren, London, 1962, p. 41.
54. M. Matsuo, C. Nozaki and Y. Jyo, *Poly. Engng Sci.* **9** (1969) 197.
55. J. Mann, R. J. Bird and G. Rooney, *Makromol. Chem.* **90** (1966) 207.
56. C. B. Bucknall, *Trans. IRI* **39** (1963) 221.
57. R. R. Durst, R. M. Griffith, A. J. Urbanic and W. J. van Essen, *ACS Div. Org. Coat. Plast. Prepr.* **34**(2) (1974) 320.
58. D. Stefan and H. L. Williams, *J. Appl. Polymer Sci.* **18** (1974) 1451.
59. N. Platzer, *ACS Poly. Prepr.* **15**(1) (1974) 28.
60. A. W. Hanson and R. L. Zimmerman, *Ind. Eng. Chem.* **49** (1957) 1803.

CHAPTER 5

VISCOELASTIC PROPERTIES

At low strains, rubber-toughened plastics behave as viscoelastic composite materials. Their moduli are determined largely by the properties of the rigid matrix, but are modified by the presence of the rubber particles. A reduction in stiffness is the inevitable penalty for adding rubber to the material, and must be weighed against the increased fracture resistance conferred by the rubber.

This reduction in stiffness is of immediate interest to the user of rubber-toughened plastics, since stiffness is often a controlling factor in the design of plastics products. In order to compensate for a low stiffness, it is necessary to increase section thicknesses, thus increasing the cost and weight of the product. For this reason, a discussion of the relationship between structure and low-strain mechanical properties is an integral part of any detailed study of rubber toughening.

In principle, it should be possible to predict the moduli of a composite material from the structure and from the properties of its components. In practice, however, there are difficulties in formulating exact equations for modulus even when both components are elastic, and the difficulties are considerably greater when one or both are viscoelastic. It is therefore necessary, in any theoretical analysis, to make certain approximations and simplifying assumptions.

One solution is to apply linear elasticity theory to the problem, and thus avoid the complications caused by viscoelasticity. This approximation is acceptable when the moduli of both matrix and rubber are varying only slowly with time, as in HIPS at room temperature, but is obviously not applicable when either component is close to its glass transition. A further sacrifice of mathematical rigour is therefore necessary in discussing dynamic mechanical loss peaks.

107

In addition to predicting low-strain properties such as shear modulus, elasticity theory provides a useful basis for the discussion of large-strain properties, including yield and fracture. Elastic stress analysis shows that the rubber particles produce high stress concentrations in the surrounding matrix, and it is now generally recognised that crazes and shear bands initiated in regions of high tensile stress are responsible for the large deformations observed in rubber-toughened plastics. Any quantitative discussion of rubber toughening should preferably include an analysis of the stress concentrations within the composite polymer in relation to the structure as the first step towards predicting rates of craze and shear band initiation.

Inhomogeneous plastic deformation due to crazing or shear mechanisms is usually first observed at tensile strains between 0·3 and 1·0 %, depending upon the polymer, temperature and time scale of the experiment. Above about 1 % strain, these plastic deformation mechanisms make a significant contribution to the stress–strain behaviour of the composite polymer, which is discussed in later chapters. The present chapter is concerned with mechanical properties at lower strains, before detectable crazing and plastic yielding has taken place.

5.1 EXPERIMENTAL OBSERVATIONS

The moduli of rubber-toughened plastics, as of most polymers, are functions of time, temperature, strain and strain history. This type of behaviour is known as *non-linear viscoelasticity*. In presenting modulus data for viscoelastic materials, the standard practice is to consider the effect of each variable separately, keeping the others constant. For example, *isochronous* curves show the relationship between modulus (or stress) and strain at a fixed time under load; and *isometric* curves show the relationship between modulus and time at a fixed strain. Similarly, dynamic mechanical data are presented as curves of modulus and loss tangent against temperature, at a fixed frequency.

Figure 5.1 compares 100 s isochronous curves for polystyrene and HIPS at 20 °C. The moduli of both materials are almost independent of strain in the small-strain region, but at a strain of about 0·5 % the moduli begin to fall, and in HIPS at strains above 1 % the decrease in modulus with strain is very marked. Some deviation from linear viscoelastic behaviour is observed in most polymers at strains of the order of 1 %:[1] in the case of HIPS, the non-linearity is due principally to crazing, which is discussed in Chapters 7

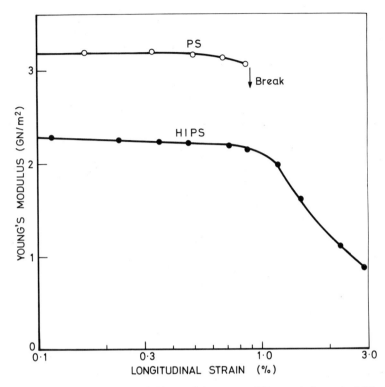

Fig. 5.1 *Isochronous curves of* 100 s *modulus against* 100 s *strain for typical PS and HIPS polymers at* 20°C.

and 8. The data suggest that both HIPS and PS can be treated as approximately linear materials, in the sense that modulus is independent of strain, in the strain region 0–0·5 %, and over the time range 0–100 s.

Figure 5.2 shows isometric modulus data obtained from short-term stress-relaxation experiments on PS and HIPS at 21 °C.[2] The viscoelasticity observed in both materials is due to a general time-dependent response of the polystyrene: crazing occurs only at longer times or higher stresses, and produces a much more marked decrease in modulus with time under load.[2] Stress relaxation in the rubber probably makes a relatively small contribution to the time-dependence of the modulus under the conditions specified, but at very short times or lower temperatures, as the rubber passes through its glass transition, the viscoelasticity of both phases must be taken into account.

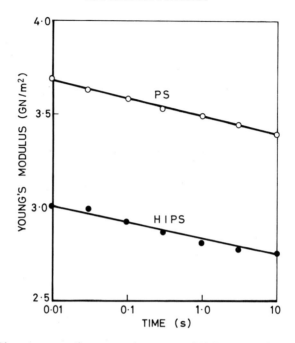

Fig. 5.2   Short-term tensile stress relaxation at 21 °C in typical PS and HIPS polymers at a strain of 0·80% (after J. A. Schmitt and H. Keskkula, ref. 2, reproduced with permission).

The effects of temperature on modulus and loss tangent are illustrated in Fig. 5.3, which compares data obtained from dynamic mechanical tests on four polymers: polystyrene, two HIPS bulk polymers and an HIPS made by blending polystyrene mechanically with rubber.[3] Between −175 and +50 °C, the modulus of polystyrene shows a steady downward trend, which becomes more marked above +50 °C, as the polymer approaches its glass transition. Conversely, the loss tangent of polystyrene rises to a peak at the glass transition, which occurs at about 100 °C. The addition of rubber produces a drop in modulus and increase in loss tangent at about −90 °C, in the glass transition region of the polybutadiene. Over the remainder of the temperature range, the properties of the HIPS polymers follow the same trends as polystyrene. The position and size of the secondary loss peak provide useful information in the analysis of rubber-toughened plastics, as discussed in Chapter 3.

The magnitude of the secondary loss peak depends not only upon the

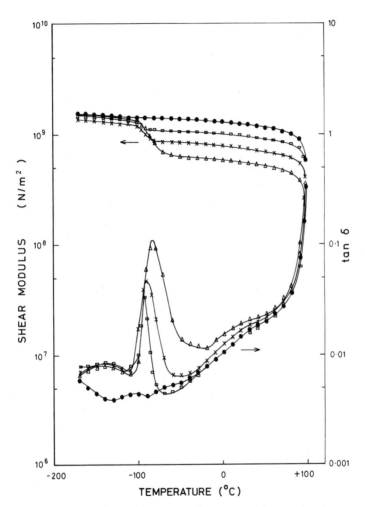

Fig. 5.3 Dynamic mechanical loss curves for styrene polymers showing secondary loss peaks in HIPS at ∼ −90°C due to glass transition of polybutadiene: (●) polystyrene containing no rubber; (□) HIPS made by blending PS mechanically with 10% rubber; (×) HIPS bulk polymer containing 5% rubber; (△) HIPS bulk polymer containing 10% rubber (from G. Cigna, ref. 3, reproduced with permission).

amount of rubber present, but also on the detailed morphology of the composite polymer. Two of the HIPS polymers illustrated in Fig. 5.3 contain 10% polybutadiene, but the loss peak in the bulk polymer is considerably larger than that in the blend. This difference is obviously connected with the difference in the volume fraction of the rubber particles in the two polymers. Polystyrene sub-inclusions greatly increase the volume fraction of the rubber phase in the bulk polymer, for the reasons discussed in Chapter 4, whereas the rubber particles in the blend consist simply of polybutadiene, and therefore occupy a much smaller volume.

The effects of rubber phase volume on the magnitude of the secondary loss peak are further illustrated in Fig. 5.4, which presents data obtained by Wagner and Robeson in experiments on HIPS polymers containing 6% polybutadiene.[4] The volume fraction of rubber particles varied from 6 to 48%. Two trends are obvious from the illustration: firstly, the size of the secondary loss peak increases rapidly with the volume fraction of the rubber phase; and, secondly, the peak tends to shift to higher temperatures as the concentration of sub-inclusions within the rubber particles increases. It is of interest to determine whether either or both of these trends can be explained simply in terms of mechanics. The upward shift in peak temperature, which can also be seen in Fig. 5.3, might be due to chemical effects such as cross-linking of the rubber.

In another experiment, Wagner and Robeson prepared two HIPS polymers containing 22% by volume of rubber particles.[4] One was a blend of polystyrene with 22% polybutadiene and 0·5% sulphur. The other was a bulk polymer containing 6% polybutadiene, the remaining 16% by volume of rubber phase consisting of polystyrene sub-inclusions. The results are shown in Fig. 5.5: the peaks are similar but not identical in area, and the loss tangent maximum is approximately 10 °C higher in the bulk polymer than in the blend. Again, it is not clear whether chemical differences are responsible for part or all of the upward shift in the loss peak.

Cross-linking of the rubber is known to shift the secondary loss maximum to higher temperatures. The effect has already been noted in Chapter 3, in connection with oxidative degradation. Another example is given in Fig. 5.6, which shows the effect of adding sulphur to an HIPS bulk polymer after polymerisation.[4] Each sample was mixed and moulded under the same conditions. The results demonstrate conclusively that cross-linking causes a very large upward shift in the glass transition of the rubber phase. This shift has a profound effect upon large-strain mechanical properties, including impact strength and elongation at break at room temperature.[4]

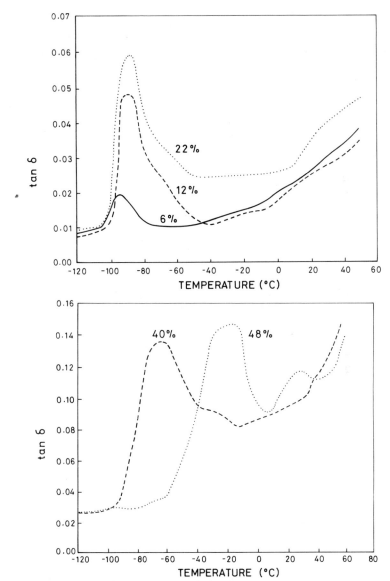

Fig. 5.4 *Dynamic loss curves for five HIPS polymers, each containing* 6%
polybutadiene, but with volume loadings of rubber particles *between* 6 *and* 48%,
*owing to differences in the concentrations of polystyrene sub-inclusions within the
composite rubber particles (from E. R. Wagner and L. M. Robeson, ref. 4,
reproduced with permission).*

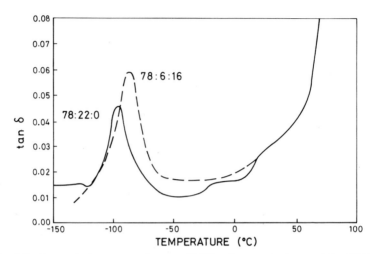

Fig. 5.5   Dynamic loss curves for two HIPS polymers, both containing 22% by volume of rubber particles, but having differing morphologies. Labels show volumetric ratios of polystyrene matrix: polybutadiene: polystyrene inclusions (from E. R. Wagner and L. M. Robeson, ref. 4, reproduced with permission).

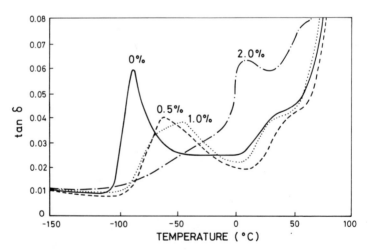

Fig. 5.6   Dynamic loss curves of HIPS polymers containing 0–2% sulphur showing the effects of cross-linking (from E. R. Wagner and L. M. Robeson, ref. 4, reproduced with permission).

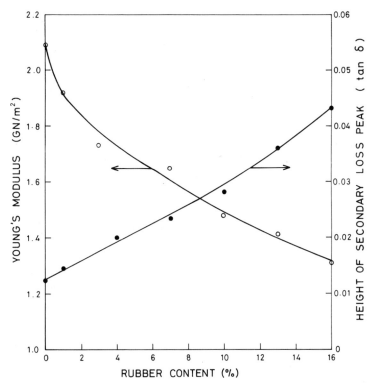

Fig. 5.7   Effects of rubber on the modulus and loss properties of poly(styrene-co-acrylonitrile-co-methyl methacrylate) (from S. Krause and L. J. Broutman, ref. 5, reproduced with permission).

It is important to remember that a peak in tan δ is accompanied by a drop in the storage shear modulus G' of the polymer, and that the factors responsible for an increase in the area of the secondary loss peak also cause a decrease in the stiffness of the composite at temperatures above the transition temperature. Figure 5.7 shows the effect of rubber content on both properties in a series of blends containing AMS (styrene–acrylonitrile–methyl methacrylate random copolymer) and grafted poly(butadiene-co-styrene).[5] The magnitude of the loss peak, as measured by the peak height, increases linearly with rubber content, whilst the modulus falls by almost 40 % on addition of 16 % rubber.

The secondary transition due to the rubber phase is sometimes obscured by a prominent secondary loss process in the matrix polymer. Examples include PVC toughened with nitrile rubbers, and epoxy resins containing

CTBN rubber. In this type of material, the rubber peak can usually be detected by careful comparison of loss curves for the toughened and untoughened polymer.[6]

5.2 MODELS FOR TOUGHENED POLYMERS

The first step in applying elasticity or viscoelasticity theory to rubber-toughened plastics is to choose a suitable physical model to represent the composite. Several models have been proposed, but there is no general agreement upon which is the most suitable. There is, of course, considerable variation in structure between materials, as demonstrated in Chapter 4, so that any model chosen must either be restricted to one class of composites, having a well-defined morphology, or be sufficiently flexible to be adapted to a range of materials of different morphologies.

Most authors have concentrated on solution- or emulsion-grafted polymers such as HIPS and ABS. Microscopy shows that in the isotropic state the standard commercial grades of these polymers contain randomly dispersed spherical rubber particles, and that the hard sub-inclusions usually found within the particles are also randomly dispersed and at least approximately spherical. In view of the importance of this class of toughened plastics, there is a strong case for restricting the discussion to this type of morphology. Any attempt to make the theory more general, for example by considering irregularly shaped rubber particles, is likely to reduce rather than enhance the value of the theory. The predictive power of available theories improves as the structure and morphology of the composite are defined more precisely.

The simplest model for HIPS is a random dispersion of polybutadiene spheres in a polystyrene matrix, as illustrated in Fig. 5.8a. There is no difficulty in assigning properties to the two components of the composite, since the moduli of polystyrene and polybutadiene are known. The chief deficiency of the model is that it does not account for the effect of polystyrene sub-inclusions on the viscoelastic properties of HIPS. This problem could be overcome by treating the whole rubber particle, including the sub-inclusions, as polybutadiene. The validity of this approach can be tested by referring to Fig. 5.5, which shows that a filled rubber particle has a similar but not identical effect on properties to that produced by a pure rubber particle of the same volume.

A second model, illustrated in Fig. 5.8b, also treats the rubber particle as an homogeneous sphere, but recognises the effects of grafting upon both the properties and volume of the rubber phase. This model could explain the upward shift in the loss maximum shown in Fig. 5.5, and account for the increase in loss tangent with increasing volume of rubber particles. The difficulty is to assign modulus values to the graft copolymer, so that the model can be used to predict the moduli of HIPS. The same objection applies to the model suggested by Turley,[7] which is illustrated in Fig. 5.8c, and is essentially a combination of the previous two: the rubber particle is

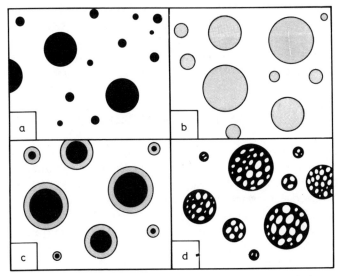

Fig. 5.8 *Models for calculating the modulus of HIPS. Black represents polybutadiene, white represents polystyrene and grey (shaded) indicates graft copolymer having properties intermediate between the two.*

treated as a two-component body, having a core of polybutadiene and a shell of graft rubber. Both the properties and the volume of the graft copolymer shell must be specified if the model is to be used to predict the properties of the composite.

These problems are avoided in the composite particle approach suggested by Bucknall.[8] The HIPS is treated as a two-phase composite containing only polystyrene and polybutadiene, with no distinct region of graft copolymer. As shown in Fig. 5.8d, the rubber particles are represented as random dispersions of polystyrene sub-inclusions in a continuous polybutadiene matrix. The volume fractions of polystyrene matrix, polybutadiene and polystyrene sub-inclusions are determined by analysis, as described in Chapter 3, and the properties assigned to the phases are the known properties of polystyrene and polybutadiene. This model has the advantage of being based on electron microscope studies of morphology, and is potentially of considerable value in predicting properties. For example, it could be used to determine whether a shift in the secondary loss maximum is to be expected purely on the basis of mechanical effects, or whether it is necessary to invoke chemical effects such as cross-linking.

There is obviously scope for further elaboration of the models by combining these principles in various ways. However, the best model is the simplest one that will represent the experimental observations adequately. The discussion will therefore be limited to the proposals represented above. Of these, the composite particle approach, model (d), has the strongest claim to consideration.

5.3 MATHEMATICAL THEORIES

In order to define the elastic properties of an isotropic solid, it is necessary to specify two independent stiffness parameters. In the present discussion, the equations are formulated in terms of the shear modulus G, which characterises the deformation of the solid at constant volume, and the bulk modulus K, which characterises the purely volumetric deformation of the solid. The problem is to calculate the moduli G_0 and K_0 of the composite solid from the moduli G_1, G_2, K_1 and K_2, and the volume fractions ϕ_1 and ϕ_2 of the components. Case (d) of Fig. 5.8 requires a double calculation: firstly, to obtain moduli for the composite rubber particles; and, secondly, to obtain moduli for the rubber-toughened polymer as a whole, using values previously calculated for the moduli of the particles. Young's modulus E and Poisson's ratio v can then be calculated if necessary, using the standard formulae of elasticity theory:

$$E = \frac{9KG}{3K + G}$$

$$v = \frac{3K - 2G}{6K + 2G}$$

(5.1)

The calculation of K_0 presents little difficulty. The difference in bulk modulus between plastics and rubbers is not very great: typical values for polystyrene and polybutadiene are $3\,GN/m^2$ and $2\,GN/m^2$ respectively. Although exact equations for K_0 are not available in the general case, any errors involved in using an approximate expression are likely to be unimportant.

The calculation of G_0 presents more formidable difficulties. The shear moduli of plastics are a thousand times greater than the shear moduli of rubbers: typical values for polystyrene and polybutadiene are $1\cdot1$ GN/m^2 and $0\cdot7$ MN/m^2 respectively. Furthermore, there is no exact expression that might be used to calculate G_0, even in the most restricted case. The basic problem is that the compositions of interest are of high concentration, and each particle therefore interacts with several of its neighbours. Rigorous mathematical treatment of this kind of many-body calculation has proved impossible in elasticity theory, as in other branches of physics. Consequently, the mathematician is forced to discard much of the available information concerning a particular system, because it cannot be handled. As an illustration of this point, it may be noted that many mathematical theories of composites do not specify which material forms the matrix and which the disperse phase. The most that can be done in a rigorous manner is to calculate bounds on the shear modulus of the composite. Approximate theories may be used, but it is important to recognise the limitations of such theories. Both alternatives are considered below.

5.3.1 Bounds on moduli

The extreme bounds on modulus are obtained by treating the composite as a set of parallel plates. When these are loaded in parallel, so that both materials are subjected to the same strain, the modulus is a maximum. When they are loaded in series, so that the stress in both materials is the same, the modulus is a minimum. The bounds on shear modulus are given by:

Parallel (*upper bound*)

$$G_0 = \phi_1 G_1 + \phi_2 G_2 \tag{5.2}$$

Series (*lower bound*)

$$\frac{1}{G_0} = \frac{\phi_1}{G_1} + \frac{\phi_2}{G_2} \tag{5.3}$$

Whilst these are rigorous bounds, they are of limited value because they are so far apart. Improved rigorous bounds, for arbitrary geometry, were obtained by Hashin and Shtrikman.[9] The results for shear modulus are

$$G_0 \, (upper) = G_2 + \phi_1 \left[\frac{1}{G_2 - G_1} + \frac{6\,(K_2 + 2G_2)\,\phi_2}{5G_2\,(3K_2 + 4G_2)} \right]^{-1} \tag{5.4}$$

and

$$G_0 \, (lower) = G_1 + \phi_2 \left[\frac{1}{G_2 - G_1} + \frac{6 \, (K_1 + 2G_1) \, \phi_1}{5G_1 \, (3K_1 + 4G_1)} \right]^{-1} \quad (5.5)$$

where $K_2 > K_1$ and $G_2 > G_1$.

Again, the bounds are widely separated, and are therefore of limited predictive value. The Hashin–Shtrikman bounds serve as a useful test of approximate theories, however, since any equation that predicts values outside the bounds over any part of the composition range must be regarded as invalid. Closer, but non-rigorous, bounds were calculated by Hashin,[10] by treating the composite as being composed of spherical, space-filling elements. Each element consisted of a spherical core of one material surrounded by a shell of the other. The model requires the introduction of infinitely small elements in order to fill the available space, and is consequently regarded as of very limited application.

5.3.2 Approximate equations

Numerous approximate equations have been proposed for the moduli of isotropic composite materials containing isotropic inclusions. Figure 5.9 shows the predictions of several of these theories for the case of a composite rubber particle containing 0–99% polystyrene sub-inclusions, and also includes the Hashin–Shtrikman upper and lower bounds. It is clear that the various theories predict widely different values for the shear modulus of the composite.

From the mathematical point of view, the most soundly based of the available approximate theories is the 'self-consistent' theory developed by Hill.[11, 12] The modulus is calculated by considering the strain within an inclusion which is surrounded by a medium having the (as yet unknown) elastic constants of the composite material. The solution for G_0 is given by

$$\left[\frac{\phi_1 K_1}{K_1 + \frac{4}{3}G_0} + \frac{\phi_2 K_2}{K_2 + \frac{4}{3}G_0} \right] + 5 \left[\frac{\phi_1 G_2}{G_0 - G_2} + \frac{\phi_2 G_1}{G_0 - G_1} \right] + 2 = 0 \quad (5.6)$$

This equation can be solved by iterative methods, or graphically, by inserting values of G_0 between G_1 and G_2 and calculating ϕ_1. As shown in Fig. 5.9, the self-consistent method predicts a sigmoidal relationship between $\log G_0$ and ϕ_2, with values close to the lower bound at low volume concentrations, and close to the upper bound at high volume concentrations. The reason is essentially that the method does not distinguish between matrix and dispersed phase. The sharp increase in modulus as the volume fraction of polystyrene approaches 0.5 effectively corresponds to a phase inversion. No provision is made for the case in which

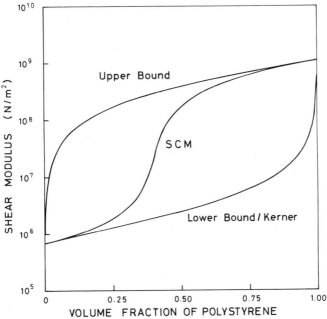

Fig. 5.9   Predictions of the self-consistent method (SCM),[11] the Kerner equation,[18] and the Hashin–Shtrikman bounds[9] for the shear moduli of composite rubber particles, using the model of Fig. 5.8d.

phase inversion does not take place as expected. The same criticism applies to the Halpin–Tsai[13–15] and Davies[16, 17] equations.

One of the most widely used of the approximate expressions is that developed by Kerner.[18] The composite is treated as a suspension of spherical inclusions, in which each inclusion is surrounded by, and bonded to, a shell of the matrix. At some distance, the composite is treated as having the properties of an homogeneous medium, as in the self-consistent method. In the intermediate region, which is of unspecified extent, the properties change gradually from those of the average composite to those of the matrix. Specific interactions between neighbouring particles are not considered. The shear modulus is given by

$$\frac{G_0}{G_m} = \frac{\dfrac{\phi_i G_i}{(7 - 5 v_m) G_m + (8 - 10 v_m) G_i} + \dfrac{\phi_m}{15 (1 - v_m)}}{\dfrac{\phi_i G_m}{(7 - 5 v_m) G_m + (8 - 10 v_m) G_i} + \dfrac{\phi_m}{15 (1 - v_m)}} \qquad (5.7)$$

where the suffixes m and i respectively refer to the matrix and the inclusion.

Kerner's theory lacks the status of the self-consistent theory from the mathematical point of view. It can be shown that eqn. (5.7) yields the Hashin–Shtrikman lower bound when the matrix has the lower shear modulus and bulk modulus in a binary composite, and the upper bound when the matrix has higher shear and bulk moduli than the inclusion. Thus, in Fig. 5.9 the shear modulus predicted by the Kerner method coincides with the lower bound.

The application of any of these theories to viscoelastic composites must be approached with caution. The interaction of a viscoelastic inclusion with a viscoelastic matrix is a problem of further complexity, which has not yet been solved in a satisfactory manner. Whilst it is tempting simply to insert complex moduli into elasticity equations, the theoretical justification for this procedure is weak, and the results of dubious value.

The theories developed for calculating the moduli of composite materials are also applicable to other properties, including dielectric permittivity,[18] specific heat, thermal conductivity and coefficient of thermal expansion.[12, 19] For example, the volume coefficient of thermal expansion β is given by[12]

$$\beta_0 = \beta_1 + \frac{(\beta_2 - \beta_1)(K_0^{-1} - K_1^{-1})}{(K_2^{-1} - K_1^{-1})} \tag{5.8}$$

5.4 APPLICATION OF THEORIES

It is clear from the foregoing discussion that mathematical theories are of limited value in relating structure to elastic and viscoelastic properties in rubber-toughened plastics. Rigorous bounds have been established, but these are very widely spaced within the composition range of principal interest. The moduli of a toughened polymer containing only 5 % by volume of pure rubber particles can be calculated with some precision, but the type of structure shown in Fig. 5.8d presents problems of a different order of difficulty. Typical rubber particles in HIPS contain 80–85 % of polystyrene sub-inclusions in a matrix of rubber, a structure with which the theories are not designed to deal. It appears probable, but by no means certain, that the self-consistent theory overestimates the shear modulus of this kind of rubber particle. On the other hand, the Kerner equation and the Hashin–Shtrikman lower bound almost certainly underestimate the modulus. If the moduli of the particle are known, the second stage of the calculation, to obtain a shear modulus for the rubber-toughened polymer, is not subject to the same degree of uncertainty. Since volume concentrations of rubber particles are rarely greater than 40 % in HIPS or ABS, the self-consistent model should give reasonably accurate predictions of G_0.

This view is supported by experimental evidence. Sigmoidal relationships between log G_0 and ϕ_2 have been observed by Sataka in blends of polystyrene with polybutadiene,[20] and by Kalfoglou and Williams in rubber-toughened epoxy resins.[21] Phase inversions occurred in both systems when the volume concentrations of the components were approximately equal.

In the field of rubber-toughened plastics, the Kerner equation has been employed more extensively than any other to represent the shear modulus of composites. Significantly, authors have found it necessary to modify the equation in various ways in order to fit experimental data.[14, 15, 22-24] The Kerner equation is also the basis of the Takayanagi treatment of viscoelastic composites containing spherical particles.[25-27] Takayanagi's model combines the series and parallel composite-element equations (5.2) and (5.3): in the first part of the calculation, material A is in parallel with material B, and the modulus of the element AB is obtained from eqn. (5.2); the composite element AB is then placed in series with another block of material A, and eqn. (5.3) is applied to calculate the modulus of the triple element ABA. Complex moduli are used in both parts of the calculation. There are two parameters in the resulting equation: the concentration ϕ_B of B in the composite ABA, and the ratio of series to parallel contributions to the total. As already indicated, the second parameter is chosen to fit the Kerner equation when the method is applied to rubber-toughened polymers.[25, 26] The Takayanagi model is essentially a curve-fitting procedure, in which role it has proved successful.[21, 28]

In order to avoid the problems arising from approximate theories, Bucknall[8] used the Hashin bound equations[10] to calculate bounds on the shear modulus of HIPS containing 5 % polybutadiene. The modulus of the composite rubber particle was first calculated as a function of polystyrene content, and the results were then used to calculate the modulus of the HIPS. The material containing 95 % matrix and 5 % rubber particles had the highest modulus. A material containing 75 % polystyrene matrix and 25 % composite rubber particles (composition 75:5:20) had a substantially lower modulus, although the use of widely spaced bounds precluded an accurate calculation of the difference between the two contrasting structures.

Bucknall and Hall extended this theoretical approach based on Hashin bounds by taking the viscoelastic properties of the phases into account.[29]

Their results are summarised in Figs. 5.10 and 5.11, which show calculated bounds on modulus and loss tangent as functions of frequency for two HIPS polymers containing poly(butadiene-*co*-styrene). Both polymers have the same rubber content, but the structures are different: the compositions can be written 95:5:0 and 75:5:20, where the ratio expresses

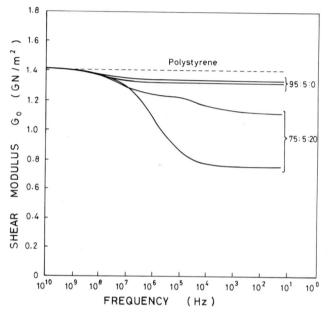

Fig. 5.10   *Bounds on shear modulus for two HIPS polymers, both containing 5% by volume of poly(butadiene-co-styrene), but having differing volumetric ratios of polystyrene matrix: polybutadiene: polystyrene inclusions, as shown (from C. B. Bucknall and M. M. Hall, ref. 29, reproduced with permission).*

the volume concentrations of polystyrene matrix: rubber: polystyrene sub-inclusions. At 20 °C, the rubber passes through a glass transition at frequencies of the order of 3 GHz. At lower frequencies, the shear modulus of HIPS is lower than that of polystyrene. Despite the uncertainty involved in using bound equations, it is clear that the composite rubber particles produce a much greater drop in modulus than the pure rubber particles. In other words, the volume of the rubber particles is more important than the shear modulus in determining the properties of the composite as a whole. The same conclusion can be drawn from the loss tangent data: the loss peak produced by pure rubber particles is much smaller than the peak due to the

composite rubber particles, although the concentration of copolymer
rubber is the same in both materials. The volume fraction of the particles is
the dominant factor. It is interesting to note that the maximum in the loss
curve shifts to lower frequencies on introducing sub-inclusions into the
rubber particles. According to the time–temperature superposition

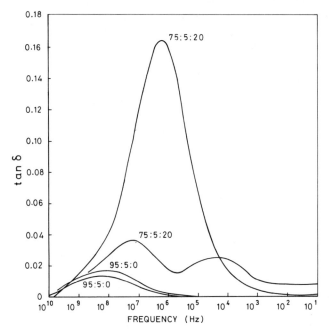

*Fig. 5.11   Bounds on loss tangent for the HIPS polymers described in Fig. 5.10
(from C. B. Bucknall and M. M. Hall, ref. 29, reproduced with permission).*

principle, this shift is equivalent to an upward shift in peak temperature at a
fixed frequency. The calculations strongly suggest that the effects illustrated
in Fig. 5.4 are largely mechanical rather than chemical in origin. The 94:6:0
HIPS in Fig. 5.4 had a secondary loss peak at tan $\delta = 0 \cdot 01$ (after subtracting
the loss tangent of the matrix), which increased in height to $0 \cdot 05$ in the
78:6:16 polymer. These figures are comparable with the calculated results
presented in Fig. 5.11.

There is obviously considerable scope for further work on
structure–modulus relationships in rubber-toughened plastics. In view of
the complexity of the mathematical problem, a simplified approach seems
desirable, and in practical terms the best solution is probably to use an

approximate expression for the shear modulus of the rubber particles, and concentrate attention upon the volume fraction of the particles. Much of the early literature on rubber-toughened plastics as composite materials is of limited value because the authors determined only rubber content, and not particle volume fraction. Bohn has shown that the dynamic mechanical properties of a range of HIPS and ABS polymers can be correlated by plotting moduli against particle volume.[30] As shown in Fig. 5.5, there are detectable differences between filled and unfilled particles of the same volume, but the errors introduced by neglecting these differences completely appear to be small compared with the uncertainties involved in using even the best of the approximate theories to calculate the modulus of the rubber-toughened polymer.

5.5 STRESS ANALYSIS

Rubber particles act as stress concentrators, since they have much lower shear moduli than the surrounding matrix material. An understanding of the inhomogeneous nature of the stress field is fundamental to any discussion of rubber toughening, which is known to be due to crazing and other yield mechanisms initiated at stress concentrations in the matrix. The first step in a quantitative treatment of toughening is to apply elasticity theory to calculate the stresses within the composite.

There are two major difficulties in attempting a theoretical stress analysis of rubber-toughened plastics. Firstly, as already discussed in this chapter, the rubber particles in many cases have a complex structure, and their moduli are consequently not known with any degree of accuracy. Secondly, the materials of practical interest have a high volume loading of rubber particles of various sizes, distributed randomly in space, so that there is a considerable overlap of stress fields between neighbouring particles. In some materials there is the additional problem that the rubber particles are spheroids or irregularly shaped, rather than simple spheres. Once again, it is necessary to apply approximate methods.

5.5.1 Isolated spherical particle

The simplest model relevant to the problem of rubber-toughened polymers is that of an isolated spherical particle embedded in an isotropic solid body, which is in a state of uniform uniaxial tension σ at all points remote from the particle. The particle is assumed to be homogeneous, isotropic and perfectly bonded to the matrix. This problem was solved by Goodier,[31] who obtained the following results, expressed in terms of the polar co-ordinates illustrated in Fig. 5.12:

$$\sigma_{rr} = 2G_m\left(\frac{2A}{r^3} - \frac{2v_m C}{(1-2v_m)r^3} + \frac{12B}{r^5}\right.$$

$$\left. + \left[-\frac{2(5-v_m)C}{(1-2v_m)r^3} + \frac{36B}{r^5}\right]\cos 2\theta\right)$$

$$\sigma_{\theta\theta} = 2G_m\left(-\frac{A}{r^3} - \frac{2v_m C}{(1-2v_m)r^3} - \frac{3B}{r^5} + \left[\frac{C}{r^3} - \frac{21B}{r^5}\right]\cos 2\theta\right) \qquad (5.9)$$

$$\sigma_{\psi\psi} = 2G_m\left(-\frac{A}{r^3} - \frac{2(1-v_m)C}{(1-2v_m)r^3} - \frac{9B}{r^5} + \left[\frac{3C}{r^3} - \frac{15B}{r^5}\right]\cos 2\theta\right)$$

$$\sigma_{r\theta} = 2G_m\left(-\frac{2(1+v_m)C}{(1-2v_m)r^3} + \frac{24B}{r^5}\right)\sin 2\theta$$

where

$$\frac{A}{R^3} = \frac{\sigma}{4G_m}\left\{\frac{\left[(1-v_m)\left(\frac{1+v_i}{1+v_m}\right) - v_i\right]G_i - (1-2v_i)G_m}{(1-2v_i)2G_m + (1+v_i)G_i}\right.$$

$$\left. - \frac{\sigma D\left[(1-2v_m)(6-5v_m)2G_m + (3+19v_i - 20v_m v_i)G_i\right]}{(1-2v_i)2G_m + (1+v_i)G_i}\right.$$

$$B = \sigma R^5 D$$

$$C = 5\sigma R^3 D(1-2v_m)$$

$$D = \frac{G_m - G_i}{8G_m\left[(7-5v_m)G_m + (8-10v_m)G_i\right]}$$

The notation for stress is explained more fully in Chapter 6.

Goodier's equations show that the maximum stress concentrations occur at the equator of the rubber particle ($r = R$, $\theta = 90°$). Figure 5.13 is a map of stress concentrations, expressed as the ratio of major principal stress to applied stress, in the neighbourhood of a polybutadiene rubber particle embedded in polystyrene. The stress concentration falls quite rapidly with distance from the surface of the particle, and the map emphasises the localised nature of the stress field. However, the importance of the stress field lies not in its extent, but in its capacity to generate crazes and shear bands.

Goodier's analysis is based on elasticity theory, which treats materials as continua. The equations are not strictly valid when the rubber particles are very small, because the scale of the stress field then approaches atomic dimensions. For example, the stress concentration factor falls from 1·92 to 1·80 over a distance of 1 nm when the particle has a diameter of 100 nm. To put these figures in perspective, it must be remembered that the diameter of

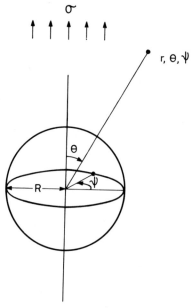

Fig. 5.12   *Spherical polar coordinates used by Goodier*[31] *for calculation of stresses around an isolated spherical inclusion.*

the polystyrene molecule is 0·6 nm, and the particle size in toughened PVC is often below 100 nm.

In the case of a spherical void, the stress concentration factor has a maximum value of 2·05 at the equator. The stress is biaxial at this point, since a normal stress cannot be generated across a free surface. The calculated maximum stress concentration factor at the surface of a rubber particle is somewhat lower, at 1·92, and the stress is triaxial, essentially because of the volume constraint represented by the bulk modulus of the rubber particle, which is comparable with that of the matrix.

Stress concentration factors are lower for composite rubber particles containing a high proportion of hard sub-inclusions. Owing to the difficulty of calculating the shear modulus of composite rubber particles, the effect of structure on stress concentrations is not known with any accuracy. Bound calculations show that the maximum stress concentration at the equator of a particle containing 80 % by volume of polystyrene sub-inclusions is between 1·54 and 1·89.

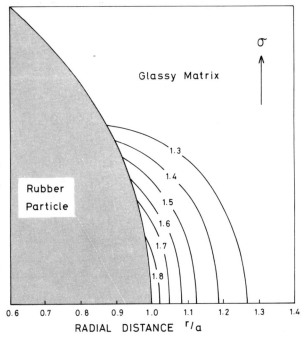

Fig. 5.13   *Contour map showing stress concentrations, calculated using Goodier's equations,*[31] *near the equator of an isolated rubber particle in HIPS under uniaxial tension.*

Goodier's equations apply to composites containing an infinitesimal concentration of particles. In most cases of practical interest, the rubber particles are too close together to be treated as isolated spheres, and a different approach is required. Analytical solutions are not available, and the stress analyst is therefore faced with the choice of using numerical methods, approximate solutions or experimental techniques.

5.5.2 Finite element analysis

The finite element method enables stress distributions to be analysed numerically where no analytical solution is available.[32, 33] Elastic properties are assigned to the materials under consideration, and the structure is divided into a large number of small elements. The problem is then solved with the aid of a computer, by finding the stress and displacement distributions that satisfy the conditions of equilibrium and continuity between elements, and at the same time satisfy the boundary conditions, which are usually stated in terms of applied forces or displacements.

In order to analyse fully the stress distribution in a typical rubber-toughened polymer, having a high volume loading of rubber particles, it is necessary to divide a representative region of the material into tetrahedral elements, and to develop a computer programme to solve the problem in three dimensions. A complete analysis of this type is laborious and expensive, and has not yet been applied to rubber-modified plastics. However, a simplified finite-element analysis has been performed by Broutman and Panizza,[34] who applied an approximate method to the case of an isotropic toughened polymer under uniaxial tension. Their technique was to draw a two-dimensional grid on the matrix material surrounding a rubber particle, and then to rotate this grid about the tensile axis, keeping the centre of the particle stationary. The unit for analysis was thus a circular cylinder having a rubber particle at its centre, and the matrix divided into axisymmetric elements. The boundary conditions were that the sum of forces acting on the ends of the cylinder equal the applied stress, the sum of radial forces acting on the cylinder be zero and the shear stresses on the cylinder be zero. The limitations of the method are recognised by the authors, but the results obtained nevertheless represent the best available information about stress concentrations at high volume loadings of rubber. At low volume loadings, the results obtained by the numerical method are in very good agreement with Goodier's analytical solution, an observation that helps to establish the validity of the approximate finite-element analysis.

Figure 5.14 shows principal stresses at the equator of the rubber particles as functions of rubber content, in a matrix having a Young's modulus of $2.75 \, GN/m^2$ and a Poisson's ratio of 0.35. It was found that variations between 7 and 21 MN/m^2 in Young's modulus of the rubber, and between 0.35 and 0.50 in Poisson's ratio made little difference to the calculated stress distributions in the matrix.

Figure 5.14 shows that with the interparticle spacings specified by the model, the stress concentration factor at the equator rises relatively slowly with rubber content up to a volume loading of about 30 %, but climbs quite rapidly between 40 and 50 %. In a real polymer, the rubber particles are of course randomly spaced, and particles are therefore much closer to their nearest neighbours than is assumed in the cylindrical-cell model on which the finite-element analysis was based. This means that the maximum stress concentrations in a rubber-toughened polymer are significantly higher, at a given rubber content, than is shown in Fig. 5.14.

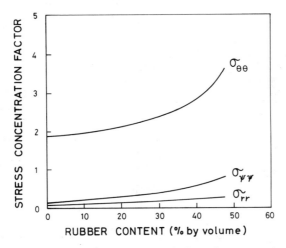

Fig. 5.14   Ratio of local principal stress to applied tensile stress at the equators of the rubber particles, calculated approximately by finite element analysis over a range of rubber volume loadings (from L. J. Broutman and G. Panizza, ref. 34, reproduced with permission).

In a more recent study, Agarwal and Broutman made a three-dimensional finite element analysis of stress distributions in composites containing spherical particles, and showed that the results were similar to those obtained by the approximate method based on axisymmetric elements.[35] They concluded that the increased accuracy of the full three-dimensional analysis was insufficient to justify the additional cost and effort.

5.5.3 Experimental methods

Standard photoelastic methods of experimental stress analysis are unsuitable for studying stress fields in rubber-toughened plastics, which are inhomogeneous in three dimensions. The photoelastic method is designed essentially for two-dimensional problems. There is, however, an alternative method of measuring maximum values of tensile stress, namely by observing craze formation. This method was adopted by Matsuo et al., who embedded 3 mm rubber spheres in polystyrene, and measured the critical applied stresses and strains required to initiate crazes at the surface of the spheres.[36] Figure 5.15 illustrates the results of experiments with two spheres. The presence of a second particle has little effect upon the crazing stress when the distance between particle centres is greater than $2 \cdot 90\, R$, where R is the radius of the particle. In other words, the stress fields do not overlap significantly unless the particles are less than $0 \cdot 9\, R$ apart.

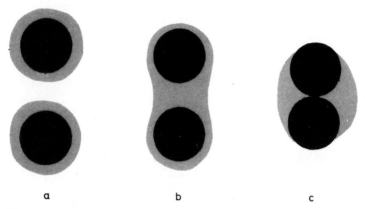

a   b   c

Fig. 5.15 Pattern of craze formation around neighbouring rubber spheres (after M. Matsuo et al., ref. 36).

Decreasing the interparticle spacing causes a rise in stress concentration in the region between the particles, so that crazes form preferentially in this region. These observations have an important bearing on the mechanism of rubber toughening, which is discussed in Chapter 7.

In order to measure stresses by the method outlined above, it is necessary to employ a criterion for craze formation. The various proposals that have been advanced are discussed in Chapter 6: experimental evidence indicates that a critical tensile strain criterion is the most satisfactory. As the minor principal stresses at the equator of the particles are small, the error involved in treating craze formation as a measure of stress rather than strain is relatively small.

In a related experiment, Oxborough and Bowden observed craze initiation in a series of HIPS polymers containing composite rubber particles at various concentrations.[37] They calculated bounds on stress concentration using Goodier's equations in combination with Hashin's bound equations, and obtained good agreement between observed and calculated crazing stress in a polymer containing 3 % by volume of rubber particles: the measured crazing stress lay midway between the bound values calculated for an isolated spherical particle of composite structure (85 % by volume of polystyrene sub-inclusions). Increasing the volume fraction of composite rubber particles to 13 % caused a drop in measured crazing stress, owing to overlap of stress fields between neighbouring rubber particles. Further reductions in crazing stress occurred at higher rubber contents. These observations form a natural link between the low-strain mechanical properties discussed in this chapter and the large-strain properties that form the subject of later chapters.

5.5.4 Thermal shrinkage stresses

Stresses are set up in rubber-toughened plastics not only by applied loads, but also by differential thermal contraction. Rubbers have higher coefficients of thermal expansion than glassy polymers, so that cooling a toughened polymer from the melt, which can be considered stress-free, produces stresses in the rubber particle and in the surrounding matrix. The magnitude of the stresses can be calculated for an isolated particle by elasticity theory.[38-40] In the case of composite rubber particles, the coefficient of expansion of the particle can be calculated from eqn. (5.8). Within the particle, thermal shrinkage produces a triaxial tensile stress given by[40]

$$\sigma_{rr} = \sigma_{\theta\theta} = \sigma_{\psi\psi} = \frac{4(\beta_p - \beta_m)(1 + v_p)G_m G_p \Delta T}{6(1 - 2v_p)G_m + 3(1 + v_p)G_p} \tag{5.10}$$

where ΔT is the temperature drop causing the thermal shrinkage stresses. Within the matrix, on the other hand, there is no hydrostatic component of stress. The stresses are purely deviatoric, as follows:

$$\sigma_{rr} = \left(\frac{R}{r}\right)^3 \left(\frac{4(\beta_p - \beta_m)(1 + v_p)G_m G_p \Delta T}{6(1 - 2v_p)G_m + 3(1 + v_p)G_p}\right) \tag{5.11}$$

$$\sigma_{\theta\theta} = \sigma_{\psi\psi} = -\frac{\sigma_{rr}}{2}$$

Thermal shrinkage stresses in the rubber particle cause a shift in the secondary loss peak to lower temperatures, owing to the pressure-dependence of T_g.[39-41] This effect can be related to the change in the free volume of the rubber under hydrostatic tension. It has been suggested that the stresses generated in the matrix by differential contraction affect the yield behaviour of toughened plastics,[40] but the evidence for this effect is slender.[41]

5.6 ORIENTATION

Moulded objects often show a high degree of molecular orientation, especially near the surface. In rubber-toughened polymers, originally spherical particles become distorted into non-spherical shapes as a result of orientation, and the foregoing calculations of elastic constants and stress distributions in isotropic materials containing spherical rubber particles are no longer relevant. The problem is not simply the shape of the particles, but also the anisotropy of stiffness in the matrix. Even in the simplest case, that of a transversely isotropic solid, five independent elastic constants are required in order to characterise stiffness properties, compared with two for an isotropic solid.

There is no analytical solution for the general case, in which the rubber particles have an ellipsoidal shape. Edwards has solved the problem of stress distributions around a prolate spheroid in an isotropic matrix,[42] and Chen has obtained expressions for the stresses around a spheroid in a transversely isotropic solid.[43] Like Goodier's equations, these solutions apply to isolated inclusions, and are therefore applicable to rubber-toughened plastics only when the volume fraction of rubber is below about 5%. Nevertheless, they form a basis for relating properties to orientation.

The effects of orientation on the properties of rubber-toughened plastics are similar to those observed in homogeneous polymers. In general, the modulus parallel to the direction of orientation increases with the level of orientation, which can be measured by observing rubber particle shape. In glassy polymers, these effects on small-strain properties are small compared with the effects upon large-strain properties such as yield and fracture, which are discussed in Chapters 8–10.

REFERENCES

1. I. V. Yannas, *J. Macromol. Sci. (Phys.)* **B6** (1972) 91.
2. J. A. Schmitt and H. Keskkula, *J. Appl. Polymer Sci.* **8** (1960) 132.
3. G. Cigna, *J. Appl. Polymer Sci.* **14** (1970) 1781.
4. E. R. Wagner and L. M. Robeson, *Rubber Chem. Technol.* **43** (1970) 1129.
5. S. Krause and L. J. Broutman, *J. Appl. Polymer Sci.* **18** (1974) 2945.
6. T. Yoshii, Ph.D. Thesis, Cranfield, England, 1975.
7. S. G. Turley, *J. Poly. Sci.* Cl (1963) 101.
8. C. B. Bucknall, *J. Materials* (ASTM) **4** (1969) 214.
9. Z. Hashin and S. Shtrikman, *J. Mech. Phys. Solids* **11** (1963) 127.
10. Z. Hashin, *J. Appl. Mech.* **29** (1962) 143.
11. R. Hill, *J. Mech. Phys. Solids* **13** (1965) 213.
12. B. Budiansky, *J. Composite Mat.* **4** (1970) 286.

13. J. C. Halpin, *J. Composite Mat.* **3** (1969) 732.
14. L. E. Nielsen, *J. Appl. Polymer Sci.* **17** (1973) 3819.
15. L. E. Nielsen, *Rheol. Acta* **13** (1974) 594.
16. W. E. A. Davies, J. Phys. (Appl. Phys.) **D4** (1971) 1176.
17. G. Allen, D. J. Blundell, M. J. Bowden, F. G. Hutchinson and G. M. Jeffs, *Faraday Spec. Discussions Chem. Soc.* **2** (1972) 127.
18. E. H. Kerner, *Proc. Phys. Soc.* **69B** (1956) 802, 808.
19. N. Laws, *J. Mech. Phys. Solids* **21** (1973) 9.
20. K. Sataka, *J. Appl. Polymer Sci.* **14** (1970) 1007.
21. N. K. Kalfoglou and H. L. Williams, *J. Appl. Polymer Sci.* **17** (1973) 1377.
22. R. A. Dickie, *J. Appl. Polymer Sci.* **17** (1973) 45.
23. R. A. Dickie and Mo-Fung Cheong, *J. Appl. Polymer Sci.* **17** (1973) 79.
24. K. D. Ziegel, *ACS Poly. Prepr.* **15**(1) (1974) 442.
25. M. Takayanagi, H. Harima and Y. Iwata, *Mem Fac. Engng Kyushu Univ.* **23** (1963) 1.
26. M. Takayanagi, S. Uemura and S. Minami, *J. Poly. Sci.* C5 (1964) 113.
27. S. Uemura and M. Takayanagi, *J. Appl. Polymer Sci.* **10** (1966) 113.
28. G. Kraus, K. W. Rollman and J. T. Gruver, *Macromolecules* **3** (1970) 92.
29. C. B. Bucknall and M. M. Hall, *J. Mater. Sci.* **6** (1971) 95.
30. L. Bohn, *ACS Poly. Prepr.* **15**(1) (1974) 323.
31. J. N. Goodier, *J. Appl. Mech.* **55** (1933) 39.
32. O. C. Zienkiewicz, *The Finite Element Method*, McGraw-Hill, New York, 1967.
33. R. T. Fenner, *Finite Element Methods for Engineers*, Macmillan, London, 1975.
34. J. J. Broutman and G. Panizza, *Int. J. Poly. Mater.* **1** (1971) 95.
35. B. D. Agarwal and L. J. Broutman, *Fibre Sci. Technol.* **7** (1974) 63.
36. M. Matsuo, T. Wang and T. K. Kwei, *J. Poly. Sci.* A2 (1972) 1085.
37. R. J. Oxborough and P. B. Bowden, *Phil. Mag.* **30** (1974) 171.
38. S. Timoshenko and J. N. Goodier, *Theory of Elasticity* (3rd edn.), McGraw-Hill, New York, 1970, Art 148.
39. T. T. Wang and L. H. Sharp, *J. Adhesion* **1** (1969) 69.
40. R. H. Beck, S. Gratch, S. Newman and K. C. Rusch, *Polymer Letters* **6** (1968) 707.
41. L. Morbitzer, K. H. Ott, H. Schuster and D. Kranz, *Angew. Makromol. Chem.* **27** (1972) 57.
42. R. H. Edwards, *J. Appl. Mech.* **18** (1951) 19.
43 W. T. Chen, *J. Appl. Mech.* **35** (1968) 770.

CHAPTER 6

DEFORMATION MECHANISMS IN GLASSY POLYMERS

Although the elastic and viscoelastic properties discussed in Chapter 5 are of considerable interest to manufacturers and users of rubber-toughened plastics, the main focus of interest is upon the large-strain properties, especially yield and fracture. It is now generally recognised that the deformation mechanisms responsible for large strains in toughened polymers are essentially the same as those observed in the homogeneous glassy polymers from which they are derived. The rubber is present as a discrete disperse phase within the glassy matrix, and cannot itself contribute directly to a large deformation: the matrix must first yield or fracture around the rubber particles. In a sense, therefore, the rubber acts as a catalyst, altering the stress distribution within the matrix, and producing a quantitative rather than a qualitative change in deformation behaviour. It follows that in order to discuss the yield and fracture of rubber-toughened plastics, it is first necessary to understand the large strain behaviour of homogeneous plastics, especially brittle glasses such as polystyrene, poly(methyl methacrylate) and epoxy resin.

At strains of the order of 0.2%, glassy polymers begin to show evidence of local plastic deformation, either by crazing or by shear band formation. The distinction between viscoelastic and plastic processes is a fine one, which is often difficult to make with any certainty. By definition, viscoelastic deformation is recoverable over a time interval, whereas plastic deformation is not. The problem is that some viscoelastic recovery processes are extremely slow, whilst many apparently plastic deformations in polymers are recoverable on heating, and are probably recoverable at room temperature over a sufficiently long time. Under these circumstances, the distinction between viscoelastic and plastic processes becomes blurred. Crazing and shear yielding are often described as plastic deformation

136

mechanisms, although they are significantly recoverable, as discussed later in this chapter.

It is convenient to classify deformation mechanisms in glassy polymers as either shear processes or cavitation processes. Shear processes include both diffuse shear yielding and localised shear band formation, and occur without loss of intermolecular cohesion in the polymer, so that they produce little, if any, change in density. Cavitation processes include crazing, void formation and fracture, and are characterised by a local loss of intermolecular cohesion, and therefore by significant local decreases in density. These two types of deformation are discussed separately below, and possible interactions between them are considered in the final section of the chapter.

6.1   SHEAR YIELDING

Shear deformation consists of a distortion of shape without significant change in volume. In crystalline materials, including both metals and plastics, shear yielding takes place by slip on specific *slip planes*, as a result of dislocation glide. Slip occurs preferentially on planes of maximum resolved shear stress. In non-crystalline polymers, large-strain deformation requires more general co-operative movement of molecular segments, and shear processes are therefore much less localised than in crystalline materials. The degree of localisation varies. In some polymers diffuse shear yielding takes place throughout the stressed region, whilst in others the yielding is localised into clearly defined shear microbands. Localised shear yielding of this type, which is due to strain-softening effects in the glassy polymer, is of particular interest in discussing rubber toughening, and is therefore considered in some detail in the following review of yielding. More extensive reviews of the subject have been published by Bowden[1] and by Ward.[2, 3]

6.1.1   Strain localisation

Strain localisation can occur for two reasons. The first is purely geometrical, and applies only to certain types of loading. The second has its origins in the properties of the material itself, and occurs under more general loading conditions.

Geometrical instabilities are often observed in tensile yield tests. The stress and strain are usually highest at one point on the parallel gauge portion of the specimen, owing to slight variations in cross-sectional area. As the test proceeds, the area is reduced preferentially at this point, causing the stress to rise further, so that a neck is formed. The true stress continues to rise whilst the nominal stress (load/original area) may remain constant. Molecular orientation within the neck gradually increases the resistance to further deformation, and thereby provides a mechanism for eventual stabilisation of the neck. Strain hardening is also observed in metals, but for different reasons.

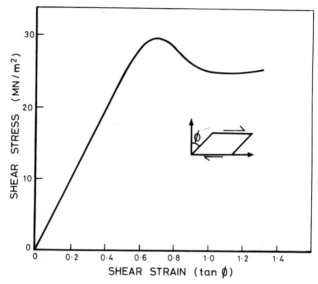

Fig. 6.1   Strain softening of PMMA during shear deformation at 70 °C. Strain rate 0.0033 s^{-1} (from S. S. Sternstein et al., ref. 4, reproduced with permission).

The second, more fundamental, cause of inhomogeneous deformation is strain softening. An example is shown in Fig. 6.1. Under an applied shear stress, PMMA yields at a shear strain of approximately 0.7, beyond which the resistance to further deformation falls with shear strain.[4] In contrast to the tensile test, there is no geometrical instability; the yield drop occurs because the properties of the polymer itself alter with strain. Strain softening is, of course, a contributory factor to the yield drop in tensile tests on many materials, in addition to the geometrical effect of necking.

Various explanations have been advanced to account for yielding and strain softening in polymers. Earlier authors tended to concentrate upon thermal softening effects, but it is now generally recognised that the adiabatic heating effect at moderate strain rates is insufficient to cause yielding, and that temperature rises occur after, rather than before, the yield point. More recent discussions have focused attention upon stress-activated flow processes, following the treatment first proposed by Eyring.[5] Robertson[6, 7] pointed out that a simple analysis based upon the Eyring equation does not explain some of the characteristic features of yielding in polymers. In particular, the stress-activated flow theory does not explain the long delay often observed between application of load and yielding in a creep test. He developed a theory of stress-induced fluidisation to account for this effect. According to this theory, the molecular conformations alter with time in the stressed polymer until they have produced a more fluid structure, and the viscosity of the material has fallen sufficiently to permit yielding. An alternative to Robertson's approach was developed by Argon,[8] who calculated the activation energy for the formation of a pair of molecular kinks, in terms of the elastic interactions with neighbouring molecules. According to this theory, local stored elastic energy is relieved by the formation of similar kinks in adjacent chains, giving an increment in shear strain. The predictions of Argon's model are in good agreement with experimental observations on the pressure, temperature and strain rate dependence of yield behaviour in glassy polymers.

A high degree of strain softening in a glassy polymer leads to the formation of *shear bands*, which are thin planar regions of high shear strain. Shear bands are initiated in regions where there are small inhomogeneities of strain due to internal or surface flaws, or to stress concentrations. Beyond the yield point, these regions have a lower resistance to deformation than the surrounding material, with the result that the strain inhomogeneity increases, and strain softening further lowers the shearing resistance of the material within the incipient band. The material at the tip of the sheared zone becomes strained and softened, causing the band to propagate along a plane of maximum resolved shear stress. The conditions necessary for shear band formation are discussed by Bowden.[9] The tendency towards localisation of shear strain into bands increases with the size of the strain inhomogeneities and with the strain softening parameter $(\partial \dot{\gamma}/\partial \gamma)_\tau$, where $\dot{\gamma}$ = strain rate, γ = shear strain and τ = shear stress. More specifically, the rate at which shear bands develop depends critically upon the ratio of the strain-rate sensitivity of the flow stress, $(\partial \tau_y/\partial \ln \dot{\gamma})_\gamma$, to the negative slope of the stress–strain curve, $- (\partial \tau/\partial \gamma)_{\dot{\gamma}}$.

The degree of strain softening, and hence the sharpness of the shear bands, depends not only upon the chemical composition of the polymer, but also upon the temperature, strain rate and thermal history of the sample. The effects of temperature and strain rate on the yield behaviour of polystyrene are illustrated in Fig. 6.2.[10] Reducing the temperature from 80°

to 75 °C produces a marked increase in both the yield stress and the magnitude of the yield drop. The trend continues at lower temperatures, and at 65 °C there is a sharp transition from diffuse yielding to shear microband formation.[10] A similar transition from homogeneous yielding to shear band formation occurs in PMMA at the much lower temperature of −131 °C.

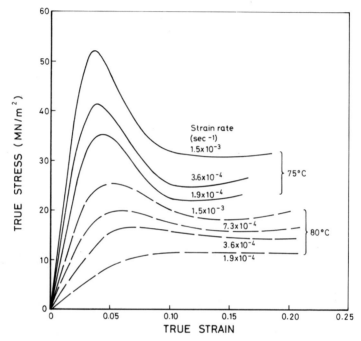

Fig. 6.2   Stress–strain curves for polystyrene in plane-strain compression (from P. B. Bowden and S. Raha, ref. 10, reproduced with permission).

Thermal history has a profound effect upon the transition from diffuse to localised yielding. The transition reported in polystyrene at 65 °C occurs in samples cooled slowly from T_g to room temperature. Air cooling shifts the transition to 55 °C, and quenching into iced water reduces it to temperatures below room temperature. Conversely, shear bands form at room temperature in PMMA that has been annealed at 95 °C for two days.[11] These observations form part of a more general pattern of behaviour known as *physical ageing:* thermal treatment below T_g is well known to alter the mechanical and thermodynamic properties of polymeric glasses.[12–14] The changes in molecular packing responsible for these physical ageing effects are not fully understood.

The relationship between the type of strain inhomogeneity developed and the constraints imposed by the surrounding material upon the deforming region is discussed by Bowden in an excellent review.[1] The criterion is that if the geometry of the formation of a strain inhomogeneity involves a restraint in any direction, the extension rate $\dot{\varepsilon}$ must be zero in that direction. This principle is illustrated for the tensile test in Fig. 6.3. In the cylindrical

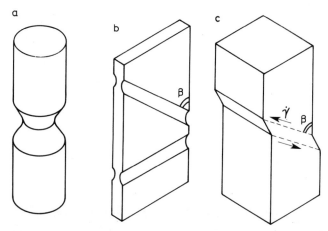

Fig. 6.3   Formation of strain inhomogeneities under tension: (a) neck in cylindrical specimen; (b) inclined neck in flat strip (c) shear band (from P. B. Bowden, ref. 1).

specimen shown in Fig. 6.3(a) there is no restraint, and the only possible form of strain inhomogeneity is a symmetrical neck of the type indicated: if the strain rate is $\dot{\varepsilon}$ in the direction of applied stress, the strain rates in the two lateral directions are both $- \dot{\varepsilon}/2$. In the flat strip specimen illustrated in Fig. 6.3(b), there is a restraint in one dimension: the extension rate must be zero along the length of the band. This requirement is met only if the angle β is 54·7°, so that $\cos 2\beta = -\frac{1}{3}$, assuming that the deformation takes place at constant volume. The angle β is greater if the material dilates in forming the band. The deformation rate tensor is the same in case (b) as in case (a).

The third case, illustrated in Fig. 6.3(c), is the one of greatest interest in a discussion of rubber toughening. Within the body of a solid specimen, the material is restrained in two dimensions. Planar plastic zones develop without lateral contraction from the sides of the specimen. Under these conditions, the only possible type of deformation at constant volume is simple shear parallel to the plane of the band; in an isotropic material the angle β is 45°. If the strain rate is $\dot{\varepsilon}$ in the direction of applied stress, the extension rates are zero and $-\dot{\varepsilon}$ respectively in the other principal directions. As in case (b), the angle β increases if the material is able to dilate in forming the band, so that volume expansion can take place normal to the plane of the band. The limiting case is represented by craze formation, which involves a large volume increase; the angle β is then 90°.

6.1.2 Structure of shear bands

Shear bands were first observed by Whitney[15] in compression experiments on polystyrene. The compression tool generated stress concentrations in the surface of the material, which gave rise to sets or packets of parallel microbands that propagated across the specimen at an angle of approximately 40° to the compression direction, as illustrated in Fig. 6.4. Later experiments showed that shear bands form in tension as well as in compression, and are produced in a wide range of glassy polymers, including PMMA, PVC, epoxy resin and amorphous poly(ethylene terephthalate).[11, 16]

Shear bands are highly birefringent, and are most clearly observed in transmitted polarised light. They are also visible as reflecting planes in

Fig. 6.4 Shear bands formed under plane strain compression in polystyrene at 60 °C viewed between crossed polars.

ordinary transmitted light at glancing incidence, owing to refractive index differences between the band and the adjacent undeformed polymer.

Light microscopy indicates a thickness of approximately 1 μm for each band,[10, 15] but under the electron microscope sections through shear bands reveal clusters of irregular lines, each approximately 0·1 μm thick, indicating that the basic structure is very fine. The origin of the electron contrast is not clear.

The shear strains within individual bands are high. Argon and co-workers obtained values of shear strain between 1·0 and 2·2 from measurements of surface scratch displacements in polystyrene,[17] and Bowden and Raha[10] calculated similar values, in the neighbourhood of 2.2, from birefringence studies. The extinction direction was at an angle of 20–23° to the plane of

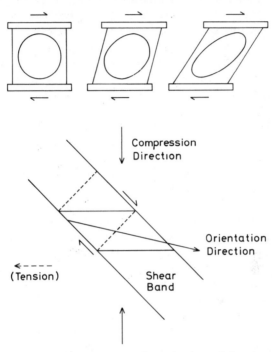

Fig. 6.5   (*Top*) *Schematic representation of simple shear deformation to large strains.* (*Bottom*) *Relationship between compression (or tension) direction, direction of molecular orientation and angle of shear band in localised shear yielding of a glassy polymer.*

the shear band. The birefringence results appear to be more reliable. Displacement measurements would be expected to underestimate the strain within the 0·1 μm thick microbands, and to give an average strain for the zone containing the set of associated microbands, which is seen as a single band in the optical microscope. The displacements involved in reaching a high homogeneous shear strain are shown schematically in Fig. 6.5. A simple shear process is the sum of a pure shear, which involves only a shape change, and a rotation. A circle inscribed on the specimen will therefore deform to an ellipse, which rotates in the manner indicated. A fuller description of large-strain shear deformation is given by Reiner.[18]

High shear strains and birefringence indicate a considerable degree of molecular orientation within the shear band. Figure 6.5 shows schematically the relationship between compression direction, band angle and orientation direction in a shear band formed at an angle of 45° to the compression axis, with a shear strain of 1·0. The diagram applies equally well to a band formed under uniaxial tension at right angles to the direction indicated for compression, i.e. horizontally in the illustration. In other words, the molecular orientation direction is at a high angle to the compression axis in a compression test, but at a low angle to the tensile axis in a tensile test. This point is of importance in considering the interaction of shear bands with crazes or cracks.

The angle between the shear band and the compression direction is usually smaller than the angle of 45° predicted by ideal plasticity theory: deviations of up to 8° have been observed.[16] Similarly, the angle between the shear band and the applied tensile stress is usually greater than 45°.[11] A contributory factor is the elastic recovery of the material; band angles are usually measured after unloading, so that they do not correspond exactly to the angle of formation. Bowden and Jukes showed that this effect accounted for the observed differences between theoretical and observed angle in a number of polymers.[16] However, the deviation in polystyrene was too large to be explained in this way.

Alternative explanations are reviewed by Kramer.[19] There is no obvious single cause. Plasticity theory predicts a band angle of 45° in isotropic materials deforming at constant volume according to a pressure-independent yield criterion. None of these conditions is satisfied for shear band formation in polystyrene. Kramer showed that even in an initially isotropic specimen there is significant anisotropy in the material ahead of the propagating shear band.[19] Furthermore, volume changes do occur during shear band formation, and the yield criterion is pressure-dependent. These points are discussed in more detail below. Their relative importance has yet to be established.

The relationship between band angle and volume effects is particularly obscure. Argon and co-workers suggested a transient local dilatation during shear band formation.[17] On the other hand, Brady and Yeh estimated the density of shear bands in polystyrene to be 0·2% *higher* than that of undeformed polymer.[11] This volume contraction is similar in magnitude to the contraction occurring as a result of ordinary homogeneous hot or cold drawing. The two views are not incompatible.

6.1.3 Shear band formation

In his study of shear band formation in polystyrene, Kramer subjected notched bars to uniaxial compression at constant strain rate, and observed shear deformation initiated at the notch, as illustrated in Fig. 6.6. Birefringence studies revealed the presence of a diffuse shear zone, which initiated at a nominal compressive strain of 1·75 % and propagated across

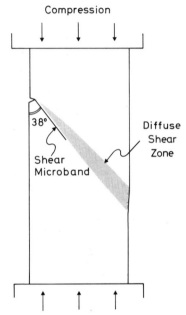

Fig. 6.6 *Drawing illustrating the extent of localised and diffuse shear yielding at the general yield point in a notched bar of polystyrene under uniaxial compression (after E. J. Kramer, ref. 20, reproduced with permission).*

the specimen at an angle of 45° to the compression direction, from the base of the notch.[19, 20] This diffuse yielding was viscoelastic rather than plastic: on removal of the stress, the material recovered almost completely within about 50 s. The strain within the diffuse shear zone was measured by a moiré fringe technique, using a grid applied to the surface of the specimen.

Shear microbands initiated at a nominal compressive strain of 2·3 %, also from the base of the notch, and propagated across the specimen at an angle of 38° to the compression direction. By comparison with diffuse shear yielding, their contribution to the total strain rate was comparatively small; a small strain over a large area had a greater effect than a large strain over a very small area. The contributions of elastic, diffuse shear and localised shear band deformation to the total strain rate are shown in Fig. 6.7 as functions of nominal strain, calculated on the assumption that strain occurs uniformly over the entire length of the specimen. The true strains at the base of the notch, where the shear processes are initiated, are of course significantly higher than the nominal strains.

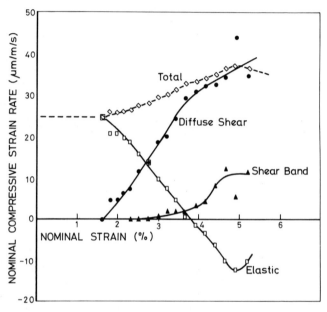

Fig. 6.7　*Contributions of shear bands, diffuse shear and elastic strain to total specimen strain in uniaxial compression of polystyrene at 25 C, as functions of strain (from E. J. Kramer, ref. 20, reproduced with permission).*

Under the imposed conditions of constant crosshead speed, the load on the material fell when the *diffuse* shear zone reached the opposite side of the specimen to the notch. This response, immediately following a load maximum, is known as *general yield*. At the general yield point, the shear band packet reached only about one-third of the way across the specimen, a fact that emphasises the relatively small contribution of shear bands to the overall deformation of the material.

Both diffuse and localised shear yielding in this experiment result from strain softening of the polystyrene. The strain inhomogeneity required by Bowden's theory to cause localisation of the strain into shear microbands is provided initially by the notch. Subsequently, as the shear band packet propagates, the stress concentrations at the tip of the packet provide the basis for maintaining the localisation of the bands. However, this localisation is not completely confined to the plane of the propagating bands. Brady and Yeh report the formation of secondary shear bands at 70–80° to the primary band packets.[14]

Kinetics

The rate of propagation of shear bands increases rapidly with both applied stress and temperature. There is some difficulty in defining precise rates of shear band propagation from direct visual observations, as the velocity initially falls and later rises. However, there is an intermediate period during which the velocity has a constant minimum value; this 'plateau velocity' is used to characterise the kinetics of shear band formation. The initial deceleration of the bands appears to be due to a reduction in stress concentration factor as the packet propagates away from the notch. The subsequent acceleration seems to be an artefact associated with reflections of light from the diffuse shear zone, which make the shear band packet look longer than it is.[19]

The relationship between shear band velocity $\dot{a}$ and applied stress σ is illustrated in Fig. 6.8. Within experimental error, ln $\dot{a}$ is linear with σ at any given temperature. At constant applied stress, ln $\dot{a}$ is linear with T^{-1}, where T is absolute temperature; this relationship is illustrated in Fig. 6.9. These observations show that shear band propagation is a stress- and temperature-activated process, which follows the Eyring flow equation, as follows:

$$\dot{\varepsilon} = 2A \exp\left(-\frac{\Delta H^*}{kT}\right) \sinh\left(\frac{\gamma V^* \sigma}{4kT}\right) \qquad (6.1)$$

where $\dot{\varepsilon}$ = strain rate, A = constant, H^* = activation enthalpy, V^* = activation volume, γ = stress concentration factor and k = Boltzmann's constant.

Eyring's treatment is an extension of the well-known Arrhenius analysis of chemical rate processes, which states that reacting molecules must pass over an energy barrier of height ΔH^* in order to transform into product molecules. The probability that a molecule will acquire sufficient thermal energy to surmount the barrier is given by the expression $\exp(-\Delta H^*/kT)$, in accordance with the Maxwell–Boltzmann law. Eyring treated viscous flow and plastic deformation in a similar way, considering each flowing molecule as passing over a symmetrical energy barrier as it squeezes past its

neighbours into a new position.[5] On application of a shear stress, there is a
net flow in one direction: the applied stress does work on the molecule,
increasing its energy when it moves in the forward direction, but reducing its
energy when it moves in the reverse direction; in effect, the height of the
energy barrier is reduced on one side and increased by the same amount on

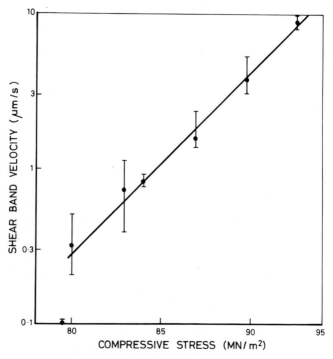

Fig. 6.8   *Eyring relationship between log (shear band velocity) and uniaxial
compressive stress in polystyrene at* 26 °C *(from E. J. Kramer, ref. 19, reproduced with
permission).*

the other side. If the area of the flow unit, which might be an atom, a
molecule or a chain segment, is given by A^*, and the distance between
equilibrium positions is λ, the work done by a shear stress τ is $\lambda A^*\tau$. More
generally, the work term is given by $\gamma\lambda A^*\tau$, where the stress concentration
factor γ takes account of the difference between the average stress and the
local stress on the flow unit. The quantity λA^*, which has the dimensions of
volume, is known as the *activation volume* V^*; it must be recognised,
however, that V^* does not necessarily represent a real physical volume. The

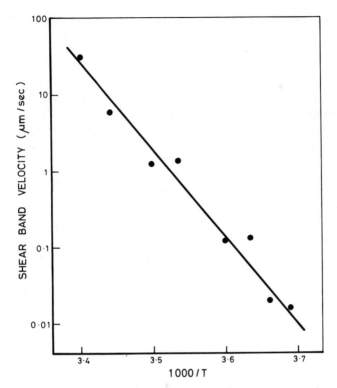

Fig. 6.9   Eyring relationship between log (shear band velocity) and reciprocal of temperature T (°K) in polystyrene under a uniaxial compressive stress of 90 MN/m² (from E. J. Kramer, ref. 19, reproduced with permission).

work done by the stress in moving the molecule from its equilibrium position, through a distance $\lambda/2$, to the top of the symmetrical energy barrier is $\gamma V^* \tau/2$. For a specimen under uniaxial tensile stress σ, the equivalent expression is $\gamma V^* \sigma/4$, since $\tau = \sigma/2$. This expression represents the amount by which the energy barrier is reduced in the forward direction and raised in the reverse direction on application of the stress. The net flow rate, which is the algebraic sum of the forward and reverse rates, is then given by

$$\dot{\varepsilon} = A \exp -\left(\frac{\Delta H^* - \gamma V^* \sigma/4}{kT}\right) - A \exp -\left(\frac{\Delta H^* + \gamma V^* \sigma/4}{kT}\right) \quad (6.2)$$

Equation (6.1) follows immediately from this equation. A more useful form of the equation is obtained by neglecting the second exponential term in eqn. (6.2), which is very small at moderate stresses, and writing

$$\dot{\varepsilon} = A \exp - \left(\frac{\Delta H^*}{kT}\right) \exp\left(\frac{\gamma V^* \sigma}{4kT}\right) \qquad (6.3)$$

This form of the equation is the basis of the logarithmic plots shown in Figs. 6.8 and 6.9; the quantities ΔH^* and γV^* are obtained directly from the slopes of the lines. The chief merit of the hyperbolic sine function in eqn. (6.1) is that it tends to zero as the stress tends to zero, whereas eqn. (6.3) suggests that the strain rate is finite in an unstressed specimen.

The effects of hydrostatic pressure on yield behaviour can be included by the use of additional terms in the Eyring equation.[11, 14] Another modification takes account of recovery effects in the deformation and flow of polymers. Recovery implies that the free energy of the unstressed material is lower in the unstrained state than in the strained state. In other words, the energy barrier is asymmetrical. Under high stress there is a net flow in the forward direction to produce a strained state, but under zero stress there is spontaneous flow in the reverse direction, so that the material recovers.

From the data presented in Figs. 6.8 and 6.9, Kramer obtained a value of 270 kJ/mol for the activation enthalpy of shear band formation in polystyrene, and 4·6 nm³ for the apparent activation volume γV^*.[19] In a more extensive study, Brady and Yeh found that ΔH^* increased from 108 kJ/mol at $-70\,°C$ to 430 kJ/mol at $89\,°C$, and that γV^* decreased from 4·6 to 3·1 nm³ over the same temperature range.[14] Thermal treatment of the polystyrene below T_g significantly affected both quantities.

6.1.4 Criteria for shear yielding

The preceding sections have focused attention upon a limited range of stressing conditions. In a more general discussion of yielding, it is necessary to consider the effects of multiaxial loading, and in particular the effects of hydrostatic pressure. A complete specification of stress requires nine separate terms, as indicated in Fig. 6.10: one normal stress and two shear stresses act on each pair of faces of the cube.

These components of stress are represented in the form of a matrix:

$$\begin{vmatrix} \sigma_{11} & \sigma_{12} & \sigma_{13} \\ \sigma_{21} & \sigma_{22} & \sigma_{23} \\ \sigma_{31} & \sigma_{32} & \sigma_{33} \end{vmatrix}$$

where σ_{ij} represents the component of stress acting in the direction j on the

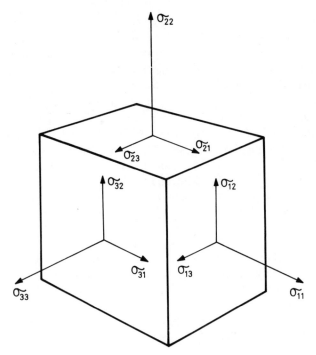

Fig. 6.10   *The nine components of the stress tensor.*

face having a normal in the i direction. The matrix can be simplified in two ways: firstly, equilibrium requires that the shear stresses be symmetrical ($\sigma_{ij} = \sigma_{ji}$); secondly, by suitable choice of axes, the shear components can be reduced to zero, so that the stress is expressed in terms of three *principal stresses*, σ_{11}, σ_{22} and σ_{33}, each acting independently. In a similar way, the strain at a point can be expressed in terms of three *principal strains*, ε_{11}, ε_{22} and ε_{33}. For a fuller description of stress as a *tensor* quantity, the reader should consult a standard text. The book by Nye[21] is especially to be recommended.

It is convenient to split the stress tensor into two terms as follows:

$$
\begin{vmatrix} \sigma_{11} & 0 & 0 \\ 0 & \sigma_{22} & 0 \\ 0 & 0 & \sigma_{33} \end{vmatrix} = \begin{vmatrix} p & 0 & 0 \\ 0 & p & 0 \\ 0 & 0 & p \end{vmatrix} + \begin{vmatrix} \sigma_{11}-p & 0 & 0 \\ 0 & \sigma_{22}-p & 0 \\ 0 & 0 & \sigma_{33}-p \end{vmatrix} \quad (6.4)
$$

Hydrostatic component   Deviatoric component

where

$$p = \tfrac{1}{3}(\sigma_{11} + \sigma_{22} + \sigma_{33}) \tag{6.5}$$

In an *isotropic* material, the effect of the hydrostatic component, which is effectively a pressure term, is to alter the volume of the material without changing its shape. The deviatoric component alters the shape of the sample without affecting its volume. This separation of the stress tensor into two components is especially useful in discussing yield behaviour.

In attempting to formulate yield criteria for polymers, the obvious starting point is the plasticity theory developed for metallic materials. The simplest criterion is due to Tresca, and states that a material will yield when the maximum shear stress within the material reaches a critical value. In terms of principal stresses, Tresca's criterion may be expressed as follows:

$$|\sigma_{11} - \sigma_{33}| = 2\tau_{\text{T}} \tag{6.6}$$

where σ_{11} and σ_{33} are maximum and minimum principal stresses, and τ_{T} is a materials constant, the critical shear stress. Tresca's criterion works well for mild steel, which yields inhomogeneously by forming Luders bands, but is less satisfactory for other metals and alloys, in which yielding is a more homogeneous process.

Homogeneous yielding of metals is better described by the von Mises criterion, which in terms of principal stresses has the form

$$(\sigma_{11} - \sigma_{22})^2 + (\sigma_{22} - \sigma_{33})^2 + (\sigma_{33} - \sigma_{11})^2 = 2\sigma_{\text{ty}}^2 \tag{6.7}$$

where σ_{ty} is another materials constant, the yield stress in uniaxial tension. In a Hookean material, *i.e.* a material in which stress and strain are linearly related, the von Mises criterion means that yielding will occur when the *shear strain energy density* reaches a critical value.

The two criteria represented by eqn. (6.6) and (6.7) are compared diagrammatically in Fig. 6.11, which shows the intersection of the yield envelopes with the plane $\sigma_{22} = 0$. The Tresca criterion corresponds to a regular hexagonal prism in principal stress space, with its axis of symmetry along the line $\sigma_{11} = \sigma_{22} = \sigma_{33}$. The von Mises yield envelope is a circular cylinder having the same axis. Fig. 6.11c is a section through both envelopes in a plane perpendicular to this axis.

Experiments show that neither the Tresca nor the von Mises criteria adequately describe the yield behaviour of polymers. Yield stresses are invariably higher in uniaxial compression than in uniaxial tension, and uniaxial tensile tests conducted in a pressure chamber show that yield stresses of polymers increase significantly with hydrostatic pressure.[22-24] However, both the Tresca and von Mises criteria are expressed in terms of

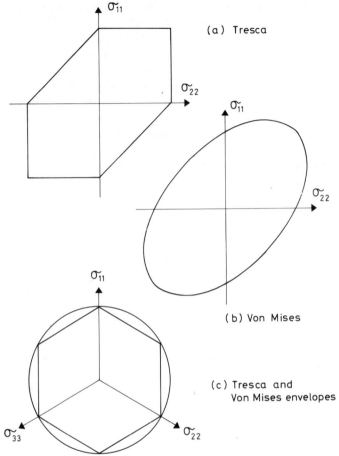

Fig. 6.11   *Cross sections of the Tresca and von Mises yield envelopes in principal stress space:* (a) *Tresca criterion in the plane* $\sigma_{33} = 0$; (b) *von Mises criterion in the plane* $\sigma_{33} = 0$; (c) *both criteria, compared in the plane normal to the* [111] *direction.*

differences between principal stresses, and therefore cannot reflect pressure effects: in both equations, the left-hand side is unaltered by substituting $(\sigma_{11} - p)$, $(\sigma_{22} - p)$ and $(\sigma_{33} - p)$ for σ_{11}, σ_{22} and σ_{33}.

Several solutions have been proposed. The first is to apply the Coulomb criterion, which is a modified version of the critical shear stress criterion, developed originally for soil mechanics.[25] The criterion may be written:

$$\tau = \tau_c + \mu_c \sigma_n \qquad (6.8)$$

where τ and σ_n are the shear stress and normal compressive stress acting on the shear plane, and τ_c and μ_c are materials constants. The quantity μ_c is analogous to a coefficient of friction, and τ_c is the *cohesion* of the material. The Coulomb criterion is appropriate for a material in which dilatation during shear flow is resisted by normal compressive forces acting on the shear plane. However, Bowden[1] has pointed out that this type of dilatation does not occur in polymers undergoing shear deformation; there is a density increase on yielding,[11] as mentioned in Section 6.1.2. Nor indeed is there evidence for dilatation in soils. In both classes of materials, the change in yield stress with pressure appears to be due to a change in the *state* of the material, caused by the *hydrostatic* component of stress, rather than to just one normal stress term.

In the light of this discussion, Bowden proposed a modified version of the Tresca criterion,[1] in the form

$$\tau = \tau_T + \mu_T p \qquad (6.9)$$

where τ_T and μ_T are again materials constants, but μ_T is no longer a 'coefficient of friction'; instead, it defines the effect of pressure, which is perhaps related to its effect upon the free volume of the material. For a uniaxial compression test, the modified Tresca criterion predicts yielding on planes at $45°$ to the compression axis, where the resolved shear stress is a maximum, whereas the Coulomb criterion predicts yielding at an angle of $(45 - \theta/2)°$, where $\tan \theta = \mu_c$: the lower angle means that the normal stress σ_n on the shear plane is reduced.

A third criterion, proposed by Sternstein and Ongchin,[26] is a modification of the von Mises yield criterion:

$$\tau_{oct} = \tfrac{1}{3}[(\sigma_{11} - \sigma_{22})^2 + (\sigma_{22} - \sigma_{33})^2 + (\sigma_{33} - \sigma_{11})^2]^{1/2} = \tau_0 - \mu_M p \qquad (6.10)$$

where τ_{oct} is the 'octahedral shear stress' and τ_0 and μ_M are materials constants. The octahedral shear stress is the stress acting on a plane having a normal that makes the same angle to all three principal stress directions: in other words, the normal defined by the equation $|\sigma_{11}| = |\sigma_{22}| = |\sigma_{33}|$. There are four lines that satisfy this equation, and it is possible to construct a regular octahedron from planes normal to these lines. As in the case of the modified Tresca criterion, the quantity μ_M can be regarded as defining the effect of pressure upon the state of the material, perhaps by altering the free volume.

On the basis of tensile yield experiments in a pressure chamber, Raghava *et al.*[27] suggested a non-linear relationship between the octahedral shear stress at yield and the pressure. The contrast with eqn. (6.10) is shown most clearly by writing the non-linear criterion as follows:

$$\tau_{oct} = (A - Bp)^{1/2} \qquad (6.11)$$

where A and B are constants. An alternative and more explicit version of the equation is:

$$(\sigma_{11} - \sigma_{22})^2 + (\sigma_{22} - \sigma_{33})^2 + (\sigma_{33} - \sigma_{11})^2$$
$$+ 2(\sigma_{ty} - \sigma_{cy})(\sigma_{11} + \sigma_{22} + \sigma_{33}) = 2\sigma_{ty}\sigma_{cy} \qquad (6.12)$$

where σ_{ty} and σ_{cy} are the tensile and compressive yield stresses at atmospheric pressure. Matsushige et al.[24] have obtained additional experimental evidence in support of this type of non-linear pressure dependence. However, the difference between the linear and non-linear modifications to von Mises criterion is small at pressures below about 150 MN/m², so that eqn. (6.10) is satisfactory for most practical purposes.

The foregoing discussion of yield criteria has been confined to the effects of stress field upon yielding; the effects of temperature and strain rate have not been specifically included. There are two ways in which these two important factors may be accomodated in a more general criterion. The first and most obvious way is to make the various materials constants functions of temperature and strain rate. The second is to modify the Eyring equation, by substituting a more complex function of stress for the shear stress. As a further complication, it must be noted that many glassy polymers fail not by shear yielding but by crazing when they are subjected to tensile stress fields. Before considering failure envelopes in more detail, it is necessary to discuss crazing and the criteria for craze formation.

6.2 CRAZING

The alternative mechanism of deformation to shear yielding is craze formation, which is both a localised yielding process and the first stage of fracture. When a tensile stress is applied to a glassy polymer, small holes form in a plane perpendicular to the stress, to produce an incipient crack. However, instead of coalescing to form a true crack, as happens in metals, the holes become stabilised by fibrils of oriented polymeric material, which span the gap and prevent it from becoming wider. The resulting yielded region consisting of an interpenetrating network of voids and polymer fibrils, is known as a *craze*. Although crazes closely resemble cracks in appearance, as illustrated in Fig. 6.12, the two types of defect differ significantly in properties.

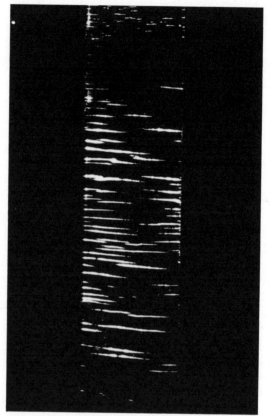

Fig. 6.12   *Crazes in a PMMA bar, viewed by reflected light.*

The distinction between cracks and crazes was made in 1949, by Sauer *et al.*,[28, 29] who demonstrated that polystyrene specimens were capable of sustaining tensile stresses of 20 MN/m^2 even when the specimens appeared to contain through-thickness cracks. On the basis of X-ray diffraction studies, they concluded that the 'cracks' contained oriented polymer molecules, which were responsible for the observed strength. The significance of this work was not generally appreciated until about 10 years later, when independent studies by Bessenov *et al.* in Russia,[30-32] and Spurr and Niegisch in the USA[33] revealed the microstructure of crazes. The subject was finally placed on a firm foundation by Kambour, who characterised the structure and properties of crazes in a classic series of papers.[34-38] Excellent reviews of crazing and fracture have been published by Kambour[39] and by Rabinowitz and Beardmore.[40]

6.2.1 Structure of crazes

The most direct method of studying the structure of crazes is transmission electron microscopy of ultrathin sections. Several different techniques have been used to prepare specimens without damaging the delicate structure of the craze during the microtoming operation. The earliest technique, used by Kambour,[35] was to introduce silver nitrate solution into the crazes, and deposit metallic silver in the voids by exposing the specimen to light; the disadvantage of this method is that the microstructure can become distorted by the growth of the silver particles at the 'development' stage.[40] In order to overcome this problem, Kambour later used liquid sulphur[41] and the lower-melting sulphur–iodine eutectic[42] to impregnate and reinforce the craze structure; the impregnants sublime away under vacuum, so that the fibrillar structure of the crazes is seen clearly in the electron microscope. The disadvantage of this method is that the specimens have to be heated to the melting point of sulphur or of the eutectic. As crazes are known to become denser on heating, some structural changes are to be expected. However, comparisons of craze impregnation in stressed and unstressed specimens suggest that these changes are small.[39] Other methods used to prepare specimens for electron microscopy include deposition of osmium in the voids, from osmium tetroxide solution; sectioning without impregnation;[43] and stretching of uncrazed sections.[44]

Figure 6.13 shows a section through a typical craze in polystyrene. Crazes generally consist of an open network of polymer fibrils between 10 and 40 nm in diameter, interspersed by voids between 10 and 20 nm in diameter. A coarser structure is sometimes observed in the central region of the craze, especially at some distance from the advancing tip of the craze.[44] The boundaries between crazes and bulk polymer are sharp to within 2 nm. The ease with which liquids penetrate the structure indicates that the voids must be interconnected. Electron microscopy shows that most crazes are between 0.1 and 1.0 μm thick, but the thickness can vary from 10 nm, at the tip of a fine craze, to 0.1 mm for a specially prepared craze.[45] In an homogeneous material, the lateral dimensions of a craze are limited only by the size of the specimen.

Further information concerning craze structure can be obtained from refractive index measurements. Kambour determined critical angles for total reflection of light at the craze boundary in a number of polymers, and hence obtained values for craze refractive index μ_c, from which he calculated

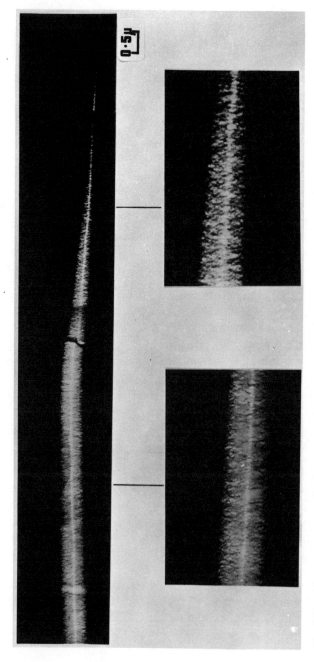

Fig. 6.13 Electron micrograph of a craze formed by applying tensile stress to a cast thin film of polystyrene. Enlargements show difference in structure between tip region and mature craze (from P. Beahan, A. Thomas and M. Bevis, J. Mater. Sci. **11** (1976) 1207, reproduced with permission).

craze densities using the Lorentz–Lorenz relationship, eqn. (3.4).[35, 36] These experiments showed that unloaded, fresh crazes contain between 40 and 60 % of polymer by volume. It is important to note that the composition of a craze is not a fixed quantity: the density increases during storage at room temperature, and decreases on application of a tensile stress. However, even after prolonged storage, the density and refractive index of crazes are low in comparison with the bulk polymer, so that light is totally reflected unless the angle of incidence is below about 10°. This explains the crack-like appearance of crazes under most conditions of illumination.

Viewed from above, under vertical illumination, crazes show interference fringes produced by reflection of light from their upper and lower boundaries.[37] If the refractive index is known, the thickness of the craze can be calculated from the order of the fringes. Noting that the brightness of the fringes often alternated near the tip of a craze, Doyle concluded that four reflecting surfaces were responsible for the interference pattern, the two additional surfaces arising from the presence of a coarse core region in the craze,[46] as previously observed by electron microscopy.[44]

The finely distributed voids and fibrils within a craze produce strong X-ray scattering at small angles.[39,47] Owing to the molecular orientation in the fibrils, interpretation of the SAXS patterns is difficult. The estimated size of the scattering elements in unstressed polystyrene crazes is between 0·9 and 1·15 nm, in reasonable agreement with electron microscope evidence.

Crazing is observed in a wide range of rigid and semi-rigid polymers. Materials in which crazes have been reported include polystyrene, poly(styrene-co-acrylonitrile), poly(methyl methacrylate), polycarbonate,[35] poly(vinyl chloride),[48] polychlorotrifluoroethylene,[49] poly(2,6-dimethyl-1,4-phenylene oxide),[50] polypropylene,[39, 51] and epoxy resins.[52] Significantly, this list contains a crystalline polymer and a thermosetting resin, as well as a number of glassy thermoplastics.

6.2.2 Craze formation

Crazes, like shear bands, are localised yielded zones formed as a consequence of strain softening. However, unlike shear yielding, crazing involves a large local increase in volume, so that the plane of the craze is not restricted to an angle of 45° to the tensile axis. Instead, crazes form at 90° to the tensile axis in an isotropic polymer. The factors governing craze formation include temperature, stress, nature of the fluid environment, molecular orientation of the sample and thermal history of the material. These factors are discussed below.

Kinetics

Craze formation can be divided into three stages: initiation, propagation and termination. In the initiation stage, an incipient craze forms through the production of voids in a region of high stress concentration. This process is repeated in the propagation stage, with the difference that the craze itself generates the stress concentration. The craze propagates outwards like a crack, and for similar reasons, becoming thicker through the incorporation of fresh material from the boundaries of the yielded zone. The process ends with the termination of the craze, which might occur simply as a result of reaching the opposite side of the specimen.

A discussion of crazing kinetics from the quantitative point of view presents considerable difficulties. The literature contains a large number of empirical observations, but it cannot be claimed that there is any fundamental understanding of the factors affecting initiation, propagation and termination, except in very broad terms. The following account aims to highlight the salient facts, as far as this can be done in the present state of knowledge.

It is generally agreed that in homogeneous polymers crazes are usually initiated either at surface flaws and scratches or at internal voids and inclusions. These defects produce the stress concentrations necessary for the initial void formation. An elastic–plastic analysis indicates that it is easier to form groups of voids than to produce a single separate void,[53] so that a co-operative process is envisaged.

On application of a tensile stress or strain to a glassy polymer, there is an induction period before crazes are observed. The period decreases with increasing stress.[54-56] As Kambour has pointed out,[39] it is extremely difficult to distinguish between initiation and propagation effects in this type of experiment; the delay in observing crazes might be due to a slow initiation step, but it might equally well be related to the time taken for a craze to grow to a size at which it is visible. In a sense, both processes could be considered as part of craze initiation.

There is no general agreement concerning the form of the relationship between rate of craze initiation and applied stress. Regel's results, which are illustrated in Fig. 6.14, indicate that the Eyring relationship (eqn. (6.3)) describes both craze formation and fracture in PMMA containing 6% dibutyl phthalate.[55] Results obtained by a number of other workers are also consistent with the Eyring model.[39] However, not all data can be interpreted on the basis of this simple model. The difficulties are illustrated in the work of Brown and Fischer, who studied the nucleation and growth of crazes in amorphous polychlorotrifluoroethylene (PCTFE) under constant applied

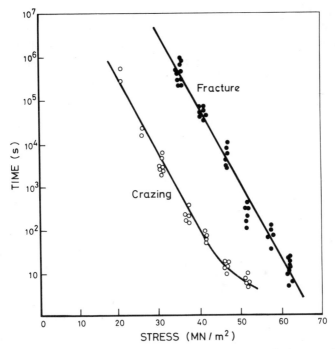

Fig. 6.14 *Time to craze and time to fracture under constant applied stress at* 25 °C *in PMMA containing* 6% *dibutyl phthalate (from V. R. Regel, ref.* 55).

stress in a liquid nitrogen environment.[49] As shown in Fig. 6.15, the number of crazes formed in unit volume of the polymer within a given time under load increased rapidly with stress, but the relationship was not of the linear form required by the Eyring treatment. Furthermore, all of the crazes appeared to be nucleated simultaneously from surface defects, and there was no evidence of time-dependent initiation of crazes. These results suggest that the stress dependence illustrated in Fig. 6.15 arises principally because smaller surface flaws are activated at higher stresses: by decreasing the critical flaw size, the number of potential crazing sites is greatly increased. The same principle forms the basis of Griffith's theory of *crack* initiation in brittle solids, which is discussed in Chapter 9.

The rate at which a craze propagates is determined by the stress concentrations in, and properties of, the material surrounding the craze tip. The stress concentration problem has been analysed by Williams *et al.* for the case of a craze growing from a stationary crack.[57, 58] By treating the craze as a plastic zone in which the stress was equal to the yield stress of the material, these authors showed that the stress concentration factor decreased as the craze propagated away from the root of the crack. Consequently, the rate of propagation decreased with time, in specimens tested at constant load.[57, 58]

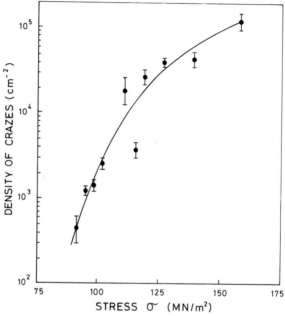

Fig. 6.15   Crazing of polychlorotrifluoroethylene (PCTFE) at 78°K. Specimens subjected to constant stress in liquid nitrogen for a period greater than the craze induction period, then removed from the test apparatus. Density of crazing is expressed as number of crazes per unit area of specimen (from N. Brown and S. Fischer, ref. 49, reproduced with permission).

Narisawa and Kondo used a photoelastic method to measure stress concentrations around crazes growing from introduced cracks in polycarbonate,[59] As predicted by the plastic zone model, they found that the stress concentration factor decreased with increasing craze length. The experiments were repeated with PMMA, but the method was not sufficiently sensitive to give satisfactory results. In another study of polycarbonate, Peterson *et al.* used the more sensitive technique of holographic interferometry to measure stress concentrations.[60] The craze was again grown from an introduced crack, but in this work the crack was machined away before beginning the experiment. The measurements showed that the stress concentration factor at the craze tip was a function of applied stress, reaching a maximum at intermediate stresses. This behaviour is clearly related to the non-linear stress–strain properties of crazes, which are described in Section 6.2.3.

The simple plastic zone model mentioned above is based upon elastic–plastic fracture mechanics (*see* Chapter 9), and takes no account of the viscoelastic nature of polymers. In a more recent treatment, Williams and Marshall refined the model by introducing two power-law relationships, one to represent stress relaxation within the craze, and the other to describe the time-dependent modulus of the bulk polymer.[57] The treatment predicts a power-law relationship between craze length and time for a craze growing under constant applied stress from a stationary crack. Experimental results for PMMA and polycarbonate were in good agreement with this prediction. This treatment of crazing as a relaxation-controlled process represents a considerable step forward in the analysis of crazing kinetics, and offers good scope for further development.

Craze growth will stop if the stress concentrations at the tip fall below the critical level for propagation. Alternatively, if the applied stress is high enough, the stress concentration will become constant, and the rate of propagation will also become constant. Both types of behaviour have been observed in PMMA immersed in alcohols.[58, 61]

One of the most interesting studies of crazing kinetics in polystyrene was made by Maxwell and Rahm, who directed a collimated beam of light into a bar subjected to constant tensile stress, and measured by means of a photometer the area of crazes produced.[62] The success of this experiment depends upon the fact that crazes act as a set of parallel mirrors. Unlike Brown and Fischer,[49] Maxwell and Rahm found that fresh crazes formed continually throughout the experiment. The rate of crazing was initially low, but after a short induction period increased to a constant value, which depended critically on applied stress and temperature, as shown in Fig. 6.16. An analysis of the published data shows that the overall rate of crazing represented by the reflecting area measurement obeys the Eyring equation, at least approximately. The apparent activation energy is 155 kJ/mol, and the apparent activation volume is $5 \cdot 6 \, nm^3$.

A similar analysis of thermally activated crazing in PMMA in the presence of ethanol–water mixture yielded an apparent activation energy of 84 kJ/mol.[61] In interpreting these data, it is important to differentiate between the true activation energy ΔH^* and the apparent activation energy $(\Delta H^* - \gamma V^* \sigma / 4)$, which is obtained by plotting log (rate of crazing) against

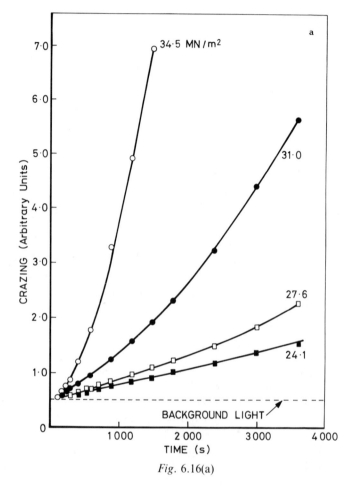

Fig. 6.16(a)

T^{-1}, for measurements made at a given stress. The figure of 155 kJ/mol for crazing in polystyrene refers to experiments at a stress of 24·2 MN/m², over the temperature range 30–70 °C. The true activation energy is estimated to be 175 kJ/mol, which may be compared with 230 kJ/mol for the activation energy of thermal bond rupture in polystyrene.[63] This comparison raises the question of whether stress-activated chain scission is a rate-determining step in craze initiation and propagation. This possibility cannot be discounted in view of the connection between crazing and cracking.

At the time of writing, there is no general agreement concerning the kinetics of crazing. Problems arise not only in the theoretical interpretation

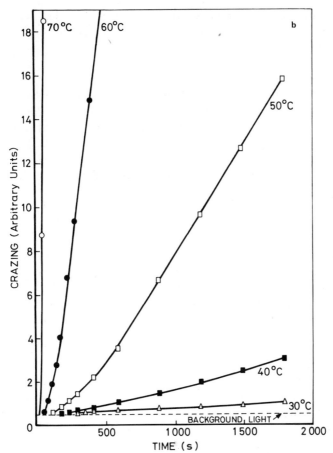

Fig. 6.16 *Kinetics of crazing in polystyrene under constant applied stress: (a) effects of stress upon rate of crazing at* 30 °C; *(b) effects of temperature upon rate of crazing under a stress of* 24·1 *MN/m². Crazing is measured as the quantity of light reflected by the crazes into a photometer (from B. Maxwell and L. F. Rahm, ref. 62, reproduced with permission).*

of the results, but also in correlating the data published by different laboratories. Many of the differences are probably due to differences in the materials used and in the conditions of the experiments. There is no intrinsic reason why solvent-crazing in polycarbonate should follow the same kinetics as crazing in polystyrene in the absence of an active liquid. Probably one of the most frequent causes of variation is the occurrence of other relaxation processes in the polymer, especially in the more ductile polymers such as polycarbonate. This aspect of the problem is discussed further in Section 6.3, which is devoted to interactions between crazes and shear bands.

Criteria for craze formation

Crazes form only under conditions of hydrostatic tension. Under hydrostatic compression, yielding occurs through shear deformation, and crazing is suppressed. This is to be expected in view of the fact that crazing is a dilatation process.

The earliest criterion for craze formation was proposed by Sternstein and Ongchin, on the basis of biaxial stressing experiments on PMMA.[26] The criterion is expressed in terms of the *stress bias* σ_b, as follows:

$$\sigma_b = |\sigma_{11} - \sigma_{22}| = A(T) + \frac{B(T)}{(\sigma_{11} + \sigma_{22})} \qquad (6.13)$$

where $A(T)$ and $B(T)$ are temperature-dependent materials parameters, and σ_{11} and σ_{22} are principal stresses. The third principal stress σ_{33} is zero. This equation is a pressure-dependent failure criterion in which the critical stress term varies with the reciprocal of the hydrostatic pressure, in contrast to the direct dependence shown in eqn. (6.9) and (6.10) for shear yielding. The form of the stress-bias equation is illustrated in Fig. 6.17, which shows the fit that is obtained with crazing data for PMMA at various temperatures.

The stress-bias criterion is a purely empirical equation, with no obvious physical significance. The stress bias represents a shear stress, but not the maximum shear stress. Furthermore, the criterion is formulated only for biaxial stress conditions. For these reasons, Bowden and Oxborough proposed the following modified version of eqn. (6.13):[64]

$$\sigma_{11} - \nu(\sigma_{22} + \sigma_{33}) = C(t, T) - \frac{D(t, T)}{(\sigma_{11} + \sigma_{22} + \sigma_{33})} \qquad (6.14)$$

where ν = Poisson's ratio, which is approximately 0·4 for a typical glassy polymer, and $C(t, T)$ and $D(t, T)$ are time- and temperature-dependent quantities. Equation (6.14) is a critical tensile strain criterion. Writing E for Young's modulus and ε_{11} for major principal strain

$$E\varepsilon_{11} = \sigma_{11} - \nu(\sigma_{22} + \sigma_{33})$$

and therefore

$$\varepsilon_{11} = \frac{1}{E}\left[C(t, T) + \frac{D(t, T)}{(\sigma_{11} + \sigma_{22} + \sigma_{33})} \right] \qquad (6.15)$$

This equation also fits the data reasonably well, so that it is difficult to compare the merits of the stress-bias and critical strain criteria purely in terms of their ability to predict the conditions for craze initiation. The main virtues of eqn. (6.15) are its generality and its relationship to a recognisable physical principle.

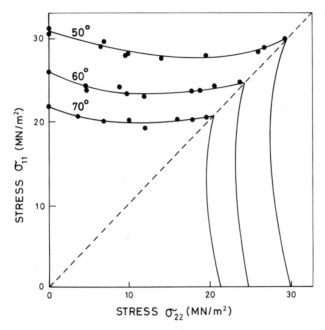

Fig. 6.17 The stress-bias criterion for crazing in PMMA at various temperatures under biaxial tension. The stress-bias curve is fitted to experimental data (from S. S. Sternstein and L. Ongchin, ref. 26, reproduced with permission).

A convenient method for comparing the various criteria that have been proposed is to subject a specimen to an inhomogeneous stress field and to observe the points at which crazes form. This approach was used by Wang *et al.*,[65] who applied uniaxial tensile stresses to polystyrene specimens containing embedded steel spheres, and found that crazes were initiated at the surface of an isolated sphere in regions of maximum strain and of maximum strain energy: both maxima occur at 37° from the 'poles' defined by the symmetry axis of the stressed sphere. Maximum dilatation occurs at 0°, maximum principal stress at 25°, and maximum shear stress at 44°. In later work with two rubber spheres embedded in polystyrene, the same authors showed that the maximum strain criterion is more satisfactory than the maximum strain energy criterion.[66] However, in neither paper did they seriously consider the stress-bias criterion.

Thus the critical strain criterion represented by eqn. (6.15) provides a firm basis for a discussion of crazing under multiaxial stress. The criterion not only fits the experimental results, but also relates the conditions for craze formation directly to the direction of the craze normal, which is the direction in which elements of the polymer are displaced in forming the craze. Significantly, this principle applies to anisotropic polymers: Beardmore and Rabinowitz showed that craze normals in hot-drawn polycarbonate coincided with principal strain axes, and deviated from the tensile stress axis by as much as 15°, depending upon the angle between the applied stress and the orientation direction.[67]

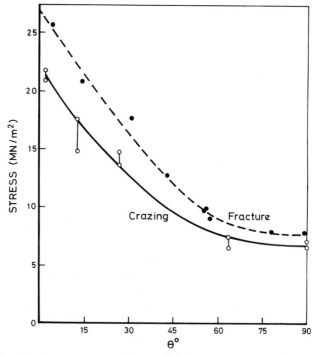

Fig. 6.18   Crazing stress and fracture stress in hot-drawn PMMA at 77°K. Material was drawn uniaxially above T_g and subsequently tested at various angles $\theta°$ to the draw direction (from P. Beardmore and S. Rabinowitz, ref. 67, reproduced with permission).

Molecular orientation

The effects of molecular orientation on craze formation have long been recognised. Figures 6.18 and 6.19 illustrate the relationships between crazing stress and orientation in tensile tests in hot-drawn PMMA.[67] In comparison with the isotropic polymer, drawn PMMA exhibits a high resistance to crazing when the stress is applied parallel to the orientation

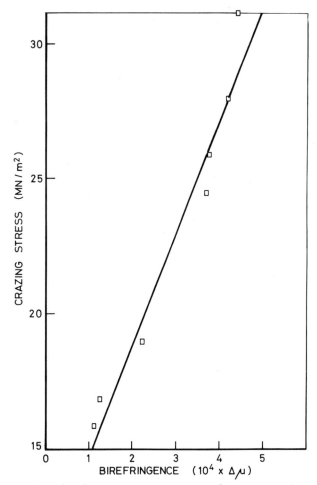

Fig. 6.19   Relationship between crazing stress and orientation (measured by birefringence) in PMMA at 77°K. Material was drawn uniaxially above T_g, and subsequently tested with tensile stress applied parallel to the draw direction (from P. Beardmore and S. Rabinowitz, ref. 67, reproduced with permission).

direction, and a low resistance to crazing at 90° to this direction. The stress required for craze initiation increases linearly with the degree of anisotropy, as measured by the birefringence of the polymer. Because of the anisotropy of modulus, the principal axes of stress and strain do not coincide in the general case, and the craze normals therefore lie at an angle to the applied stress direction, as mentioned above.

The shape of the crazes also changes with the direction of the applied stress. A stress applied parallel to the orientation direction produces large numbers of short, thin crazes, whereas a stress applied at 90° to the orientation direction produces a few long, thick crazes.[67] These observations show that the propagation behaviour of crazes in glassy polymers is affected by orientation in the same way as the initiation behaviour.

The effects of orientation on craze propagation are most clearly demonstrated in tensile tests on injection-moulded bars. Crazes form readily in the interior of the bar, where the orientation is low, and propagate outwards towards the surface, where the high orientation along the length of the bar brings them to a halt.[68] A similar interaction between crazes and shear bands is discussed in Section 6.3.

The most obvious explanation of these observations is that oriented fibrils of the type seen in crazes are unable to form when the bulk polymer is already oriented. In effect, the existing orientation pre-empts the crazing process. An alternative approach, due to Sternstein and Rosenthal,[69] is based on the idea that crazing, like other failure processes in polymers, is promoted by intrinsic flaws in the material. These flaws are represented by a spectrum of ellipses, which in an isotropic polymer are randomly oriented in space. On the basis of this model, hot drawing is treated as deforming and rotating the ellipses, thus producing a change in the pattern of stress concentrations. In this way the flaw spectrum model provides a quantitative relationship between orientation and failure stress, including crazing stress, at any angle to the orientation direction. As the authors point out, the model offers considerable scope for further development: the initial analysis neglects effects due to molecular orientation, and treats only two-dimensional flaws.

Molecular weight

Although molecular weight has a profound effect upon the fracture behaviour of polymers, it has a negligible effect upon craze formation. These points are clearly demonstrated in Fig. 6.20, which presents data obtained by Fellers and Kee in experiments on polystyrene.[70] The experiments consisted of injection moulding tensile bars at high temperatures and low flow rates in order to minimise orientation, and observing craze formation in these bars under uniaxial tension by Maxwell and Rahm's light-reflection method. When the molecular weight was below 70 000, the polymer was very brittle, and stable crazes did not form.

The morphology of the crazes also varied with the number average molecular weight $\overline{M}_n$. At $\overline{M}_n$ values of about 80 000, the crazes were short,

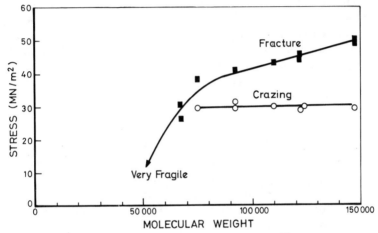

Fig. 6.20   Effect of number average molecular weight $\overline{M}_n$ upon crazing and fracture stresses in polystyrene at 25 °C. Strain rate $1.7 \times 10^{-4} s^{-1}$ (from J. F. Fellers and B. F. Kee, ref. 70, reproduced with permission).

thick, irregular in shape and few in number. At much higher values of $\overline{M}_n$, they were long, thin, planar and numerous.

Solvent crazing

The critical strain for craze formation is greatly reduced in the presence of liquids having solubility parameters close to that of the polymer. This phenomenon is known as *solvent crazing*, although in general the liquids are non-solvents for the polymer: the essential features of solvent crazing are that the liquids have no significant effect on the unstressed polymer, and that the low stresses involved have no significant effect in the absence of the liquid. However, it is not necessary that an external stress be applied to the polymer; internal stresses due to molecular orientation are a primary cause of solvent crazing.

The work of Kambour et al.,[71] Andrews et al.[72] and other workers has shown that solvent crazing is due to the plasticisation of the polymer by the liquid or vapour environment. A local reduction in T_g at a point of high stress concentration facilitates craze formation, and the void channels within the craze provide a ready path for penetration of the active fluid to the stressed tip region of the advancing craze. In oriented polymers, the plasticisation also generates internal stresses, by allowing the extended molecules to recover towards their equilibrium configurations.

In practical terms, solvent crazing is a very widespread and important phenomenon, encompassing the action of butter on moulded polystyrene, interactions between various oils and polycarbonate, which severely restrict the use of that polymer in car engine compartments, and even the effects of finger grease on the strength of polymeric glasses. At very low temperatures, liquid nitrogen and other gases near their condensation temperatures act as crazing agents.[49, 73, 74]

6.2.3 Properties of crazes
Because of their structure, crazes have very distinctive properties, differing considerably from those of the bulk polymer. Some progress has been made in studying these properties, but the investigator is faced with two major problems. The first is experimental: crazes are extremely thin, and measurements of many properties are therefore difficult to make. The second is more fundamental: crazes are continually changing their structure. Under stress, they relax and extend, and the fibrils begin to break, allowing the voids to coalesce. In the unloaded condition, they tend to close up, and return to the bulk state. Consequently, the properties depend markedly upon the mechanical and thermal history of the sample. There is no definitive state to which the measurements of craze properties can be related, except the bulk state.

Stress–strain properties
The first and most extensive study of the stress–strain properties of crazes was made by Kambour and Kopp, who subjected a bar of polycarbonate to incremental tensile loading under ethanol, and produced a craze 100 μm in thickness and occupying the complete cross-section of the bar.[45] The bar was held under load for seven months in order to allow the ethanol to vaporise away before making measurements. The crazed specimen was then loaded in tension at a constant cross-head speed, and the thickness of the craze was measured by direct observation under a microscope. The observations confirmed that extension was due entirely to deformation of

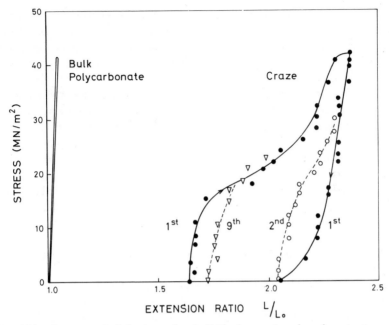

Fig. 6.21   Stress–strain behaviour of an individual craze in polycarbonate at room temperature: (●) first cycle—specimen unloaded for 7 weeks prior to test; (○) second cycle—load reapplied immediately on completion of first cycle; (▽) ninth cycle—specimen unloaded for 4 weeks prior to test. Bulk polymer is reference state in calculating extension ratios (from R. P. Kambour and R. W. Kopp, ref. 45, reproduced with permission).

material already within the craze at the beginning of the test, and not to incorporation of fresh material from the surrounding bulk polymer.

Results from these experiments are presented in Fig. 6.21. The extension ratio quoted in this figure is the ratio of the current craze thickness to the thickness of the original layer of bulk polymer from which it was formed. The value of 1·6 for the extension ratio of the fully relaxed craze was calculated from refractive index measurements. Initially, the craze was rather stiff, exhibiting a modulus comparable with that of the bulk polymer, but at a stress of about 10 MN/m^2 the craze began to yield, as the fibrils stretched, and the cross-sectional area of load-bearing polymer consequently decreased. At an extension ratio of about 1·8 the curve began to rise again, as molecular orientation caused the fibrils to strain-harden. On unloading, only part of the strain was recovered within the time scale of the experiment. A substantial degree of molecular orientation was retained by the craze fibrils, and immediate reloading gave a steep stress–strain curve, with a small hysteresis loop. During storage for a month under zero load, the craze relaxed, and its properties became similar to those originally observed: a repeat of the cyclic loading test gave a large mechanical hysteresis loop.

Craze deformation in polycarbonate is clearly a complex viscoelastic process, characterised by high hysteresis and slow recovery. Similar behaviour is observed in polystyrene.[75] Recovery occurs because the unloaded crazes have a higher free energy than the bulk polymer, and is probably entropic in origin. The process is hindered by the existence of free energy barriers, and takes place relatively slowly at room temperature. Heating has the effect of increasing the thermal energy available for surmounting the energy barriers, and reducing the height of the barriers, so that the rate of recovery increases with temperature, and is very rapid above T_g.

Sauer *et al.*[28] noted that crazes could be made to disappear under compression, but reappeared in the same place when reloaded in tension. This experiment showed that compression does not restore the polymer to the bulk conformational state, but simply increases the density of the craze to the point at which its refractive index approaches that of the bulk polymer. Frustrated internal reflection might also be a contributory factor.

Energy of formation
Figure 6.21 emphasises the fact that only a small fraction of the energy expended in forming a craze is stored as elastic strain energy. The remainder is dissipated. The stresses required for craze formation in PS and PMMA are of the order of $40 \, MN/m^2$, so that the work done in producing a craze approximately $1·5 \, \mu m$ thick is $40 \, J/m^2$, assuming that the original layer of bulk polymer was $0·5 \, \mu m$ thick.

This capacity to dissipate energy is an important factor in the fracture resistance of both homogeneous and rubber-toughened polymers. There is a substantial body of evidence that crazing precedes crack formation in these materials unless the molecular weight is extremely low or the polymer is very highly oriented. It is therefore significant that fracture mechanics studies show the energies absorbed in forming cracks in PMMA to be between 120 and $650 \, J/m^2$, depending upon the method of measurement. These figures suggest that the minimum energy for fracture is the energy required to form and break a single craze. The fracture surface energy of polystyrene is between 250 and $1000 \, J/m^2$, the higher values reflecting the fact that multiple crazing is usually observed at the crack tip in this polymer.

Craze strength

Since crazing is an integral part of the fracture process in glassy polymers, the factors affecting craze strength are of great importance. The breakdown of a craze to form a crack begins slowly, with the appearance of slightly expanded voids which coalesce to produce a cavity equal in thickness to the craze itself.[76, 77] At some critical stage, this cavity begins to extend along the centre of the craze, as illustrated in Fig. 6.22, meeting secondary crack centres which generate parabolic markings on the fracture surface. The crack accelerates, and the fracture process becomes catastrophic. Rapid

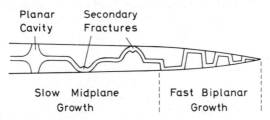

Fig. 6.22   *Stages in the formation of a crack by failure of a craze (from J. Murray and D. Hull, ref. 77, reproduced with permission).*

failure in thinner regions of the craze gives adiabatic heating. Similar effects are observed when a crack propagates into a previously uncrazed polymer: a craze is formed ahead of the advancing crack tip, and secondary cracks are generated by fracture within this craze layer. As the main crack overtakes the secondary crack, a parabolic step is produced on the fracture surface.[78]

In many cases, Murray and Hull were able to identify the regions of craze breakdown with impurity particles.[77] Internal failure of the craze due to rupture of the fibrils, whether associated with foreign particles or not, appears to be characteristic of fracture at room temperature. At low temperature, cracks initiate at the surface of the tensile specimen, where the craze is not under triaxial tension and therefore retracts inwards, producing a small depression. In addition to this effect, crazes give rise to surface steps about $0\cdot1\ \mu m$ high, which act as stress concentrators. Beardmore and Johnson concluded that these surface steps were responsible for crack initiation in PMMA at $78\,^{\circ}K$.[79] This view is questioned by Kambour, who points out that crazing and fracture might be affected by the nitrogen or oxygen environment, since both gases are mild stress crazing agents at cryogenic temperatures.[39]

The stability of crazes under stress is clearly due to molecular entanglements. Figure 6.20 shows that stable crazes do not form when the molecular weight is below the critical length for chain entanglement,[70] and that the material is very brittle in consequence. The strength of the polymer rises steadily as the molecular weight is increased from this critical value.

Kinetics of fracture

Fracture of craze fibrils is a rate process, dependent upon the stress, temperature and detailed structure of the crazes. In the case of solvent crazing, the nature of the environment is also important. Little is known about the precise mechanisms by which the fibrils fail; presumably viscoelastic stress relaxation, plastic flow and chemical bond rupture all contribute to the eventual fracture process. The influence of molecular weight is evidence of flow processes, and kinetic data indicate that bond rupture is also involved.

Regel's studies of crazing and fracture of PMMA, which are illustrated in Fig. 6.14, show that both processes exhibit the same Eyring dependence upon stress,[55] and therefore suggest that both might be controlled by the same rate-determining step. A more extensive study of fracture kinetics by Zhurkov and co-workers[80-82] confirms that the Eyring relationship is obeyed, and clearly identifies the rate-determining step in fracture as rupture of main-chain carbon–carbon bonds, although it must be noted that many of Zhurkov's experiments were performed on highly oriented semi-crystalline polymers. Graphs of log (fracture time) against applied stress for these materials, for isotropic PMMA and for a number of metals all showed the same characteristics: linear relationships were obtained, and in each case the lines for a given material at different temperatures converged at a rupture time of 10^{-13} s, which corresponds to the period of molecular vibrations. In other words, if the stress could be raised sufficiently high, thermal energy would make no contribution to the rate, which would be controlled purely by the frequency with which the flow units approached the energy barrier; under these conditions of high stress, the effective activation energy would be zero.

Activation energies obtained from these experiments correspond closely with activation energies of thermal degradation for a wide range of polymers, including PS, PMMA, PVC, PP and PTFE.[80] Additional evidence for bond rupture comes from infra-red spectroscopy. Under tensile stress, the absorption peaks for carbon–carbon skeletal vibrations shift to lower frequencies, so that the distribution of stress over individual bonds can be measured.[80-82] Observed distributions are continuous up to a certain stress, and then fall abruptly, indicating that the bonds are unable to support higher stresses, and that a small fraction of highly loaded bonds breaks under tension. This experiment provides a direct measurement of the stress concentration factor γ in the Eyring equation.

Thus there is a case for considering main-chain bond rupture as the rate-determining step in both craze formation and fracture of polymers. This statement does not imply that bond rupture is the principal mechanism in either case. Physical chemists will recognise that the rate-determining step is only a small part, albeit a crucial one, of a sequence of events. Once a few highly stressed bonds have broken, viscoelastic and plastic flow processes can follow.

Optical properties

A craze behaves optically as a thin, homogeneous, transparent layer of low refractive index because the voids are small. If the voids were larger, they would scatter light, and the crazes would lose their property of mirror-like reflection. Some opacity has been observed in solvent-crazed PMMA, PC and PPO as a result of fracture of the fibrils and consequent coarsening of the craze structure.[39]

The fracture of a craze leaves a thin layer of deformed polymer on the fracture surface. Critical angle measurements show that this layer has the same refractive index as the unstressed craze. In PMMA, which forms a relatively flat fracture surface, the broken craze layer produces strong interference colours, owing to reflections from the upper and lower surfaces. If the refractive index of the craze is known, the thickness can be calculated from the fringe colour.[37]

Porosity

Crazes readily imbibe liquids, indicating that the voids consist of interconnecting channels. The dominant driving force is capillary action resulting from the small void diameter.[39] In some applications, this porosity is a distinct disadvantage. Nicolais et al. used the property to advantage in measuring craze formation and recovery in PS and PMMA: crazing greatly increased the permeability of these polymers to water vapour, but the porosity decreased on heating, as the crazes closed up.[83] The recovery was so rapid at T_g that permeability measurements could be used to determine the position of the glass transition.

6.3 INTERACTIONS BETWEEN CRAZES AND SHEAR BANDS

The preceding discussion has considered crazing and shear deformation separately. However, under certain stress conditions both mechanisms operate simultaneously, and interactions between them become possible. These interactions are of fundamental importance in the discussion of toughness and fracture resistance, especially as the stress conditions required for simultaneous crazing and shear yielding in many cases include uniaxial tension.

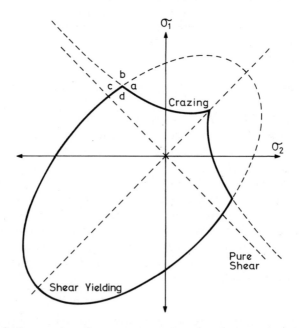

Fig. 6.23   Failure envelope (heavy continuous line) for PMMA under biaxial stress at room temperature showing four types of response to applied stress: (a) crazing; (b) crazing with shear yielding; (c) shear yielding; (d) elastic deformation (from S. Sternstein and L. Ongchin, ref. 26, reproduced with permission).

Figure 6.23 shows a complete failure envelope for PMMA under biaxial stress.[26] In the first quadrant, under biaxial tension, the polymer fails by crazing as predicted by the critical tensile strain criterion, and subsequently fractures in a brittle manner at a stress well below that predicted for shear yielding. In the third quadrant, under biaxial compression, the polymer yields by shear deformation according to the modified von Mises criterion; no crazing is possible. In the second and fourth quadrants, however, the two envelopes intersect, and there are conditions under which both crazing and shear yielding can occur simultaneously. In order to achieve this situation, eqns. (6.10) and (6.15) must both be satisfied, and the stress must be below the level required to cause fracture. Sternstein and Myers have confirmed the validity of this analysis experimentally.[84]

The chief significance of the intersection of the two failure envelopes is that it marks the transition from brittle to ductile fracture in the polymer. It is in this region that interactions between crazes and shear bands are to be expected, especially in the area marked (a) in the diagram. Here the polymer is above the crazing stress and just below the shear yield stress. As crazes form, their growth will be affected by the presence of shear zones, which in PMMA are usually diffuse, but in polystyrene are more likely to take the form of shear microbands. As Kramer's work has shown, shear zones are formed well before general yield in glassy polymers.[19, 20]

Interactions between crazes and shear zones might take several forms. One possibility is that the craze will meet an existing shear band and come to a halt, owing to the high molecular orientation within the band. Another possibility is that shear bands will initiate from the highly stressed tip of the craze, again bringing it to a halt. There is evidence for both types of interaction,[11, 39] although systematic investigation has been lacking, and a number of points remain unresolved. In addition to these effects, kinetic studies of craze initiation and propagation suggest a link between rates of crazing and the general segmental relaxation of the polymer: observed decreases in rate with time under load might well be due to changes in the structure of the glassy polymer resulting from these relaxations.

The relative positions of the crazing and shear yielding envelopes are affected by a number of factors, including the magnitude of the third principal stress, temperature, strain rate and liquid environments. Each of these factors produces a shift in the brittle–ductile transition which may be sufficient to alter the stress quadrant in which the transition occurs. Hydrostatic pressure raises the crazing envelope to a greater extent than the shear envelope, increasing the ductility of the polymer essentially by suppressing the dilatational mechanism.[84] Conversely, active liquids lower the crazing stress in relation to the shear yield stress, so that a polymer that is normally ductile under uniaxial tension, such as polycarbonate or polysulphone, becomes brittle. Similar arguments apply to the effects of temperature and strain rate: increasing temperature generally promotes shear yielding at the expense of crazing, whilst increasing strain rate usually results in a more brittle response from the polymer. Chapter 7 considers another important factor affecting the ductile–brittle transition—the addition of rubber particles.

REFERENCES

1. P. B. Bowden, in *The Physics of Glassy Polymers*, R. N. Haward (ed.), Applied Science, London, 1973, p. 279.
2. I. M. Ward, *Mechanical Properties of Solid Polymers*, Wiley, New York, 1971.
3. I. M. Ward, *J. Mater. Sci.* **6** (1971) 1397.
4. S. S. Sternstein, L. Ongchin and A. Silverman, *Appl. Polymer Symp.* **7** (1968) 175.
5. H. Eyring, *J. Chem. Phys.* **4** (1936) 283.
6. R. E. Robertson, *Appl. Polymer Symp.* **7** (1968) 201.
7. R. P. Kambour and R. E. Robertson, in *Polymer Science*, A. D. Jenkins (ed.) North Holland, Amsterdam, 1972, p. 687.
8. A. S. Argon, *Phil. Mag.* **28** (1973) 839.
9. P. B. Bowden, *Phil. Mag.* **22** (1970) 455.
10. P. B. Bowden and S. Raha, *Phil. Mag.* **22** (1970) 463.
11. T. E. Brady and G. S. Y. Yeh, *J. Appl. Phys.* **42** (1971) 4622.
12. S. E. B. Petrie, *J. Poly. Sci.* A2, **10** (1972) 1255.
13. R. M. Mininni, R. S. Moore, J. R. Flick and S. E. B. Petrie, *J. Macromol. Sci., Phys.* **B8** (1973) 343.
14. T. E. Brady and G. S. Y. Yeh, *J. Macromol. Sci., Phys.* **B9** (1974) 659.
15. W. Whitney, *J. Appl. Phys.* **34** (1963) 3633.
16. P. B. Bowden and J. A. Jukes, *J. Mater. Sci.* **3** (1968) 183.
17. A. S. Argon, R. D. Andrews, J. A. Godrick and W. Whitney, *J. Appl. Phys.* **39** (1968) 1899.
18. M. Reiner in *Rheology*, Vol. 1, F. R. Eirich (ed.), Academic Press, New York, 1956, p. 10.
19. E. J. Kramer, *J. Poly. Sci. (Phys.)* **13** (1975) 509.
20. E. J. Kramer, *J. Macromol. Sci., Phys.* **B10** (1974) 191.
21. J. F. Nye, *Physical Properties of Crystals*, Clarendon Press, Oxford, 1957.
22. L. Holliday, J. Mann, G. Pogany, H. D. Pugh and D. A. Green, *Nature* **202** (1964) 381.
23. A. W. Christiansen, E. Baer and S. V. Radcliffe, *Phil. Mag.* **24** (1971) 451.
24. K. Matsushige, S. V. Radcliffe and E. Baer, *J. Mater. Sci.* **10** (1975) 833.
25. C. A. Coulomb, *Mem. Math. et Phys.* **7** (1773) 343.
26. S. S. Sternstein and L. Ongchin, *ACS Poly. Prepr.* **10**(2) (1969) 1117.
27. R. Raghava, R. M. Caddell and G. S. Y. Yeh, *J. Mater. Sci.* **8** (1973) 225.
28. J. A. Sauer, J. Marin and C. C. Hsiao, *J. Appl. Phys.* **20** (1949) 507.
29. C. C. Hsiao and J. A. Sauer, *J. Appl. Phys.* **21** (1950) 1071.
30. M. I. Bessenov and E. V. Kuvshinskii, *Vys. Soed.* **1** (1959) 1561.
31. M. I. Bessenov and E. V. Kuvshinskii, *Sov. Phys. Solid State* **1** (1960) 1321.
32. G. A. Lebedev and E. V. Kuvshinskii, *Sov. Phys. Solid State* **3** (1962) 1947.
33. O. K. Spurr and W. D. Niegisch, *J. Appl. Polymer Sci.* **6** (1962) 585.
34. R. P. Kambour, *Nature* **195** (1962) 1299.
35. R. P. Kambour, *Polymer* **5** (1964) 143.
36. R. P. Kambour, *J. Poly Sci.* A2 (1964) 4159.
37. R. P. Kambour, *J. Poly. Sci.* A2, **4** (1966) 349.
38. R. P. Kambour, *J. Poly. Sci.* A3 (1965) 1713.
39. R. P. Kambour, *J. Poly. Sci.* D7 (1973) 1.
40. S. Rabinowitz and P. Beardmore, *CRC Crit. Revs Macromol. Sci.* **1** (1972) 1.
41. R. P. Kambour and A. S. Holik, *J. Poly. Sci.* A2, **7** (1969) 1393.
42. R. P. Kambour and R. R. Russell, *Polymer* **12** (1973) 237.
43. P. Beahan, M. Bevis and D. Hull, *Phil. Mag.* **24** (1971) 1267.
44. P. Beahan, M. Bevis and D. Hull, *J. Mater. Sci.* **8** (1973) 162.
45. R. P. Kambour and R. W. Kopp, *J. Poly. Sci.* A2, **7** (1969) 183.
46. M. J. Doyle, *J. Poly Sci. (Phys.)* **13** (1975) 2429.
47. D. G. LeGrand, R. P. Kambour and W. R. Haaf, *J. Poly. Sci.* A2, **10** (1972) 1565.
48. M. Matsuo, C. Nozaki and Y. Jyo, *Poly. Engng Sci.* **9** (1969) 197.

49. N. Brown and S. Fischer, *J. Poly. Sci. (Phys.)* **13** (1975) 1315.
50. G. A. Bernier and R. P. Kambour, *Macromolecules* **1** (1968) 393.
51. H. G. Olf and A. Peterlin, *J. Poly. Sci. (Phys.)* **12** (1974) 2209.
52. J. Lilley and D. G. Holloway, *Phil. Mag.* **28** (1973) 215.
53. R. N. Haward and D. R. J. Owen, *J. Mater. Sci.* **8** (1973) 1136.
54. E. E. Ziegler and W. E. Brown, *Plast. Technol.* **1** (1955) 341.
55. V. R. Regel, *Sov. Phys. Tech. Phys.* **1** (1956) 353.
56. Y. Sato, *Kobunshi Kagaku* **23** (1966) 69.
57. J. G. Williams and G. P. Marshall, *Proc. Roy. Soc.* **A342** (1975) 55.
58. G. P. Marshall, L. E. Culver and J. G. Williams, *Proc. Roy. Soc.* **A319** (1970) 165.
59. I. Narisawa and T. Kondo, *J. Poly. Sci. (Phys.)* **11** (1973) 223.
60. T. L. Peterson, D. G. Ast and E. J. Kramer, *J. Appl. Phys.* **45** (1974) 4220.
61. S. S. Sternstein and K. J. Sims, *ACS Poly. Prepr.* **5**(2) (1964) 422.
62. B. Maxwell and L. F. Rahm, *Ind. Eng. Chem.* **41** (1949) 1988.
63. S. Madorsky, *J. Res. Nat. Bur. Stand.* **62** (1959) 219.
64. R. J. Oxborough and P. B. Bowden, *Phil. Mag.* **28** (1973) 547.
65. T. T. Wang, M. Matsuo and T. K. Kwei, *J. Appl. Phys.* **42** (1971) 4188.
66. M. Matsuo, T. T. Wang and T. K. Kwei, *J. Poly. Sci.* A2, **10** (1972) 1085.
67. P. Beardmore and S. Rabinowitz, *J. Mater. Sci.* **10** (1975) 1763.
68. R. N. Haward, B. M. Murphy and E. F. T. White, in *Fracture* 1969, Chapman and Hall, London, 1969, p. 519.
69. S. S. Sternstein and J. Rosenthal, *ACS. Adv. Chem. Ser.* (1976), in press.
70. J. F. Fellers and B. F. Kee, *J. Appl. Polymer Sci.* **18** (1974) 2355.
71. R. P. Kambour, C. L. Gruner and E. E. Romagosa, *J. Poly. Sci. (Phys.)* **11** (1973) 1879.
72. E. H. Andrews, G. M. Levy and J. Willis, *J. Mater Sci.* **8** (1973) 1000.
73. H. G. Olf and A. Peterlin, *J. Coll. Inter. Sci.* **47** (1974) 628.
74. N. Brown, *J. Poly. Sci. (Phys.)* **11** (1973) 2099.
75. J. Hoare and D. Hull, *Phil. Mag.* **26** (1972) 443.
76. M. Bevis and D. Hull, *J. Mater. Sci.* **5** (1970) 983.
77. J. Murray and D. Hull, *Polymer* **10** (1969) 451.
78. R. N. Haward, in *The Physics of Glassy Polymers*, R. N. Haward (ed.), Applied Science, London, 1973, p. 340.
79. P. Beardmore and T. L. Johnson, *Phil. Mag.* **23** (1971) 1119.
80. S. N. Zhurkov and E. E. Tomashevsky, in *The Physical Basis of Yield and Fracture*, A. C. Stickland (ed.), Institute of Physics, London, 1966, p. 200.
81. S. N. Zhurkov and V. E. Korsukov, *J. Poly. Sci. (Phys.)* **12** (1974) 385.
82. S. N. Zhurkov, V. A. Zakrevsky, V. E. Korsukov and V. S. Kuksenko, *J. Poly. Sci. (Phys.)* **10** (1972) 1509.
83. L. Nicolais, E. Drioli and C. Migliaresi, *J. Appl. Polymer Sci.* **19** (1975) 1999.
84. S. S. Sternstein and F. A. Myers, *J. Macromol. Sci. (Phys.)* **B8** (1973) 539.

CHAPTER 7

MECHANISMS OF RUBBER TOUGHENING

How do small quantities of rubber produce such dramatic increases in the fracture resistance of brittle plastics? What is the mechanism of rubber toughening? The answer to these fundamental questions is far from obvious. The first satisfactory theory of toughening was advanced in 1964, almost 40 years after Ostromislensky's original discovery, and 15 years after the commercial introduction of HIPS. Since 1964, the subject has developed to the stage at which quantitative theories can be constructed, but there are still many outstanding questions concerning the relationship between structure and fracture resistance. This chapter presents some of the qualitative and quantitative theories of rubber toughening, and discusses the problems that remain unresolved.

7.1 STATEMENT OF THE PROBLEM

The basic problem can be illustrated by comparing the behaviour of polystyrene and HIPS in two standard fracture tests, the tensile test and the notched Izod test. Figure 7.1 illustrates the differences in stress–strain properties between the two materials in a tensile test at constant strain rate. The polystyrene extends almost linearly to a strain of 2%, breaking in a brittle manner under a stress of 45 MN/m². Fracture is preceded by crazing, which is first observed at a strain of 1·5% and a stress of 35 MN/m². The initial stress–strain response of the HIPS is also approximately linear, although the modulus is lower owing to the presence of the rubber. However, at a stress of 12·5 MN/m² the specimen begins to whiten, and thereafter the stress–strain properties are completely different from those of polystyrene. There is a yield point at 15·5 MN/m², followed by a load drop to 14 MN/m². The stress then rises very slowly until the specimen breaks at a stress of 16 MN/m² and a strain of 40 %. The *stress-whitening* observed at a strain of 1 % becomes more intense as the test proceeds, and it is noticeable that there is no sign of necking: the cross-sectional area of the specimen changes very little between the yield point and fracture.

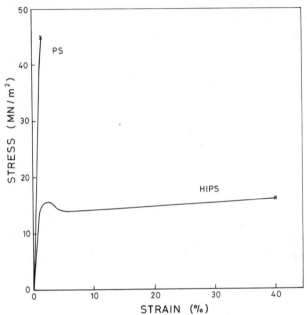

Fig. 7.1   *Tensile stress–strain curves for PS and HIPS at 20°C.*

Other rubber-toughened plastics exhibit similar, but not identical, stress–strain behaviour. In most ABS polymers, stress-whitening and yielding are followed by localised necking. The neck becomes intensely white, and the specimen breaks at a lower strain than a comparable HIPS because the stress is increased in the neck region. HIPS behaves in a similar way at temperatures above about 60°C. These observations, including the differences between materials, must be explained by the theory of toughening.

Whilst tensile properties are of considerable interest to users of rubber-toughened plastics, the main reason for choosing these materials is that they offer high impact resistance. Figure 7.2 compares notched Izod impact strengths for polystyrene and HIPS over a wide range of temperatures, and again emphasises the differences in mechanical properties between the two related polymers. The impact strength of PS remains constant at the low value of 0·16 J/cm notch over the whole of the temperature range, and the fracture surface has the broken appearance typical of a brittle solid. The impact strength of HIPS is also low at low temperatures, but begins to rise at about −90 °C, near the glass transition of the polybutadiene rubber. The

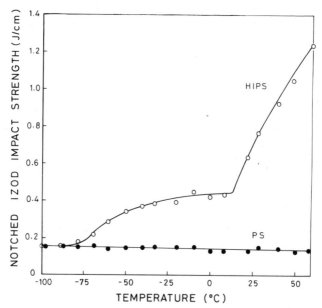

Fig. 7.2   *Relationship between Izod impact strength and temperature for PS and HIPS.*

rise is accompanied by the appearance of a whitened zone near the base of the notch, which increases steadily in extent with increasing temperature. At 12 °C there is a further steep rise in impact strength, and the fracture surface becomes fully whitened. These observations correlate well with the results of the tensile test. Stress-whitening is obviously connected with an energy-absorbing yield process in HIPS, which makes an increasing contribution to impact strength as the temperature rises.

   In addition to explaining these general characteristics of rubber-modified polymers, a theory of toughening must account for the effects of structure on fracture resistance, including the effects of rubber–matrix adhesion, rubber particle size and relaxation behaviour of the rubber. The importance

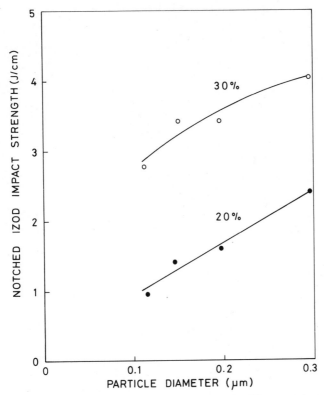

Fig. 7.3   *Relationship between Izod impact strength and rubber particle size in ABS emulsion polymers containing* 20% *and* 30% *polybutadiene (from C. F. Parsons and E. L. Suck, ref.* 1, *reproduced with permission).*

of adhesion between the phases has been emphasised in Chapters 2 and 3: a weakly bonded rubber particle is mechanically equivalent to a void, and the evidence indicates that voids are much less effective than rubber particles in toughening brittle plastics. Similarly, very small particles are inefficient compared with the same volume of larger particles, as illustrated in Fig. 7.3.[1] Increasing the rubber particle size from 0·1 to 0·3 μm produces a substantial rise in the impact strength of ABS. The same kind of relationship has been observed in other polymers, and it is generally agreed within the industry that there is an optimum particle size for each type of matrix: polystyrene requires rubber particles over 1 μm in diameter, whereas particles less than 0·1 μm in diameter appear to be adequate in PVC. However, not all of the evidence is completely reliable, since in many experiments the change in

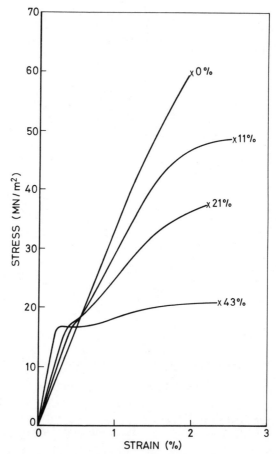

Fig. 7.4   Tensile stress–strain curves at 24 °C for SAN containing various volume concentrations of glass beads. Strain rate $2 \cdot 2 \times 10^{-3}\,s^{-1}$ *(from R. E. Lavengood et al., ref. 3, reproduced with permission).*

particle size is accompanied by changes in other important parameters. This problem is apparent especially in studies on toughened polymers made by the bulk process: an increase in stirrer speed, for example, reduces both the average diameter and the volume fraction of the rubber particles in HIPS and related polymers, with the result that the modulus rises as the impact strength falls.[2]

A comparison of rubber particles with other types of spherical fillers is illuminating. Figure 7.2 shows that polybutadiene loses its toughening properties below the glass transition at about $-90\,^\circ$C, thus demonstrating that the rubbery characteristics of the dispersed particles have at least some bearing on the fracture resistance of the composite. An even more striking effect is shown in Fig. 7.4, which presents stress–strain curves obtained by Lavengood et al. in experiments on SAN.[3] The polymer was blended on a heated two-roll mill with spherical glass beads between 12 and 36 μm in diameter, and the blends were subjected to tensile tests. The similarities and differences between Figs. 7.4 and 7.1 are most interesting. Glass beads, like rubber particles, induce yielding in styrene-based polymers. Furthermore, the yielding is accompanied by stress-whitening, and the elongation at break is higher in the composite than in the homogeneous polymer. Only the magnitude of the effect is different: the glass-filled blends fracture at a strain of about 3 %, whereas the rubber-toughened polymers reach strains of 40 % or more.

These experiments show that rubber particles are not unique in their capacity to induce yielding in brittle glassy polymers. Other types of stress concentrators, including particles with a higher modulus than the matrix, can also cause yielding. The distinctive feature of rubber-toughened polymers is their post-yield behaviour. The problem can therefore be resolved into two separate questions: what are the mechanisms of yielding and energy absorption in filled glassy polymers, and why does fracture occur later in rubber-toughened plastics than in composites containing other types of filler particles?

To sum up, the theory of toughening must explain the following phenomena in materials based largely upon brittle glassy polymers:

(i)   yielding;
(ii) high elongation at break;
(iii) high energy to break in impact;
(iv) stress-whitening.

The theory should also account for the influence of the following factors on toughness:

(a)   rubber–matrix adhesion;
(b)   particle size and size distribution;
(c)   rubber content;
(d)   relaxation behaviour of rubber;
(e)   composition of matrix;
(f)   temperature.

Finally, the theory should not be designed simply to explain previous observations, but should have some predictive value. The first step is to establish a qualitative understanding of the causes of toughening, but the ultimate aim must be to develop a quantitative treatment, which can account in detail for the observed effects, and indicate how a better balance of properties might be achieved in any particular class of rubber-modified polymers.

7.2 QUALITATIVE THEORIES OF TOUGHENING

7.2.1 Early theories

One of the first hypotheses advanced to explain rubber toughening was that the rubber absorbed impact energy by mechanical damping. Buchdahl and Nielsen had observed the secondary loss peak due to the rubber,[4] and it was known that a number of other tough polymers also had prominent secondary loss peaks below room temperature. These observations appeared to be inter-related. However, whilst damping might explain some of the energy absorption in impact, it did not account for stress-whitening or large strain deformation. Obviously, some other mechanism was responsible for these effects.

The first theory of toughening was published in 1956 by Merz et al.[5] The basic idea was that the rubber particles held together the opposite faces of a propagating crack, so that the energy absorbed in impact was the sum of the energy to fracture the glassy matrix and the work to break the rubber particles. In order to explain tensile yielding, it was necessary to postulate the formation of a large number of microcracks, each spanned by a rubber particle and separated from the neighbouring microcrack by a layer of polystyrene. Large tensile deformation could then occur through the opening of the microcracks, extension of the rubber particles and buckling of the polystyrene layers.

This theory accounted for a number of experimental observations. Scattering of light from microcracks explained stress-whitening. Opening of the microcracks provided a mechanism for large strain deformation, and the bridging role of the rubber particles required that the particles be elastomeric and well bonded to the matrix. Merz et al. demonstrated that the density of HIPS fell by 8 % as a result of a tensile test, thus confirming that large strains were achieved by void formation within the polymer. The chief weakness of the theory was that it concentrated attention upon the rubber rather than on the matrix. There were significant differences in fracture behaviour between toughened polystyrene and toughened PVC, which could not adequately be explained by the microcrack theory.

7.2.2 Multiple crazing theory

The multiple crazing theory, advanced by Bucknall and Smith in 1965,[6] was a development of the Merz–Claver–Baer microcrack theory. The important new feature was that stress-whitening was attributed not to cracks but to crazes. The development followed logically from the work of Kambour, who had shown that crazing almost invariably preceded fracture in glassy polymers,[7] and was the mechanism responsible for the high fracture surface energies measured by Berry in experiments on PS and PMMA.[8] The properties of crazes were discussed in detail in Chapter 6. By emphasising the role of the matrix polymer in deformation and energy absorption, the multiple crazing theory resolved a number of the difficulties inherent in the earlier microcrack theory, and also stimulated research in this hitherto neglected branch of polymer science.

The basis of the theory is that rubber particles both initiate and control craze growth. Under tensile stress, crazes are initiated at points of maximum principal strain, which are usually near the equators of rubber particles, and propagate outwards, again following planes of maximum principal strain. Craze growth is terminated when the stress concentration at the tip falls below the critical level for propagation, or when a large rubber particle or other obstacle is encountered. The result is a large number of small crazes, in contrast with the small number of large crazes formed in the same polymer in the absence of rubber particles. In fracture mechanics terms, the flaw size is reduced from several millimetres to a few microns, or less. Consequently, the material can reach a much higher strain energy density before fracture (*see* Chapter 9). Dense crazing throughout a comparatively large volume of material accounts for the high energy absorption in tensile and impact tests.

The first experimental evidence for multiple crazing was obtained by optical microscopy.[6] A thin section of HIPS was attached by adhesive tape to a stretching device, and mounted between two glass cover slips, using K_2HgI_4 in glycerol as an immersion oil, as described in Chapter 3. Photomicrographs were then taken between crossed polars and under phase contrast conditions at increasing tensile strains. Figure 7.5 shows the results obtained at an early stage of the experiment. Under polarised light, the crazes are seen as bright birefringent bands about $50\,\mu m$ long, lying perpendicular to the applied stress; the remainder of the specimen is dark. Under negative phase contrast conditions, both the rubber particles and the crazes appear dark against the background of the uncrazed polystyrene,

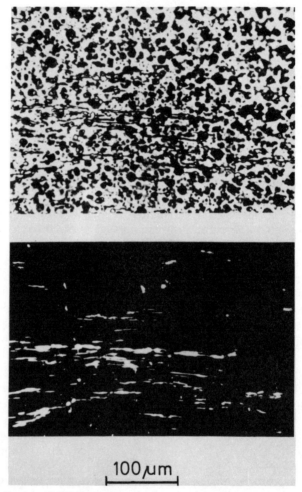

Fig. 7.5 Section of HIPS viewed (top) by negative phase contrast and (bottom) between crossed polars, showing crazes formed by stretching the section on the stage of the optical microscope (from C. B. Bucknall and R. R. Smith, ref. 6).

which has a higher refractive index, and therefore appears brighter in the microscope.

Metallographic techniques have also been used to study crazing in rubber-modified plastics. Moskowitz and Turner polished toughened PMMA with emery paper and alumina, coated the polished surface with a 20 nm layer of gold, and observed craze formation under stress by means of reflection microscopy.[9] Polished surfaces may also be prepared by microtoming the specimen.[10] Crazes formed by applying stress before polishing may be examined by etching: the solutions listed in Chapter 3 as etches for rubber particles are also effective in etching crazes and shear bands.[10]

Later work on crazing in toughened plastics has been based mainly on electron microscopy. The first published electron microscope study was made by Matsuo, who applied replica techniques to crazed HIPS and ABS.[11] This work revealed that the spherical rubber particles in ABS became spheroidal as a result of craze formation in the adjacent SAN matrix material, and that crazes tended to branch in rubber-toughened plastics. More recent electron microscope studies, based mainly on transmission electron microscopy of ultrathin sections, have confirmed that crazes are formed in HIPS,[12-14] ABS,[14] toughened PVC,[15] toughened PMMA[9] and toughened epoxy resin.[16] Figure 7.6 shows crazing in an ultrathin section of HIPS, and illustrates the way in which crazes run from one rubber particle to the next.

The multiple crazing theory is well founded on experimental evidence, and successfully explains the impact and tensile properties of HIPS, including stress-whitening, decreases in density and elongation without lateral contraction. The effects of rubber content, particle size and size distribution, temperature and rubber–matrix adhesion can all be understood in terms of the theory, and will be discussed in detail later. However, there are some observations that do not fit into the relatively simple treatment presented above. In particular, both ABS and toughened PVC exhibit marked necking in tensile yielding experiments, and in the case of PVC-based materials this necking often occurs without detectable stress-whitening. In order to explain these observations, it is necessary to postulate that shear yielding mechanisms also contribute to the tensile deformation of rubber-toughened plastics.

7.2.3 Shear yielding theories

The suggestion that rubber toughening might be due to shear yielding in the matrix was made by Newman and Strella.[17, 18] Their theory was based on optical microscope studies of rubber particle distortion in ABS tensile specimens. Recognising that yielding must take place in the matrix, Newman and Strella attributed the deformation to a local reduction in the T_g of the SAN as a result of triaxial tension. Both mechanical effects and differential thermal contraction are known to produce triaxial stresses in the neighbourhood of rubber particles (*see* Chapter 5).

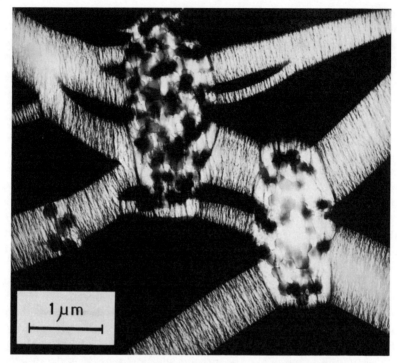

Fig. 7.6   Crazes connecting rubber particles in a cast thin film of HIPS, which was strained in tension, stained with osmium tetroxide and examined in the transmission electron microscope (from P. Beahan, A. Thomas and M. Bevis, J. Mater. Sci. **11** (1976) 1207, *reproduced with permission).*

Although shear yielding undoubtedly contributes to toughening in some rubber-modified plastics, the detailed mechanism outlined above is not supported by the evidence now available. The failure envelope studies described in Chapter 6 show that triaxial tension promotes crazing and brittle fracture rather than shear yielding, and that shear deformation takes place well below T_g even in non-dilatational stress fields. Rubber particles initiate shear deformation by producing local increases in the octahedral shear stress, rather than by modifying the relaxation behaviour of the matrix polymer.

A further difficulty of shear yielding theories is that they do not explain stress-whitening, density changes, elongation without necking and other characteristics of rubber-toughened plastics. Only multiple crazing can account in a satisfactory manner for these effects. The currently accepted view is that crazing is the principal mechanism of toughening, but that shear yielding also contributes in some rubber-toughened plastics, especially those based on the more ductile plastics such as PVC. Interactions between crazes and shear bands appear to be important in these polymers.

7.2.4 Crazing with shear yielding

By recognising that crazing and shear yielding occur simultaneously in most rubber-toughened plastics, it is possible to resolve the problems inherent in the simple crazing theory of toughening. The differences in tensile behaviour between HIPS and ABS, for example, are due to differences in the contribution of the two mechanisms to the overall deformation. In HIPS, crazing dominates, and there is little evidence of shear yielding. In ABS, on the other hand, crazing and shear yielding proceed simultaneously, so that the specimen exhibits both stress-whitening and necking. At higher temperatures, HIPS itself necks under tensile stress, an effect that can be explained in a similar way. The creep experiments described in the latter part of this chapter confirm that the mechanism of deformation changes with temperature, strain rate, composition of matrix, orientation and other factors, and the theme is further developed in Chapter 8.

Figure 7.7 is a scanning electron micrograph showing simultaneous crazing and shear band formation in a rubber-toughened polymer.[19] The specimen is a blend of HIPS with poly(2,6-dimethyl-1,4-phenylene oxide) which was deformed in uniaxial tension, polished by microtoming and etched with a mixture of chromic and phosphoric acids. The hemispherical holes are produced by etching of the rubber particles; the parallel fissures lying normal to the stress axis are formed by attack on crazes; and the lightly etched bands at 45° to the stress axis are shear bands. Many of the crazes run from rubber particles, as expected, but some appear to be unconnected with particles, probably because they were initiated at points above or below the plane of the micrograph. The shear bands also tend to run between rubber particles, and the micrograph generally supports the view that both crazes and shear bands initiate at stress concentrations produced by the rubber. In this material, few of the crazes terminate at a neighbouring rubber particle. The crazes are relatively short, and appear to terminate at shear bands.

Interactions between crazes and shear bands were discussed in Chapter 6. The orientation within a shear band is roughly parallel to the applied tensile stress, and therefore normal to the plane of the crazes. For this reason, shear bands would be expected to act as obstacles to craze propagation. Evidence from electron microscopy and from creep studies is consistent with the view that shear bands control craze size in this way, and thus help to increase toughness: in fracture mechanics terms, the intrinsic flaw size of the material is reduced. The two contrasting mechanisms of deformation in rubber-toughened plastics are not merely simultaneous, but synergistic.

X-ray scattering experiments support the electron microscope observations. Initially isotropic ABS specimens show clear evidence of

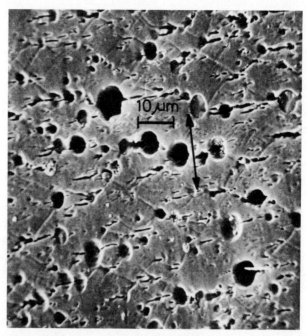

Fig. 7.7   Crazes and shear bands in a HIPS/PPO blend. Specimen was subjected to a tensile strain (direction arrowed), polished by microtoming, etched with chromic acid, coated with Au/Pd and viewed in the scanning electron microscope.

molecular orientation in WAXS (wide angle X-ray scattering) after being deformed to high strains in tension, whereas HIPS specimens show no such orientation in the same experiment.[20] The orientation within the shear yield zones is retained on unloading, whilst the orientation within crazes is to a large extent lost as a result of viscoelastic recovery. The small angle X-ray scattering (SAXS) results shown in Fig. 7.8 again emphasise the difference in tensile yield behaviour between HIPS and ABS.[21] The scattering intensity is low for the HIPS specimen in the unstrained state, but increases by a factor of 100 on straining the material by 8 %. Because of its smaller rubber particles, ABS produces more intense scattering in the unstrained state. However, the increase in scattering on yielding is much less marked in ABS than in HIPS. These differences reflect the lower contribution of crazing to the deformation of ABS, since it is the void–fibril structure of the crazes that produces the small-angle scattering. SAXS techniques are potentially of great value in studying deformation mechanisms in toughened plastics, as

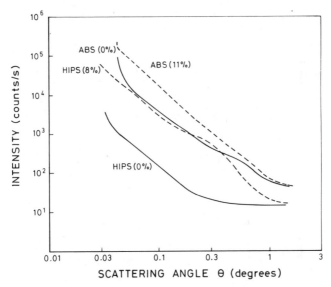

Fig. 7.8   *Small angle X-ray scattering curves showing the effects of applying strains of 8 % and 11 % to HIPS and ABS respectively (from L. E. Nielsen et al., ref. 21, reproduced with permission).*

demonstrated by Nielsen and co-workers,[21] and there is a good case for more extensive use of the method in the future.

7.3   VOLUMETRIC STRAIN MEASUREMENTS

Since crazes consist of approximately 50 % by volume of voids, multiple crazing produces a significant increase in the volume of a toughened polymer. Furthermore, the magnitude of the volume change is directly proportional to the amount of craze matter formed. This principle was applied by Bucknall and Clayton in the first quantitative study of crazing in rubber-modified plastics.[22, 23]

The apparatus used in these quantitative experiments was developed by Darlington and Saunders,[24, 25] and is illustrated in Fig. 7.9. The specimen is a standard dumb-bell creep specimen of rectangular cross-section, with a parallel gauge portion approximately 40 mm long. The lower grip is fixed rigidly to the base of the apparatus, whilst the upper grip is connected through a linear bearing to the lever loading arm. The linear bearing

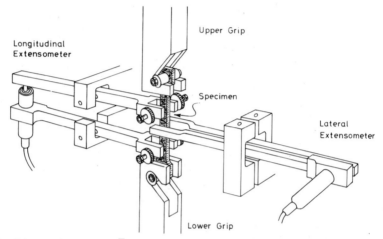

Fig. 7.9   Darlington–Saunders creep apparatus for simultaneous measurement of longitudinal and lateral strains in polymers.

eliminates twisting of the specimen, which would produce spurious lateral strain readings. The extensometers are also designed on the lever-loading principle, and are balanced about the pivot point, so that they do not subject the specimen to additional load. Four small pins attached to the longitudinal extensometer make light contact with the specimen, whilst a light spring holds the lateral extensometer in contact with the flat faces of the specimen. Strains are recorded by high-accuracy linear capacitance transducers. Where necessary, an additional lateral extensometer can be accommodated to measure the third principal strain.

The strain measurements are valid only when the deformation is uniform throughout the gauge portion. Some polymers, including HIPS, can reach high strains without necking, but most rubber-toughened plastics begin to neck at strains between 3 and 10%, at which stage the test must be terminated.

This sensitive creep technique is unsuitable for measurements at high rates of strain. Instead, it is necessary to use alternative methods to follow lateral strains. Cessna[26] and Fenelon and Wilson[27] used high-speed cinematography in experiments on ABS specimens subjected to rapid tensile elongation. Any non-uniformity in the deformation is recorded by the camera in this type of experiment.

For isotropic or transversely isotropic specimens it is sufficient to measure only the longitudinal strain and one lateral strain, since the two lateral strains are equal. The volume strain ΔV is then given by

$$\Delta V = \frac{V}{V_0} - 1 = \lambda_1 \lambda_2 \lambda_3 - 1 = (1 + e_3)(1 + e_1)^2 - 1 \qquad (7.1)$$

where V is the current volume of the gauge portion at time t, V_0 is the original volume, λ_1, λ_2 and λ_3 are the extension ratios in the thickness, width and longitudinal directions, and e_1, e_2 and e_3 are the corresponding strains.

As explained in Chapter 6, a tensile stress is equivalent to a deviatoric (shear) stress plus a hydrostatic tension. On application of the stress, the hydrostatic component produces an immediate increase in volume, $\Delta V(0)$, which can be regarded as the elastic volume strain. Writing σ for the tensile stress, and K for bulk modulus, $\Delta V(0)$ is given by

$$\Delta V(0) = \frac{\sigma}{3K} \qquad (7.2)$$

Goldbach and Rehage measured the time-dependence of K in polystyrene at 93 °C,[28] and showed that the effect was relatively small. They applied a pressure of $4 \cdot 5 \, \text{MN/m}^2$ to the specimen, and observed a volume change of $0 \cdot 08 \%$ over a period of 10^5 s following the release of the pressure. Volume changes of similar magnitude are observed in other polymers.[29] A hydrostatic tension of $4 \cdot 5 \, \text{MN/m}^2$ corresponds to (the hydrostatic component of) a tensile stress of $13 \cdot 5 \, \text{MN/m}^2$, approximately half the yield stress of a typical HIPS, which means that viscoelastic volume changes in this polymer are unlikely to exceed $0 \cdot 2 \%$ in creep tests. Indeed, volume relaxations at 20 °C are probably much smaller than at 93 °C.

On the basis of this evidence, it may be concluded that the large volume strains observed in HIPS and other rubber-toughened plastics are due to the elastic volume response $\Delta V(0)$, which is instantaneous, plus crazing, which is a time-dependent process. Any time-dependent bulk response of the polymer will be neglected in the remainder of the discussion, and the volume strains will be analysed on the assumption that

$$\Delta V(t) = \Delta V(0) + \Delta V \, (\text{crazing}) \qquad (7.3)$$

Since crazes are formed in planes approximately normal to the applied tensile stress, crazing makes a negligible contribution to the lateral strain. One consequence has already been noted: beyond the yield point, HIPS extends to large strains without significant change in cross-sectional area. The stress remains almost constant throughout the tensile test, and shear yielding makes very little contribution to deformation, so that e_1 is small and approximately constant: a typical value of e_1 in a creep or tensile test would be -0.005, *i.e.* a lateral contraction of 0.5%. Under these conditions, differentiation of eqn. (7.1) yields:

$$\left(\frac{\partial \Delta V}{\partial e_3}\right)_{e_1} = (1 + e_1)^2 \approx 1 \tag{7.4}$$

Thus a graph of ΔV against e_3 will have unit slope when the deformation is entirely due to crazing. This type of relationship is usually observed in tests on isotropic HIPS. More generally, for small strains, eqn. (7.1) expands to

$$\Delta V = e_3 + 2e_1 \tag{7.5}$$

or

$$e_3 = \Delta V - 2e_1$$

The latter version of the equation emphasises the fact that the extension e_3 is the sum of two terms: crazing, represented by ΔV, and shear yielding, measured by $-2e_1$ (the negative sign occurs because e_1 is a contraction). Thus the creep test provides quantitative information about the contributions of both mechanisms. The current contributions of crazing and shear yielding to creep at time t are given by $d\Delta V/de_3$ and $-2de_1/de_3$ respectively, so that the balance between the two mechanisms can be determined by plotting ΔV against e_3 and measuring the slope. Zero slope means that there is no crazing, and that creep is entirely due to shear processes. Conversely, unit slope means that there is no shear yielding, and that creep is entirely due to crazing, as already stated. It is important to note that the slope characterises the mechanism at a single point during the creep test. In some polymers, the slope changes with strain, as the balance between crazing and shear yielding alters.

The lateral contraction, like the volume strain, can be separated into an initial elastic response and a time-dependent term. However, the distinction is much less clear than that between elastic volume change and dilatation due to crazing, and it is preferable to include short-term viscoelastic response and long-term creep in the same quantity.

The foregoing principles are illustrated in Figs. 7.10–7.11, which present creep and recovery data for HIPS and toughened PVC. The specimens were unloaded at 5% extension, so that the approximation to small strains is valid. In the HIPS experiment, the first measurements were made 10 s after loading: the recorded values of e_3, e_1 and ΔV can be regarded as instantaneous values for practical purposes. During the remainder of the period under load, the lateral strain e_1 changed very little. The deformation

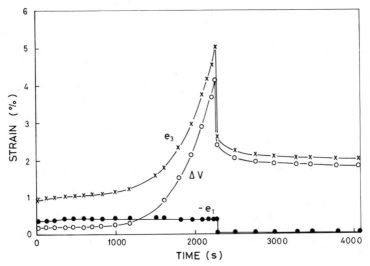

Fig. 7.10   *Tensile creep and recovery of HIPS at 20° C under a stress of* 19·7 *MN/m².*

was dominated by crazing, which began slowly, but accelerated after 1000 s, and became relatively rapid. On unloading, the recovery was also dominated by the response of the crazes, which made a large initial recovery, followed by a slower time-dependent response. The creep of toughened PVC is in complete contrast. The change in volume with time is small, and shear processes dominate the creep and recovery stages of the test. Furthermore, the total rate of deformation decreases with time, whereas the rate increased in the HIPS experiment. The difference in deformation mechanism between the two polymers is shown in Fig. 7.12. Both give a linear relationship between ΔV and e_3, but the slope of the line is 0·95 for HIPS, compared with 0·08 for toughened PVC. It must therefore be concluded that crazing is responsible for 95% of the time-dependent part of the creep in HIPS, and for 8% of the creep in the PVC, the remainder in each case being due to shear yielding.

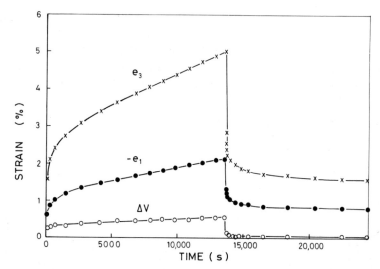

Fig. 7.11   Tensile creep and recovery of toughened PVC at 20°C under a stress of 36·0 MN/m². Polymer made by blending PVC with 5% ABS concentrate.

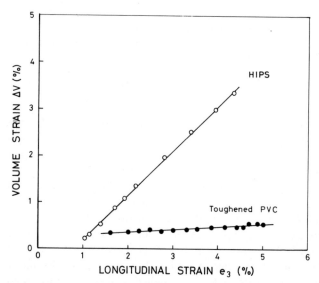

Fig. 7.12   Relationship between ΔV and e₃ during tensile creep of toughened PS and PVC at 20°C. Slope of 0·95 for HIPS under a stress of 22·7 MN/m² indicates craze-dominated deformation. Slope of 0·08 for toughened PVC under a stress of 36·0 MN/m² indicates shear-dominated creep.

The volumetric strain experiment described above provides information not only about the mechanisms of toughening, but also about the *kinetics* of crazing and shear yielding. Thus a quantitative study of rubber toughening is possible. Applications of the methods and typical results are discussed in the latter part of this chapter and in Chapter 8.

7.4 KINETICS OF CRAZING

The deformation kinetics of rubber-toughened plastics is a complex subject, involving at least two mechanisms, neither of which is properly understood in homogeneous glassy polymers. The problems encountered in characterising the kinetics of crazing and shear band formation in glassy polymers were discussed in Chapter 6. In rubber-toughened plastics, the presence of rubber particles complicates the problem further: stress fields become inhomogeneous, and phase boundaries are presented as barriers to propagating bands of both types. These considerations obviously limit the extent to which quantitative models can be developed. Nevertheless, it is useful to examine simple models based on volumetric strain data. More satisfactory models can then be developed as additional experimental evidence becomes available. Creep studies indicate that there are two important cases to consider: crazing without interaction, and crazing modified by interaction with shear bands.

7.4.1 Crazing without interaction
The relationship between ΔV and t shown in Fig. 7.13 is typical of HIPS polymers at room temperature, and is also observed in some ABS polymers. There is an initial 'induction period' τ_i during which the rate of crazing is low, a steady acceleration and finally a period in which the rate of crazing is constant. The reciprocal induction period, τ_i^{-1}, and the maximum rate of crazing, $d\Delta V/dt$, are convenient parameters for describing the kinetics of crazing. There are several possible explanations of this pattern of behaviour. One possibility is that the crazes form simultaneously after an induction period, and that each propagates at a rate that increases to a constant value. On balance, it appears more probable that the increasing rate of crazing reflects the growing number of crazes present in the stressed specimen.

Bucknall and Clayton proposed a simple model for creep of HIPS under constant load, based on this explanation.[23] The model involves three assumptions, each of which can be justified using published data on homogeneous glassy polymers, as follows:

(a) The number of crazes, N, formed in unit volume of the polymer is proportional to the time t under load:

$$N = k_i t \qquad (7.6)$$

where k_i is the rate coefficient for craze initiation.

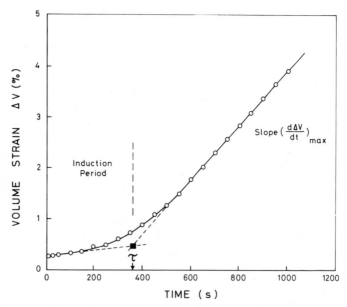

Fig. 7.13   Volume changes during creep of HIPS at $20\,^\circ C$ under a stress of $23{\cdot}35\,MN/m^2$ illustrating definition of induction time τ and of maximum rate of crazing.

(b) The craze tip advances unidirectionally at a constant rate, its thickness and width remaining unchanged:

$$\frac{dr}{dt} = k_p \qquad (7.7)$$

where r is the length of the craze, and k_p is the rate coefficient for craze propagation.

(c) On average, crazes stop growing when they reach a certain length r^* where r^* is the mean distance between neighbouring rubber particles in the plane of craze formation.

On the basis of these simple assumptions concerning initiation, propagation and termination, the kinetics of crazing can be developed in two stages: one relating to the initial period, before the crazes are large enough to terminate; and the second relating to the later period, when termination becomes a controlling factor.

Initial period
At time t during the initial period under load, the volume increase v produced by a craze initiated at time ζ is given by

$$v = ar = ak_p(t - \zeta) \qquad (7.8)$$

where a is the cross-sectional area of the craze.

Since the fraction of crazes formed in a time interval $d\zeta$ is $d\zeta/t$, the average volume contribution $\bar{v}$ per craze is given by

$$\bar{v} = \int_{\zeta=0}^{t} ak_p(t - \zeta)\frac{d\zeta}{t} = \frac{a}{2}k_p t$$

The total volume strain ΔV is obtained by combining this result with eqn. (7.6) as follows:

$$\Delta V = N\bar{v} = \frac{a}{2}k_i k_p t^2 \qquad (7.9)$$

This equation describes the creep of HIPS during the initial induction period. Agreement with experimental results is reasonably good.[23]

Steady state
At a later stage in the creep experiment, the rate of crazing becomes constant, *i.e.* d$\Delta V/$dt is constant. The condition for this steady state can be deduced from eqn. (7.6):

$$\Delta V = N\bar{v} = k_i \bar{v} t$$

$$\frac{d\Delta V}{dt} = k_i \bar{v} + k_i t\frac{d\bar{v}}{dt} \qquad (7.10)$$

Steady state conditions require that d$\bar{v}/$dt be negligible, and therefore that $\bar{v}$ be effectively constant, so that

$$\left(\frac{d\Delta V}{dt}\right)_{max} = k_i \bar{v} \qquad (7.11)$$

Equation (7.11) would be expected to apply to the later stages of the creep test, when most of the crazes present have already reached their maximum permitted size, and terminated at the surface of a neighbouring rubber particle, leaving a relatively small number of crazes propagating. The average volume of a craze would then be governed by the size of the average craze at termination:

$$\lim_{t \to \infty} \bar{v} = ar^*$$

Substituting for $\bar{v}$ in eqn. (7.11), and integrating, gives the following equation for the linear portion of the creep curve:

$$\Delta V = k_i ar^*(t - \tau_i) \quad \text{for} \quad t > \tau_i \qquad (7.12)$$

where τ_i is the induction period. The end of the induction period marks the time at which crazes first begin to terminate. From eqn. (7.7), the value of τ_i is given by

$$\tau_i = r^*/k_p \qquad (7.13)$$

Thus the two rate quantities defined by the linear portion of the creep curve measure the two rate coefficients for crazing. The slope $(\mathrm{d}\,\Delta V/\mathrm{d}t)_{max}$ measures k_i, as shown by eqn. (7.11), and the reciprocal induction period τ_i^{-1} is proportional to k_p, as shown by eqn. (7.13). Observed variations in rates of crazing with stress and temperature are consistent with the Eyring activated-flow treatment of creep.[23] In other words, the rate coefficients are exponential functions of stress and reciprocal temperature.

Discussion of model

The model presented above gives good agreement with experimental data for isotropic HIPS, in which crazing is virtually the only mechanism of creep, and also for some types of ABS, in which both crazing and shear deformation occur simultaneously, but without significant interaction. As explained earlier, it is necessary to make a number of simplifying assumptions in constructing the model, and there is obviously scope for the development of more sophisticated models as our understanding of the kinetics of crazing improves.

The assumption that crazes are initiated at constant rate derives from observations on polystyrene, which show that crazes are nucleated at intervals throughout the period under load, rather than being nucleated simultaneously, or at the end of an induction period. A constant rate of initiation is to be expected in a specimen held at constant stress, unless the number of initiation sites becomes depleted. A possible refinement of the model is to set a limit on the number of sites available for craze initiation, and to allow for the decrease in available sites with time. A further refinement would be to make k_i a decreasing function of time, to take account of relaxation effects in the matrix polymer. A more radical modification would be to treat the initiation sites as having a range of activation free energies, so that initiation takes place initially at the small number of sites with low activation energies, and begins at less favourable sites at a later stage of the test.

Similar problems arise in relation to craze propagation. Constant rates of propagation are observed in glassy polymers, as described in Chapter 6, so that there is some justification for assumption (b). However, there is no evidence that rates of propagation are constant in HIPS or any other toughened polymer. Indeed, some variation in rate with craze length is to be expected in view of the inhomogeneity of the stress field.

There is also evidence from model experiments that crazes will grow unidirectionally from one rubber particle to the next. Propagation of this type was observed by Matsuo et al. in their experiments with two embedded rubber spheres.[30] Again, other growth patterns can be justified. Growth in two dimensions is obviously possible, and growth in a third direction, by craze thickening, cannot be excluded.

Termination by rubber particles is a more basic assumption of the model. If rubber particles do not control craze growth, then it is difficult to see how they differ from simple stress raisers such as glass beads, which are ineffective as toughening agents, as shown in Section 7.1. The unidirectional propagation model leads naturally to the conclusion that a craze ceases to grow, or is at least slowed down, when it reaches the neighbouring rubber particle towards which it is propagating. The fate of a craze growing in two dimensions is less obvious, although termination mechanisms related to the complex stress fields in toughened plastics could doubtless be devised. The role of the rubber in craze termination is discussed in more detail in Section 7.5. In the more ductile polymers, there is an alternative termination mechanism, interaction with a shear band.

7.4.2 Crazing–shear interaction

Toughened PVC (PVC/ABS blend) and toughened poly(phenylene oxide) (HIPS/PPO blend) are examples of materials in which shear bands appear to form simultaneously, and interact, with crazes. Whereas the simple model for crazing in HIPS predicts an increase in rate with time under load, these polymers show a decrease in rate with time. Such decreases are observed only when shear mechanisms play a major role in the deformation.

The characteristics shown in Figs. 7.11 and 7.12 by the toughened PVC specimen are typical of the materials studied to date: when significant shear deformation takes place, the rates of both shear and crazing processes decrease with time in a similar manner, with the result that the balance between the mechanisms remains unchanged. Referring to Fig. 7.12, the slope obtained in a graph of ΔV against e_3 is constant. This observation supports the view that there is some interaction between crazes and shear bands.

The early stages of shear deformation in toughened PVC and toughened polypropylene can usually be fitted to a power law relationship of the form[31]

$$e_1(0) - e_1(t) = k_s t^n \qquad (7.14)$$

where k_s is the rate coefficient for shear creep. When the exponent n has the value $\frac{1}{3}$, eqn. (7.14) corresponds to the Andrade creep law.[32] Shear deformation approximating to the Andrade law has been observed in both PVC- and PP-based toughened polymers. During the later stages of shear deformation, the creep rate increases as the cross-sectional area of the specimen falls and the true stress therefore increases. As explained earlier, volumetric creep measurements are terminated at low tensile strains, so that this effect is relatively small in most experiments.

In at least two rubber-modified polymer systems, toughened PVC and HIPS/PPO blends, craze formation follows the same kinetics as shear deformation, the rate decreasing with time as shown in eqn. (7.14) with the same exponent n, so that the mechanisms do not change in relative importance with time. According to the theory outlined in Section 7.2.4, the rate of initiation of crazes is not affected. The shear bands affect the rate of crazing by providing an additional mechanism of termination, either by acting as barriers to craze growth or by propagating from the tip of a growing craze and so dissipating the strain energy in the tip region. In both cases, the result is the same: shorter crazes are formed.

7.5 STRUCTURE–PROPERTY RELATIONSHIPS

The preceding discussion answers many of the questions posed at the beginning of the chapter. Stress-whitening, yielding and tough fracture are explained in general terms, and the differences between families of rubber-toughened plastics are accounted for. Both matrix and rubber particles are seen to play a part in the phenomenon of toughening.

Important points of detail remain unresolved. In particular, the well-known particle size effect has not yet been discussed, and other properties of the rubber particles require further examination.

7.5.1 Rubber particle size

There are two possible reasons why small rubber particles might be less effective than larger particles in toughening a glassy polymer. The first possibility is that the smaller particles are inefficient in *initiating* crazes. The second possibility is that small particles are ineffective in *terminating* crazes.

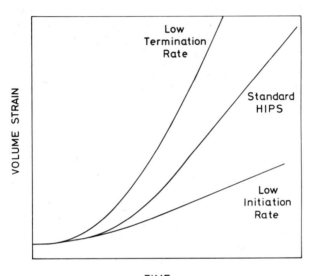

TIME

Fig. 7.14   Suggested relationships between volume strain and time during creep of HIPS according to the model proposed in Section 7.4.1. The middle curve is taken from experimental observations on HIPS. The top and bottom curves show the predicted effects of reducing one of the rate coefficients for crazing in this polymer, whilst in each case keeping the other two rate coefficients constant.

Available evidence suggests that the second explanation accounts for the critical particle size effect noted in Section 7.1. When the particle size is extremely small, the rubber-modified polymer acts as a compatible blend; but critical particle sizes are large compared with the scale of dispersion at which pseudo-compatibility is exhibited, so that some other explanation must be sought for the effect.

Volumetric creep tests help to resolve the problem by providing direct evidence concerning rates of craze formation. Figure 7.14 shows schematically how crazing kinetics are affected by (a) poor initiation and (b) poor termination, at a given rubber content. If initiation rates are low, the overall rate of crazing will also be low, so that creep will be slower than in a normal polymer. If, on the other hand, it is only the termination mechanism that is affected by particle size, the rate of crazing will be higher than in a normal polymer: instead of levelling out at a constant rate at the end of the 'induction period', the material will continue to creep at an increasing rate, as predicted by eqn. (7.9). Both the number of growing crazes and the average craze size will continue to increase with time up to the point of failure. Poor termination means that very large crazes are formed, and the elongation at break is therefore low (see Chapter 9).

In experiments on HIPS containing small rubber particles, Bucknall, Clayton and Keast showed that rates of crazing were high, and continued to increase throughout the test,[19] as indicated in the upper curve of Fig. 7.14. Despite the high rate of crazing and low yield stress, impact strengths and elongation at break were low. These results offer strong support for the view that poor termination is the critical factor. Small rubber particles appear to be unable to stop crazes initiated by neighbouring particles. In this respects, they function in a similar manner to glass beads, and with similar effects upon properties.

A similar argument could be applied to shear band formation, which does not, of course, affect fracture properties in the same way as crazing. If large rubber particles terminate shear bands as well as crazes, then a reduction in rubber particle size should result in an increase in the rate of shear deformation. There is some evidence for this effect in the work of Sultan and McGarry, who studied failure envelopes in rubber-toughened epoxy resins.[33] Failure of unmodified epoxy resin and of resin containing 40 nm rubber particles followed a modified von Mises envelope. By contrast, resin containing $1 \cdot 2 \mu m$ particles stress-whitened in the tensile quadrant, and failed according to a Sternstein–Ongchin relationship (see Chapter 6).

A large particle size is less important when crazing is accompanied by shear band formation. The shear bands provide an alternative mechanism of craze termination, thus reducing the requirement for large particles to control craze growth. It is significant that the critical particle size for toughening is smaller in ductile polymers such as PVC than in relatively brittle polymers such as polystyrene. In the more ductile polymers, the rubber particles must be large enough to initiate crazes and shear bands, but need not necessarily participate directly in craze termination. In experiments on HIPS/PPO blends, Bucknall and co-workers showed that a high fracture resistance could be achieved in a blended PS/PPO matrix using rubber particles that were too small to toughen polystyrene itself.[19]

7.5.2 Rubber–matrix adhesion

The craze termination mechanism also fails when the bond between the rubber and the matrix is weak. Instead of stabilising the craze, a weakly bonded rubber particle is pulled away from the matrix, leaving a hole from which the craze can propagate further, and from which breakdown of the craze to form a crack is probable. The effect is similar to that observed with glass beads. When there is good adhesion between the rubber and the surrounding matrix, fracture surfaces reveal rubber particles that have fractured into two halves along the equatorial plane.[34] Clearly, in this type of toughened polymer the rubber particle is a load-bearing component, and high stresses are transmitted to the centre of the particle. At first sight, the load-bearing capacity of the rubber might appear to be small, as the shear modulus and Young's modulus are much lower than those of the matrix. However, the bulk modulus is comparable with that of the matrix, so that the particle is capable of supporting large hydrostatic stresses, provided the adhesion to the matrix is strong.

The effects of adhesion and of particle size upon fracture resistance in HIPS are well illustrated in Fig. 7.15, which presents results obtained by Durst et al. in experiments using block copolymers.[35] Polystyrene was blended in solution with SBS-type styrene–butadiene block copolymers of various compositions, to give a series of HIPS polymers all containing 20 % by weight of polybutadiene. In a second experiment, polybutadiene was also added, again keeping the total polybutadiene concentration at 20 %. The results show a number of interesting features. Impact strengths are low when the styrene content of the block copolymer is low, because the interfacial adhesion between the rubber and matrix is poor. Increasing the styrene–butadiene ratio in the block copolymer dramatically increases the impact strength, which reaches 500 J/m when the ratio is 50:50. At higher styrene contents, the block copolymer becomes much less effective, because the surfactant properties of the block copolymer lead to a reduction in particle size below the critical level for toughening. Addition of polybutadiene, to form a ternary blend, increases the particle size and shifts the drop in impact strength to the right. In order to achieve high impact strengths in melt-blended mixtures, it is found necessary to use block copolymers of high molecular weight, otherwise, the rubber particle size is reduced below 1 μm during melt blending, and toughness is lost.

The impact strengths obtained in the best of the block copolymer blends are five times higher than those usually recorded in HIPS polymers, although the volume fraction of the rubber phase is comparable with, and indeed lower than, the volume fraction of rubber particles in a typical commercial HIPS. There is no reason to suspect the interfacial adhesion or the particle size in ordinary HIPS, and it must therefore be concluded that

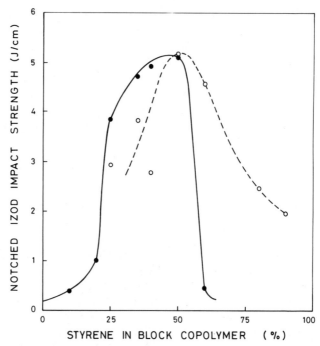

Fig. 7.15   Izod impact strength of HIPS made by melt blending: (●) binary blends of polystyrene with poly(butadiene-b-styrene); (○) ternary blends of polystyrene with poly(butadiene-b-styrene) and polybutadiene. All blends contain 20% by weight of polybutadiene chains (from R. R. Durst et al., ref. 35, reproduced with permission).

the difference is due to the composition of the rubber particle. The block copolymer particles approximate to pure rubber particles, whereas a typical HIPS contains composite rubber particles, consisting largely of polystyrene sub-inclusions. On the basis of this evidence, composite rubber particles are far less effective in toughening polystyrene than a similar volume of pure rubber particles. There is obviously considerable scope for improving the properties of HIPS by preparing structures similar to those of emulsion ABS, but with larger rubber particles. Block copolymers offer one method of achieving this desirable result.

REFERENCES

1. C. F. Parsons and E. L. Suck, *ACS Adv. Chem. Ser.* **99** (1971) 340.
2. M. Baer, *J. Appl. Polymer Sci.* **16** (1972) 1109.

3. R. E. Lavengood, L. Nicolais and M. Narkis, *J. Appl. Polymer Sci.* **17** (1973) 1173.
4. R. Buchdahl and L. E. Nielsen, *J. Appl. Phys.* **21** (1950) 482.
5. E. H. Merz, G. C. Claver and M. Baer, *J. Poly. Sci.* **22** (1956) 325.
6. C. B. Bucknall and R. R. Smith, *Polymer* **6** (1965) 437.
7. R. P. Kambour, *Nature* **195** (1962) 1299.
8. J. P. Berry, *J. Poly. Sci.* **50** (1961) 107.
9. H. D. Moskowitz and D. T. Turner, *J. Appl. Polymer Sci.* **18** (1974) 143.
10. C. B. Bucknall, I. C. Drinkwater and W. E. Keast, *Polymer* **13** (1972) 115.
11. M. Matsuo, *Polymer* **7** (1966) 421.
12. R. P. Kambour and R. R. Russell, *Polymer* **12** (1971) 237.
13. R. J. Seward, *J. Appl. Polymer Sci.* **14** (1970) 852.
14. G. Michler, K. Gruber, G. Pohl and G. Kaestner, *Plaste u Kaut* **20** (1973) 756.
15. M. Matsuo, *Poly. Engng Sci.* **9** (1969) 206.
16. T. Yoshii, Ph.D. Thesis, Cranfield, England, 1975.
17. S. Newman and S. Strella, *J. Appl. Polymer Sci.* **9** (1965) 2297.
18. S. Strella, *J. Poly. Sci.* A2, **3** (1966) 527.
19. C. B. Bucknall, D. Clayton and W. E. Keast, *J. Mater. Sci.* **7** (1972) 1443.
20. R. N. Haward, J. Mann and G. Pogany, *Brit. Poly. J.* **2** (1970) 209.
21. L. E. Nielsen, D. J. Dahm, P. A. Berger, V. S. Murty and J. L. Kardos, *J. Poly. Sci. (Phys.)* **12** (1974) 1239.
22. C. B. Bucknall and D. Clayton, *Nature (Phys. Sci.)* **231** (1971) 107.
23. C. B. Bucknall and D. Clayton, *J. Mater. Sci.* **7** (1972) 202.
24. M. W. Darlington and D. W. Saunders, *J. Phys.* **E3** (1970) 511.
25. M. W. Darlington and D. W. Saunders, in *Structure and Properties of Oriented Polymers*, I. M. Ward (ed.), Applied Science, London, 1975, p. 326.
26. L. C. Cessna, *Poly. Engng Sci.* **14** (1974) 696.
27. P. J. Fenelon and J. R. Wilson, *ACS Div. Org. Coat. Plast. Prepr.* **34**(2) (1974) 326.
28. G. Goldbach and G. Rehage, *J. Poly. Sci.* C16 (1967) 2289.
29. J. D. Ferry, *Viscoelastic Properties of Polymers*, 2nd edn., Wiley, New York, 1970, Chapter 18.
30. M. Matsuo, T. Wang and T. W. Kwei, *J. Poly. Sci.* A2, **10** (1972) 1085.
31. C. J. Page and C. B. Bucknall, unpublished results.
32. E. N. da C. Andrade, *Proc. Roy. Soc.* **A84** (1910) 1.
33. J. N. Sultan and F. J. McGarry, *Poly. Engng Sci.* **13** (1973) 29.
34. J. A. Manson and R. W. Hertzberg, *J. Poly. Sci. (Phys.)* **11** (1973) 2483.
35. R. R. Durst, R. M. Griffith, A. J. Urbanic and W. J. van Essen, *ACS Div. Org. Coat. Plast. Prepr.* **34**(2) (1974) 320.

CHAPTER 8

DEFORMATION AND YIELDING

The theory of rubber toughening presented in Chapter 7 provides a basis for the discussion of the general deformation behaviour of rubber-modified plastics at large strains. The present chapter develops the themes of multiple crazing and interaction with shear bands, by considering the mechanical properties of HIPS, ABS and other toughened plastics, as displayed in creep, tensile, multiaxial and fatigue tests.

8.1   CREEP

Tensile creep experiments provide information in a convenient form for analysis. A fixed load is applied throughout the test, and the stress is therefore constant, provided the cross-sectional area of the specimen does not change appreciably. By holding the stress constant, it is possible to study changes in deformation behaviour with time and temperature in a systematic way. This is most important in rubber-toughened plastics, which respond in a complex manner to applied loads. The factors affecting creep behaviour are discussed below.

8.1.1 Stress
Both crazing and shear deformation are accelerated by an increase in stress (*see* Chapter 6). When both mechanisms are operating simultaneously, and especially when there is significant interaction between the two, the relationship between creep rate and applied stress can be quite complicated. In discussing the effects of stress, it is therefore preferable to begin with a relatively simple system, in which there is essentially only one deformation mechanism.

212

The effects of stress upon creep rates in isotropic HIPS at 20 °C are illustrated in Fig. 8.1.[1] The dominant mechanism of deformation is craze formation, which is characterised by the two rate quantities: the maximum rate of crazing $(d \Delta V/dt)_{max}$, and the reciprocal induction period τ_i^{-1}. According to the model proposed in Chapter 7, these quantities are proportional to k_i and k_p respectively, the rate coefficients for initiation and

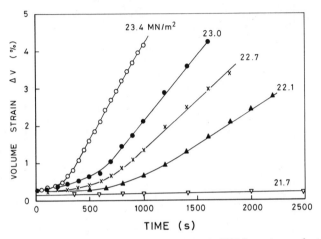

Fig. 8.1 Volume changes due to crazing in isotropic HIPS specimens during tensile creep at 20 °C.

propagation of crazes. Both increase rapidly with stress, following the Eyring equation (6.3):

$$\log_e \frac{\dot{\varepsilon}}{A} = \frac{\gamma V^* \sigma}{4kT} - \frac{\Delta H^*}{kT} \qquad (6.3)$$

The two rate quantities are plotted on a logarithmic scale against applied stress σ in Fig. 8.2. Two parallel lines are obtained, from which the apparent activation energies of the processes can be calculated: the slopes yield a value of 20 nm³ for γV^*. As the stress analyses discussed in Chapter 5 indicate a value of about 2·5 for the stress concentration factor γ at the equator of a rubber particle in a typical HIPS, the activation volume V^* for polystyrene is estimated to be approximately 8 nm³. The similarity between the two lines suggests that the same activation volume and stress concentration factor apply to both initiation and propagation processes. In any comparison of results, it should be noted that some authors use a slightly different form of the Eyring equation from that given above: the

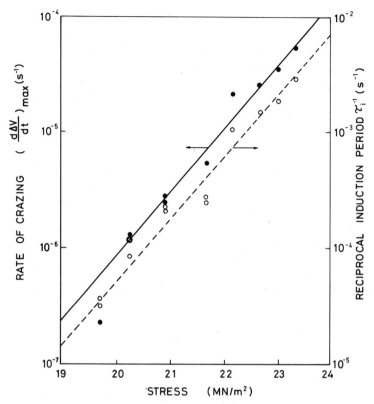

Fig. 8.2 Relationship between log (rate) and applied stress in isotropic HIPS creep specimens at 20°C showing application of the Eyring equation to craze formation.

figure 4 is omitted from the denominator of the stress term, and the activation volumes obtained are correspondingly lower.

The analysis is less straightforward when crazing is accompanied by shear deformation, especially when there is strong interaction between the two processes. However, in ABS polymers, the interactions appear to be relatively weak, perhaps because the shear deformation is of the diffuse type. If the SAN matrix forms diffuse shear zones rather than shear microbands, then craze growth will not be affected to the same extent. Shear microbands are effective barriers because they consist of highly oriented polymer molecules, in contrast to the diffuse shear zones. At present, there is no direct information concerning the mechanism of shear deformation in ABS, and the reason for the low degree of interaction remains a matter for

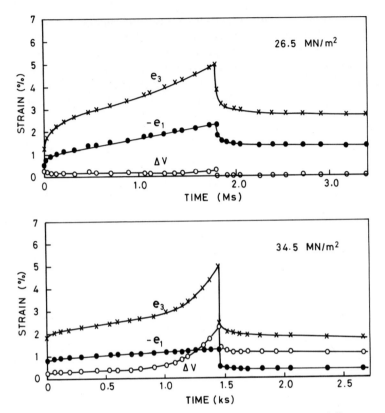

Fig. 8.3   Tensile creep and recovery curves for ABS at 20°C at two different stress levels.

speculation. Whatever the reason, it appears to be possible to treat crazing and shear deformation as independent processes in ABS polymers, and to analyse the contribution of each process to the creep of the material separately.

Creep and recovery curves for ABS are presented in Fig. 8.3.[2] At the relatively low stress of 26·5 MN/m^2 there is little evidence of crazing: creep takes place predominantly by a shear process, at almost constant volume. The rate decreases with time. At the higher stress of 34·5 MN/m^2, shear deformation again dominates the creep process during the first 800 s under load, but the creep rate then increases as a result of accelerating craze formation. Increasing the stress reduces the time and strain at which crazing becomes apparent, with the result that crazing plays a larger part in the creep of the polymer. This point is clearly illustrated in Fig. 8.4, which shows the relationship between volume and longitudinal strains over the stress range 26·5–34·5 MN/m². At all five stresses, the initial creep is dominated by shear yielding, but the curves begin to diverge at strains of the order of 2·5 %. Beyond that strain, the slopes indicate a change of deformation mechanism, from 100 % shear yielding at 26·5 MN/m² to 85 % crazing at 34·5 MN/m².

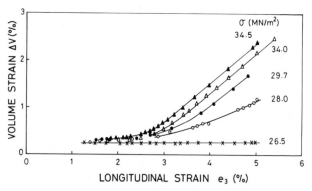

Fig. 8.4   Relationship between ΔV and e_3 for ABS at $20\,^\circ C$ showing change in the mechanism of tensile creep with stress and strain.

This change reflects the differences in kinetics between the two mechanisms of creep, and suggests that shear yielding is unlikely to make a significant contribution to the initial deformation of ABS polymers at high rates of strain. The work of Fenelon and Wilson supports this conclusion.[3]

There is some evidence for a similar change in mechanism with stress in other polymers. Reid found that the creep of HIPS at low stresses followed an Andrade creep law, suggesting deformation predominantly by shear yielding.[4] At the other end of the scale, toughened PVC blends exhibit stress-whitening in tests conducted at high rates of strain, whilst showing little or no volume change in a slow creep test. As the kinetics of crazing are basically different from those of shear yielding, with different rate coefficients, there is every reason to expect a change in the relative importance of the two mechanisms with stress. This principle is expressed in the following equation, which has been applied by Reid[4] to creep in HIPS, and by Bergen to creep in ABS[5]:

$$\varepsilon = \varepsilon_0 + A(\sigma, T)t^{1/3} + B(\sigma, T)t \qquad (8.1)$$

where ε_0 is the initial elastic strain, and $A(\sigma, T)$ and $B(\sigma, T)$ are Eyring rate coefficients. The obvious interpretation of this equation is that there are two independent contributions to creep: the term in $t^{1/3}$ can be identified with shear yielding according to the Andrade relationship, and the linear term in t can then be related to craze formation at a constant rate. This type of equation is not applicable when there is strong interaction between the two mechanisms. As explained in Chapter 7, crazing has been observed to follow an Andrade relationship in HIPS/PPO blends.[6] The simplest explanation of this observation is that crazing kinetics are controlled by shear band formation, which itself is proportional to $t^{1/3}$. Under these conditions, the contributions of the two mechanisms would not be expected to change relative to each other with increasing stress.

8.1.2 Temperature

The creep behaviour of a typical ABS polymer over a range of temperatures is illustrated in Fig. 8.5, which relates $\log J(t)$ to $\log t$, where $J(t)$ is compliance.[7] As noted at 20 °C, the creep curves were initially quite steep, passed through a plateau region, and then rose sharply. Moore and Gieniewski found that there were two separate contributions to creep strain, both of which were temperature and stress activated, following the Eyring

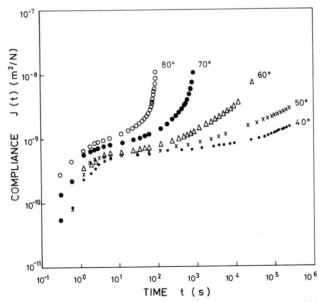

Fig. 8.5 *Tensile creep curves for ABS over the temperature range 40–80 °C under a stress of 17·8 MN/m² (from R. S. Moore and C. Gieniewski, ref. 7, reproduced with permission).*

equation.[7, 8] The dominant term at low temperatures had an activation enthalpy ΔH^* of 43 kJ/mole and an activation volume γV^* (defined by eqn. (6.3)) of 13 nm^3. Between 60 and 85 °C, creep was dominated by a second process, with $\Delta H^* = 550$ kJ/mole and $\gamma V^* = 256$ nm^3. These figures refer to creep rates characterised by the 12-s isochronous creep compliance $J(12)$. The first term represents the initial portion of the creep curves, and the second term is responsible for the sharp increase in creep rate at longer times. Moore and Gieniewski suggested that the second term arose from a glass transition in the ABS, the temperature of which decreased with stress. However, in the light of more recent work on the creep of ABS, it appears more likely that the second term is due to craze formation. Without volumetric or other information, it is impossible to identify the crazing mechanism positively, but crazing is known to occur in ABS at temperatures between 40° and 80 °C, and would be expected to cause an increase in creep rate of the type described by Moore and Gieniewski.

8.1.3 Matrix composition

The structure and composition of the matrix play an important part in determining the relative contributions of crazing and shear yielding to the creep of a rubber-toughened polymer. In order to demonstrate this principle, Bucknall et al. made a series of ternary blends containing HIPS, PS and PPO.[6] The compositions are shown in Table 8.1. Each blend

TABLE 8.1
COMPOSITION OF HIPS/PS/PPO BLENDS
(Percentage by weight)

Blend	HIPS	PS	PPO
A	50	50	0
B	50	$37\frac{1}{2}$	$12\frac{1}{2}$
C	50	25	25
D	50	$12\frac{1}{2}$	$37\frac{1}{2}$
E	50	0	50

contained 50 % of the same HIPS polymer, so that rubber particle size, size distribution, structure and composition were identical throughout the series. Only the composition of the matrix varied. The HIPS chosen for the experiment contained rather small rubber particles, and therefore gave a relatively low elongation at break before blending.

Figure 8.6 shows how creep characteristics alter with matrix composition. Curves presented in 8.6(a) are typical of HIPS polymers: the strain increases

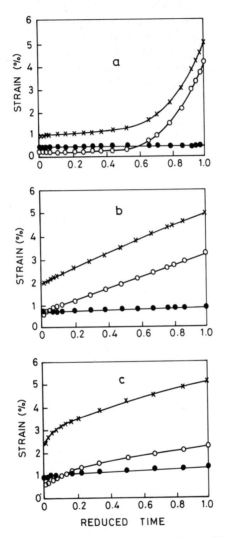

Fig. 8.6   *Tensile creep curves for HIPS/PS/PPO blends:* (a) 50:50:0 *blend;* (b)
50:25:25; (c) 50:0:50. (○) *volume strain* ΔV; (●) *lateral strain* $-e_1$; (×)
longitudinal strain e_3.

slowly during the initial stages of the test, then accelerates as a result of craze formation; shear yielding makes a negligible contribution. On replacing some of the polystyrene in the matrix with PPO, the curves change in shape. Creep curve (b) is straight, and curve (c) has a decreasing slope. The increasing ductility of the matrix leads to a larger contribution from shear yielding, and a correspondingly smaller contribution from crazing. In contrast to ABS, the curves of volume strain against time change shape from

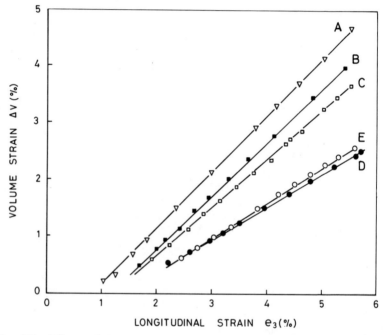

Fig. 8.7 *Effects of matrix composition on mechanism of tensile creep in HIPS/PS/PPO blends at 20°C. Compositions as shown in Table 8.1.*

one blend to the next as a result of increasing shear yielding, suggesting that shear band formation exerts a controlling influence over crazing kinetics. Electron microscope evidence in support of this view was given in Chapter 7. Figure 8.7 shows how the mechanisms alter in relative importance with matrix composition: crazing accounts for all of the time-dependent creep in blend A, but only 60 % of the observed creep in blend E. The slopes do not change with strain, another indication of interaction or co-operation between crazing and shear yielding.

Another interesting example of the effects of matrix composition occurs in rubber-toughened epoxy resins. Unmodified epoxy resin deforms by a shear mechanism, the rate of which decreases with increasing cross-linking. On adding rubber, a toughened product is obtained (*see* Chapter 4), which shows an increase in volume during creep tests, in addition to shear deformation.[9] As in rubber-toughened thermoplastics, the volume

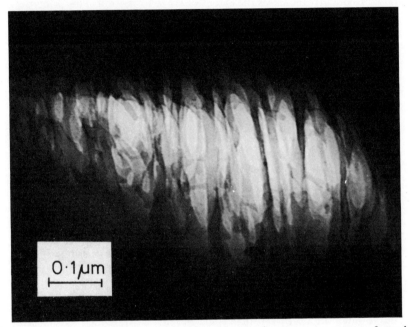

Fig. 8.8   *Ultrathin section of toughened epoxy resin showing craze structure formed by applying tension (vertical direction) on the stage of the electron microscope (from T. Yoshii, ref. 9, reproduced with permission).*

change is due to crazing. Figure 8.8 is an electron micrograph of an ultrathin section of CTBN-modified epoxy resin which has been stretched on the stage of the electron microscope; the fibrillar structure is characteristic of crazing.[9] The rate of crazing increases with the volume fraction of rubber particles, but is virtually independent of the degree of cross-linking of the resin, over the normal range of cross-link densities. The contribution of crazing to the total creep of the resin therefore increases with the amount of rubber added, the degree of phase separation and the degree of cross-linking

of the resin. Rates of shear deformation increase with volume fraction of
rubber, but decrease with cross-linking. Thus the composition of the epoxy
resin matrix affects the balance of mechanisms in two ways: the direct effect
of cross-linking upon mechanisms has already been mentioned; and there is
an indirect effect arising from the varying levels of rubber phase separation
in resins of different formulations. Adding Bisphenol A to the epoxy resin
increases the volume fraction of rubber particles, as discussed in Chapter 4,
and consequently increases the extent of craze formation. This principle is
illustrated in Fig. 8.9.

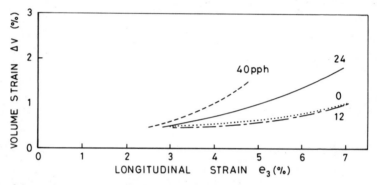

*Fig. 8.9   Relationship between ΔV and e_3 for toughened epoxy resin at 20 °C showing
effect of adding 0–40 pph Bisphenol A on the mechanism of tensile creep (from T.
Yoshii, ref. 9, reproduced with permission).*

8.1.4 Rubber content

The volume fraction of rubber particles in the polymer affects creep
behaviour in several different ways. The principal effect of increasing the
rubber content is to raise the level of stress concentration in the
neighbourhood of the equator of each rubber particle. Not only do the
stress fields begin to overlap, as described in Chapter 5, but also the volume
fraction of the load-bearing matrix is reduced. The resulting increase in the
stress concentration factor γ in the Eyring equation (6.3) produces a large
rise in creep rate, which depends exponentially upon γ. A higher rubber
content also means a larger number of rubber particles, and hence a larger
number of sites for craze initiation. However, this increase is offset by the
reduced spacing between rubber particles: larger numbers of crazes may
form, but each terminates at a neighbouring particle at a correspondingly
earlier stage of growth, so that there is little or no net gain in rate of crazing.

Figures 8.10 and 8.11 compare data obtained in creep experiments on two ASA polymers, containing different concentrations of acrylate rubber in an SAN matrix.[10] Crazing is characterised by $\log(d\,\Delta V/dt)_{max}$ and $\log \tau_i^{-1}$, both of which are linear with applied stress σ. As in the case of HIPS, Eyring plots for each material yield parallel lines for the two rate parameters. Increasing the nominal rubber content from 30% to 40% produces a substantial increase in creep rate: the rate of crazing is approximately 10 times higher in the ASA polymer having the higher rubber content. There is a similar increase in the rate of shear deformation, defined by $-(de_1/dt)_{min}$, and representing the rate of lateral contraction during the later stages of the creep test.

These results are consistent with the interpretation given above. The differences in creep behaviour between the two ASA polymers can be explained simply by assuming that the stress concentration factor γ is 15% higher in the polymer containing 40% rubber than in the polymer containing 30% rubber. In other words, the two materials give a single curve for each rate parameter when the rate parameters are plotted against $\gamma\sigma$ rather than σ.

8.1.5 Orientation

Uniaxial hot drawing produces two changes in the structure of a rubber-toughened polymer. The rubber particles change shape, and the matrix becomes oriented, probably in a non-uniform manner. Both factors act in the same direction, reducing creep rates parallel to the draw direction. Stress concentration factors are lower for spheroidal particles than for spherical particles, and a comparatively small decrease in γ is sufficient to lower the creep rate substantially, as already stated in the preceding section. Molecular orientation parallel to the draw direction further reduces the creep rate.

Shear yielding and crazing are affected differently by the drawing process. In the case of uniaxial drawing, the result is usually to suppress crazing. Tests parallel to the draw direction show that rates of both shear yielding and crazing are reduced, but the net result is that shear deformation becomes relatively more important. This point is illustrated in Fig. 8.12, which presents data from creep experiments on an ABS suspension polymer.[11] Creep is dominated by crazing in the isotropic polymer, but the contribution of shear yielding increases with draw ratio. The behaviour observed at a draw ratio of 1·47 is particularly interesting: the mechanism of creep varies with stress and with strain, as in the case of the isotropic ABS emulsion polymer discussed in Section 8.1.1.

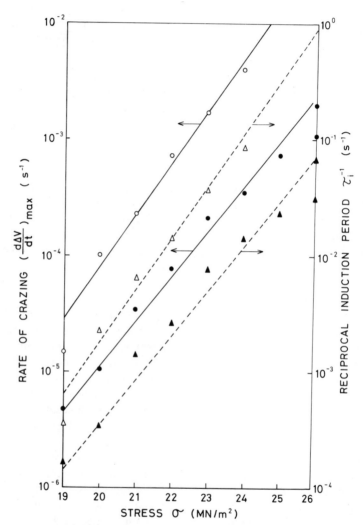

Fig. 8.10   *Relationship between log (creep rate) and applied stress in ASA specimens showing application of the Eyring equation to craze formation:* ($\bigcirc$) *rate of crazing in ASA containing 40% rubber;* ($\triangle$) τ_i^{-1} *for ASA containing 40% rubber;* ($\bullet$) *rate of crazing in ASA containing 30% rubber;* ($\blacktriangle$) τ_i^{-1} *for ASA containing 30% rubber* (*from C. B. Bucknall* et al., *ref.* 10).

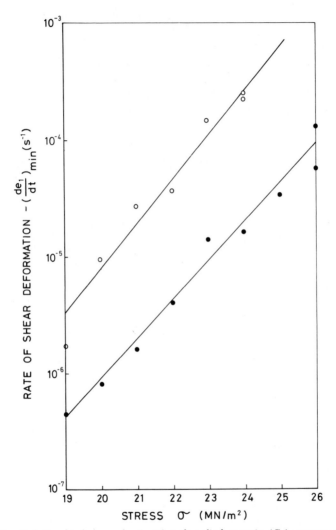

Fig. 8.11 *Relationship between log (rate) and applied stress in ASA creep specimens showing application of the Eyring equation to shear deformation in ASA polymers containing:* (◯) 40% *rubber;* (●) 30% *rubber* (*from C. B. Bucknall et al., ref.* 10).

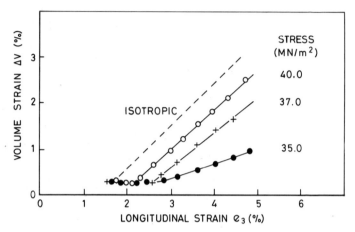

Fig. 8.12   *Relationship between* ΔV *and* e_3 *in hot-drawn ABS bulk-suspension polymer showing change in mechanism of tensile creep with stress and strain at* 20°C. *Specimens drawn uniaxially to draw ratio of* 1·47 *and subjected to creep tests parallel to the draw direction (from C. B. Bucknall et al., ref.* 11).

8.2   TENSILE TESTING

Standard tensile tests are conducted not at constant stress, as in creep experiments, but at a constant rate of grip separation. To a first approximation, this type of tensile test can be treated as a constant strain-rate experiment.

The stress–strain behaviour of HIPS in a standard tensile test was illustrated in Fig. 7.1. The first part of the curve is approximately linear, and reflects the elastic response of the polymer at stresses below those required for craze formation. Crazing becomes significant at relatively high stresses: the material begins to stress-whiten visibly, and the stress–strain curve becomes non-linear. The stress continues to rise, producing a rapid increase in the rate of crazing, until the *upper yield point* is reached. The stress then falls rather steeply, levelling out at the *lower yield point*. The remainder of the curve is comparatively flat: the stress rises gradually with increasing strain until the specimen breaks.

The *yield drop* which is observed between the upper and lower yield points is not due simply to a reduction in cross-sectional area of the specimen, as in some other materials. The lateral contraction of the specimen between the beginning of craze formation and the lower yield point is only about 0·5 %

in HIPS, so that the difference between true stress and nominal stress is only about 1 %. A graph of true stress against strain would therefore show a yield drop similar but not identical to that illustrated in Fig. 7.1. The cause of the drop is strain-softening of the HIPS due to craze formation.

The upper and lower yield stresses represent two conditions under which the rate of crazing matches the imposed strain rate. During the early stages of deformation, a relatively high stress is necessary to produce the required rate of crazing because the number of crazes present is comparatively small. The upper yield stress therefore corresponds to the early stages of the creep test, before the end of the induction period, when the strain rate produced by a given stress is low. Referring to eqn. (7.10), the number of crazes N and the average volume of a craze $\bar{v}$ are both relatively small at this stage of the deformation. As the number and size of the crazes increase, the stress required to produce a given strain rate falls, until the stress–strain curve levels out at the lower yield point. This relatively flat portion of the curve corresponds to the linear portion of the creep curve, beyond the induction period, when the rate of crazing is a maximum. The later part of the tensile test approximates to a creep test, as the stress varies only slowly. Experiments on ABS show that the upper yield stress σ_y increases linearly with log $\dot{\varepsilon}$, where $\dot{\varepsilon}$ is the imposed strain rate.[12, 13] By applying the Eyring equation in this way, Truss and Chadwick showed that the apparent activation volume γV^* (defined according to eqn. (6.3)) was $10 \, nm^3$ for ABS containing SAN of low molecular weight, and $12 \cdot 7 \, nm^3$ for ABS of higher matrix molecular weight.[13]

A true yield drop is not observed in materials that begin to strain harden at an early stage of deformation. The PPO/HIPS blend E defined in Table 8.1 shows a continuous fall in strain rate under constant applied stress in the creep test (see Fig. 8.6). The corollary of this statement is that the stress required to maintain a constant strain rate must increase continuously throughout the tensile test, as shown in Fig. 8.13.[6] Blends A and B, which contain a high proportion of polystyrene in the matrix, exhibit a yield drop, but blends C, D and E do not. These differences reflect the differences in creep behaviour.

Figure 8.13 also shows that the yield stress increases with PPO content, although the change in matrix composition brings a second deformation mechanism into play. Furthermore, the addition of PPO results in a large increase in elongation at break. The reason for both effects is that shear bands hinder craze growth: higher stresses are needed to produce a given rate of crazing; and shorter crazes are formed, so that fracture occurs at a much higher strain energy density (see Chapter 9). It should be noted,

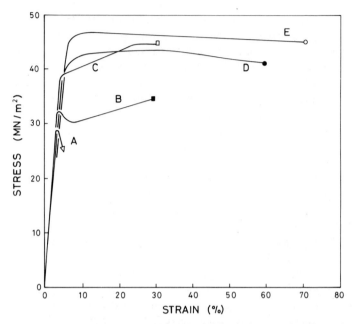

Fig. 8.13   Tensile test results for HIPS/PS/PPO blends showing the effects of matrix composition on properties. Test temperature $20\,^{\circ}C$; *strain rate* $4\cdot0 \times 10^{-4}\,s^{-1}$. *Compositions as shown in Table 8.1.*

however, that the HIPS used in the blending experiments contained very small rubber particles. Shear bands provide a craze termination mechanism which compensates for the poor termination capacity of the small particles (*see* Chapter 7).

Mechanism of tensile deformation
On the basis of creep experiments, there can be little doubt that tensile deformation of isotropic HIPS occurs largely by craze formation. Volume changes in strained HIPS were first noted by Merz *et al.*,[14] but their measurements were made on unloaded specimens, in which there was substantial recovery of the volume strain. More recent studies have concentrated upon measuring volume changes during testing, especially in ABS.[3, 13, 15, 16] Liquid displacement dilatometry is suitable for tensile tests at low strain rates: high-speed cinematography is necessary under tensile impact conditions.

These studies have shown that volume changes occur in ABS both during ordinary tensile tests and during tensile impact. Crazing dominates the deformation at high strain rates, especially in the early stages of the test. Shear yielding becomes apparent in the tensile impact test at strains of 10 % or more,[3, 15, 16] perhaps as a result of delayed necking. At high strains, tensile deformation of ABS polymers tends to be non-uniform along the gauge portion, and the response becomes more difficult to analyse without additional information. In some ABS polymers, elongations at break are low at low strain rates as a result of shear yielding followed by necking rupture, but increase at high strain rates because the specimen is able to reach high strains largely by crazing, without substantial decrease in cross-sectional area.

Temperature
Figure 8.14 presents results obtained by Truss and Chadwick in experiments on ABS polymer over a wide range of temperatures.[13] The version of the Eyring equation (6.3) quoted in Section 8.1.1 predicts that a family of parallel straight lines should be obtained by plotting σ_y/T against $\log \dot\varepsilon$, and the experimental results confirm this prediction. Activation volumes calculated from these results have already been quoted. The Eyring equation also predicts that σ_y/T at any given strain rate should increase linearly with T^{-1}, the slope of the line being given by $\Delta H^*/k$. Arrhenius plots of this kind yield values of 164 kJ/mole and 257 kJ/mole respectively for ABS polymers of low and high matrix molecular weight. Volumetric studies show that these figures refer to the activation energy of crazing in the ABS.

Orientation
The effects of orientation upon yielding in HIPS are illustrated in Fig. 8.15.[17] Isotropic HIPS sheets were drawn uniaxially at various strain rates and at various temperatures above T_g, and subjected to tensile tests at 0°, 45° and 90° to the draw direction. The correlation between yield stress and natural draw ratio (L/L_0) was poor because relaxation during drawing varied with temperature and strain rate. However, it was found possible to correlate the data by using rubber particle shape to characterise orientation. Specimens were sectioned parallel to the draw direction, and etched as described in Chapter 3. The axial ratio of the elliptical rubber particles was measured using optical and electron microscopy. The correlation between yield stress and particle shape suggests that the orientation of the matrix is uniquely related to particle shape in this series of experiments. Etching thus provides a rapid method for studying orientation patterns which are of direct relevance to mechanical properties.

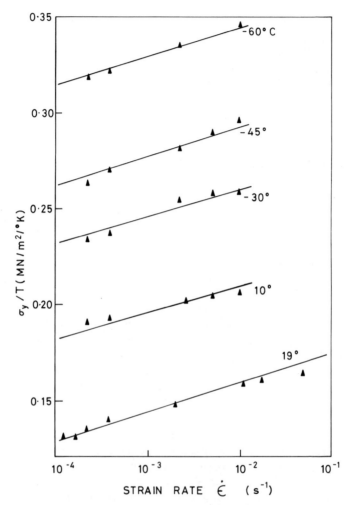

Fig. 8.14   *Relationship between σ_y/T and $\log \dot{\epsilon}$ for ABS subjected to uniaxial tension at constant strain rate over a range of temperatures showing application of the Eyring equation to yield behaviour (from R. W. Truss and G. A. Chadwick, ref. 13, reproduced with permission).*

Grancio *et al.* compared hot drawing with cold rolling in ABS polymers.[18, 19] Whereas hot drawing produces spheroidal rubber particles with smooth outlines, cold rolling deforms the particles into more irregular shapes, with sharp, jagged outlines, probably as a result of localised shear yielding under compression in the SAN matrix. A further factor is the

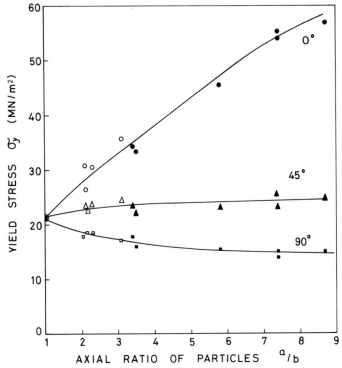

Fig. 8.15   *Tensile yield stress of uniaxially hot-drawn HIPS, tested at* 0°, 45° *and* 90° *to the draw direction, plotted against axial ratio of elliptical rubber particles. Test temperature* 21 °C; *strain rate* $1\cdot66 \times 10^{-3}$ s^{-1}. (○,△, □) *drawn at* 110 °C; (●, ▲,■) *drawn at* 95 °C (*from L. J. Evans* et al., *ref.* 17).

rigidity of the SAN sub-inclusions at temperatures below T_g. During hot drawing, the sub-inclusions are elongated into spheroidal shapes similar to those of the parent particles; but in cold rolling the sub-inclusions retain their spherical shape. Both types of orientation process result in a reduction in the extent of stress-whitening in the tensile test, and an increase in yield and fracture stress.

8.3   MULTIAXIAL LOADING

The first part of this chapter has been concerned with yielding and deformation under uniaxial tension, which is of obvious importance. However, there is also a practical interest in yielding under multiaxial loading, including deformation in compression. This interest arises partly from work on cold forming of ABS sheet: the requirement for successful cold forming is that the polymer should yield without tearing, and should not recover immediately upon removal from the press. Yielding by craze formation is therefore undesirable, since crazes exhibit very large recovery

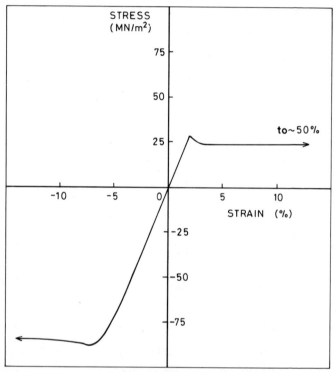

Fig. 8.16   Stress–strain curves for compression-moulded HIPS in uniaxial tension and compression at room temperature. Strain rate $10^{-3} s^{-1}$ (from R. J. Oxborough and P. B. Bowden, ref. 20, reproduced with permission).

on unloading. Crazing is also undesirable for other reasons: crazes are permeable, and act as nuclei for subsequent cracking, especially in the presence of active liquids.

Stress–strain curves in tension and compression are compared in Fig. 8.16 for HIPS containing 13 % rubber.[20] The yield stress in compression is over three times higher than the yield stress in tension, partly because the modified von Mises envelope is displaced towards the compression quadrant (*see* Chapter 6), but mainly because the critical stress for craze formation is well below the shear yield stress in polystyrene. Crazing is suppressed in compression, and the yield stress is consequently high.

Biglione *et al.* used a pressure chamber to study yielding in polystyrene and HIPS in the compression–compression quadrant.[21] Specimens were subjected to pressures of up to 600 MN/m^2, and an axial tensile stress was

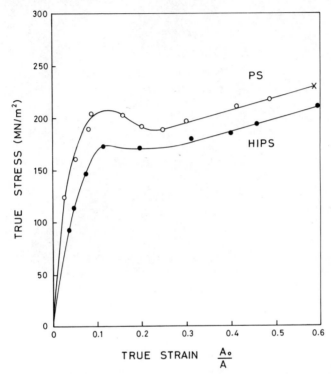

Fig. 8.17   Yielding of PS and HIPS under a superimposed hydrostatic pressure of 600 MN/m². Curves show additional tensile stresses applied to specimens in the pressure chamber (from G. Biglione et al., ref. 21, reproduced with permission).

then superimposed upon the compressive stresses transmitted by the fluid in the chamber. This superimposed variable tension, which is recorded in Fig. 8.17, had the effect of reducing the compressive stress in the axial direction, which nevertheless remained compressive in tests conducted at high pressures. For example, in tests on polystyrene at a pressure of 600 MN/m^2, the principal stresses acting on the specimen at the yield point were $(-390, -600, -600) \text{ MN/m}^2$, where the negative sign indicates compression. The experiments showed that both polystyrene and HIPS yield under pressure,

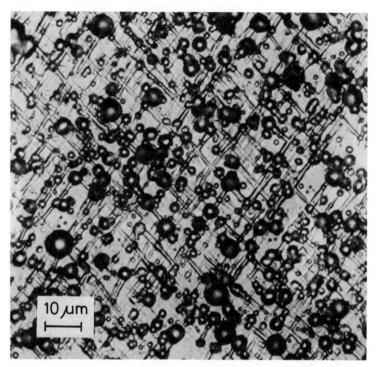

Fig. 8.18 Shear bands formed in a HIPS/PPO blend under plane strain compression. Specimen polished by microtoming, etched with chromic acid and viewed by reflected light.

and subsequently extend to relatively high elongations. HIPS yields at lower stresses than polystyrene, but the difference between the two polymers is much smaller than in tension. In the pressure chamber experiment, the rubber particles act as stress concentrators, but initiate shear bands rather than crazes. The failure envelope studies described in Chapter 6 show that crazes are not formed under hydrostatic compression.

Figure 8.18 is an optical micrograph of shear bands formed in an HIPS/PPO blend under plane strain compression.[22] In order to achieve plane strain conditions, pressure was applied to a strip of the polymer by means of parallel rectangular dies.[23] Yielding occurred in the central portion of the strip, which was between the dies, with constraint imposed by the polymer on either side of the dies. The strip was sectioned to produce a block with a polished surface, as described in Chapter 3, and etched in a mixture of chromic and phosphoric acids. The pattern of shear bands initiated by the rubber particles and running at an angle of 45° to the compression direction is clearly visible in the etched sample. Oxborough and Bowden used the plane strain compression test, followed by etching, to study yielding of HIPS in the tension–compression quadrant of biaxial stress: a tensile stress was applied along the length of the strip specimen, whilst a constant compressive stress was maintained in the perpendicular direction by means of the dies.[20] This work demonstrated the transition in yield mechanism from multiple crazing in uniaxial tension to shear band formation in uniaxial compression. The shape of the failure envelope reflected this transition, but the exact position of the transition was difficult to determine, probably because both mechanisms operate simultaneously over part of the stress quadrant.

Another technique for studying yielding under multiaxial loading is to apply tensile stress to strips containing oblique grooves. The stress conditions within the groove are determined by the angle between the groove and the tensile axis of the specimen.[24] Stress–strain curves in the principal stress directions are obtained by measuring the load–displacement relationship at the groove, and the direction of relative motion of the material on either side of the groove. Lee used this technique to study yielding of HIPS under multiaxial loading,[25] and concluded that the strain rate tensor was related to the stress tensor, as predicted by plasticity theory. The analysis is complicated in the case of rubber-toughened plastics, since yielding cannot be assumed to take place at constant volume, so that plasticity theory is not applicable under all conditions of stress. The grooved strip method is perhaps best suited to testing conducted in a pressure chamber, under conditions that suppress craze formation.

The mechanisms of yielding determine whether a polymer can be processed by cold forming. Tensile stresses are almost inevitably generated during forming, and materials such as isotropic HIPS, which craze in the tension–tension quadrant, are unsuitable subjects for the process. ABS emulsion polymers are more suitable, because they can be deformed to high strains without crazing if the strain rate is sufficiently low, but some problems still remain. Raising the forming temperature is one method by which cold drawing performance may be improved. An alternative is to subject the ABS sheet to biaxial cold rolling before drawing.[26] The resulting biaxial orientation suppresses craze formation, so that drawing can take place by shear yielding. Thus the cold forming behaviour of rubber-toughened plastics correlates well with volumetric strain data obtained in creep tests: a low slope in the plot of volume against elongation indicates good drawing performance.

8.4   STRAIN DAMAGE

Crazes differ significantly in properties from the bulk polymer that gave rise to them. Chapter 6 refers to the low modulus and high porosity of crazes, and these properties are very apparent when crazes are formed in large numbers, as in the tensile deformation of HIPS. The term *strain damage* is not too strong to use in describing the effects of craze formation. Some aspects of strain damage are considered below.

Shear bands are much closer than crazes in their mechanical properties to the undeformed bulk polymer. Nevertheless, shear band formation also results in a change in properties, since the process involves strain softening, and therefore a reduction in modulus. Permeability is affected very little, if at all, by shear band formation.

The properties of a rubber-toughened polymer beyond the viscoelastic region therefore depend critically upon the mechanisms and extent of yielding. For some applications, *e.g.* containers for carbonated drinks, porosity due to crazing is a serious problem in itself. The permeability of ABS and other rubber-toughened plastics increases dramatically as a result of multiple crazing.[26, 27] It is therefore necessary to suppress crazing by suitable choice of material and of processing conditions: biaxially oriented polymers are less susceptible to crazing, as already explained. Similarly, a reduction in modulus due to multiple crazing is undesirable in many load-bearing applications.

8.4.1 Modulus changes
The modulus of a craze decreases with increasing strain, because the cross-sectional area of the craze fibrils decreases as the craze structure opens up under applied tensile stress. In other words, the properties of a craze are functions of its density. This marked non-linearity in stress–strain behaviour is shown very clearly in materials such as HIPS, which deform principally by craze formation. Figure 8.19a shows the relationship between modulus and tensile strain, after 100 s under load, in an HIPS tensile creep specimen. Before the creep test, the polymer is relatively stiff, and the modulus falls only slowly with strain. After being extended to 5 % strain in a creep test, the same specimen exhibits very different properties: the modulus is much lower, especially at higher strains, reflecting the lower modulus of

the crazes formed during the creep test. These results, obtained in experiments on blend A of the series listed in Table 8.1, are in marked contrast to those obtained in an identical experiment on the HIPS/PPO blend E. Blend E, which deformed by a combination of crazing and shear yielding, shows a much smaller reduction in modulus as a result of extension in the creep test (Fig. 8.19b). These changes in properties refer only to tests in the direction of the tensile creep strain: properties in the two lateral directions are little affected by the presence of crazes.[6]

Another striking demonstration of the effects of crazing upon properties is shown in Fig. 8.20.[28] The cyclic stress–strain curves emphasise the progressive decrease in modulus resulting from an increasing number of crazes. Considerable recovery takes place on unloading, as the crazes close

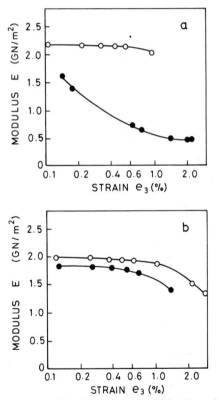

Fig. 8.19   Effects of creep to 5% elongation upon 100 s Young's modulus at 20 °C: (a) in 50/50 HIPS/PS blend; (b) in 50/50 HIPS/PPO blend. Modulus measured ($\bigcirc$) before and ($\bullet$) after long-term creep test.

up, a process usually described as *craze healing*. This craze healing underlines the fact that crazing is essentially as much a viscoelastic as a plastic yielding process. Viscoelasticity is also apparent in the large hysteresis loops, which increase linearly in size with increasing strain as the test progresses.[29] The similarities between the cyclic stress–strain behaviour of HIPS and the properties of crazes can be seen by comparing Fig. 8.20 with Fig. 6.21.

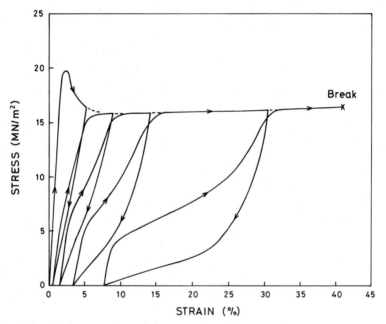

Fig. 8.20 *Cyclic stress–strain behaviour of HIPS at 21 °C showing hysteresis loops due to crazes, which increase in number during the test.*

8.4.2 Fatigue

The cyclic stress–strain experiment described above is essentially a fatigue test. Figure 8.21 presents results obtained by Beardmore and Rabinowitz in a more conventional fatigue test on ABS, in which specimens were cycled between fixed strain limits in tension and compression.[30] During the initial stages of the test, peak stresses in both tension and compression fell rapidly, before levelling out. This strain softening effect is probably due to some form of shear yielding of the SAN matrix, promoted by the stress concentrations around the rubber particles. At a later stage of the test, the peak tensile stress fell sharply, whilst the peak compressive stress remained almost constant. This second change in properties is almost certainly due to craze formation. Under tension, the craze structure opens up, and the low modulus of the crazes becomes apparent. Under compression, however, neither crazes nor cracks affect the compliance of the specimen, and the peak stress therefore remains unchanged. After relatively few cycles, the crazes break down to form cracks of critical dimensions, and the specimen

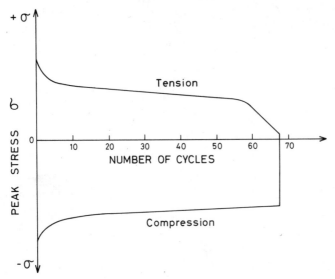

Fig. 8.21   Fatigue of ABS between fixed strain limits at room temperature showing decreases in peak stress with number of cycles (from P. Beardmore and S. Rabinowitz, ref. 30, reproduced with permission).

fractures. The observed changes in peak stress are accompanied by changes in the hysteresis loop, which becomes larger and alters in shape, reflecting the variations in compliance in both halves of the strain cycle.

An increase in the size of the hysteresis loop means that the specimen dissipates more heat per cycle. At frequencies of the order of 1 Hz, temperatures can rise quite rapidly during fatigue, owing to the low thermal conductivity of polymers. Furthermore, the heat generation is localised within crazes and shear yield zones, so that these regions are particularly likely to be affected. An understanding of the mechanisms of deformation, and of the resulting thermal effects, is of obvious importance in any study of fatigue in rubber-toughened plastics.

8.4.3 Recovery

Crazes exhibit substantial recovery immediately after unloading, and continue to 'heal' on subsequent storage. The extent of the recovery is apparent in Fig. 8.20. There are several ways in which recovery can be characterised. Perhaps the most informative is the volumetric strain measurement used for studying creep mechanisms. Figures 7.10 and 8.3 show that the volume strain due to crazing in HIPS and ABS falls on

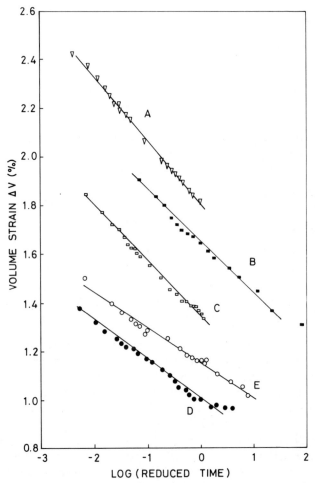

Fig. 8.22 *Recovery of HIPS/PS/PPO blends at 20 °C from tensile creep to 5%
strain (from C. B. Bucknall et al., ref. 6, reproduced with permission).*

unloading the specimen, and continues to fall slowly with time. Figure 8.22 illustrates the relationship between volume strain and log (time) during recovery in ternary blends of HIPS with PS and PPO (*see* Table 8.1): the curves are almost linear, and it is obvious that a very long time must elapse before the crazes can even approach the density of the bulk polymer.[6] Heating accelerates recovery, but it is necessary to heat the polymer above the glass transition of the matrix in order to restore the material to its original state.

Very rapid recovery is observed when a highly strained rubber-toughened polymer breaks.[15, 16, 31] Strain energy stored in the crazes is suddenly released, and it is possible that sufficient heat is generated to raise the temperature above T_g. Cessna has suggested that rapid recovery is responsible for the apparent absence of crazes in toughened PVC impact specimens.[15]

As the volume of the crazed specimen decreases, and the crazes tend towards the density of the bulk polymer, the properties of the rubber-toughened polymer change accordingly. The initial modulus of the specimen rises, and the yield stress of the crazed material, represented in Fig. 8.20 by a knee in the stress–strain curve, slowly increases. Measurements of these properties show that the yield stress increases linearly with log (time) over a range of temperatures below T_g.[6] None of these changes is permanent, however: on reloading, the crazes reopen, and the modulus and yield stress fall to the level of the freshly crazed sample.

REFERENCES

1. C. B. Bucknall and D. Clayton, *J. Mater. Sci.* **7** (1972) 202.
2. C. B. Bucknall and I. C. Drinkwater, *J. Mater. Sci.* **8** (1973) 1800.
3. P. J. Fenelon and J. R. Wilson, *ACS Div. Org. Coat. Plast. Prepr.* **34**(2) (1974) 326.
4. D. R. Reid, *Brit. Plast.* **32** (1959) 460.
5. R. L. Bergen, *SPE J.* **23** (1967) 57.
6. C. B. Bucknall, D. Clayton and W. E. Keast, *J. Mater. Sci.* **8** (1973) 514.
7. R. S. Moore and C. Gieniewski, *Poly. Engng Sci.* **9** (1969) 190.
8. R. S. Moore and C. Gieniewski, *Macromolecules* **1** (1968) 540.
9. T. Yoshii, Ph.D. Thesis, Cranfield, England, 1975.
10. C. B. Bucknall, C. J. Page and V. O. Young, *ACS Adv. Chem. Ser.* (1976) in press.
11. C. B. Bucknall, C. J. Page and V. O. Young, *ACS Div. Org. Coat. Plast. Prepr.* **34**(2) (1974) 288.
12. J. N. Sultan and F. J. McGarry, *Poly. Engng Sci.* **14** (1974) 282.
13. R. W. Truss and G. A. Chadwick, *J. Mater. Sci.* **11** (1976) 111.
14. E. H. Merz, G. C. Claver and M. Baer, *J. Poly. Sci.* **22** (1956) 325.
15. L. Cessna, *ACS Poly. Prepr.* **15**(1) (1974) 229.
16. L. Cessna, *Poly. Engng Sci.* **14** (1974) 696.

17. L. J. Evans, C. B. Bucknall and M. M. Hall, *Plast. Polym.* **39** (1971) 118.
18. M. R. Grancio, *Poly. Engng Sci.* **12** (1972) 213.
19. M. R. Grancio, A. A. Bibeau and G. C. Claver, *Poly. Engng Sci.* **12** (1972) 450.
20. R. J. Oxborough and P. B. Bowden, *Phil. Mag.* **30** (1974) 171.
21. G. Biglione, E. Baer and S. V. Radcliffe, in *Fracture* 1969, P. L. Pratt (ed.), Chapman and Hall, London, 1969, p. 503.
22. C. B. Bucknall, I. C. Drinkwater and W. E. Keast, *Polymer* **13** (1972) 115.
23. R. J. Oxborough and P. B. Bowden, *Phil. Mag.* **28** (1973) 547.
24. R. Hill, *J. Mech. Phys. Solids* **1** (1953) 271.
25. D. Lee, *J. Mater. Sci.* **10** (1975) 661.
26. L. Nicolais, E. Drioli and R. F. Landel, *Polymer* **14** (1973) 21.
27. E. Drioli, L. Nicolais and A. Ciferri, *J. Poly. Sci.* (*Chem.*) **11** (1973) 3327.
28. C. B. Bucknall, *Brit. Plast.* **40**(12) (1967) 84.
29. C. B. Bucknall, *J. Materials* (ASTM) **4** (1969) 214.
30. P. Beardmore and S. Rabinowitz, *Appl. Polymer Symp.* **24** (1974) 25.
31. K. Takahashi, *J. Poly. Sci.* (*Phys.*) **12** (1974) 1697.

CHAPTER 9

FRACTURE MECHANICS

Whilst viscoelastic properties and yield characteristics are of undoubted importance, the main interest in the mechanical properties of rubber-toughened plastics centres upon fracture behaviour, especially under impact loading. Rubber-modified polymers are formulated primarily to overcome problems of brittle fracture encountered in the parent polymer. This point is emphasised in the terminology of the subject, which features terms such as 'high-impact' and 'toughened' to describe materials containing added rubber.

The toughness of these materials is easy to demonstrate but difficult to quantify in a satisfactory manner. Unlike stiffness and yield strength, fracture resistance is highly dependent upon geometry, and it is therefore extremely difficult to employ standard test results as design data to predict the fracture behaviour of a component in service. For this reason, many manufacturers prefer to rely on component testing as the primary method for assessing the quality of mouldings. The problem is, of course, not specific to the plastics industry, but applies to all structural materials.

These difficulties have stimulated a widespread interest in fracture mechanics, which aims to develop a more fundamental understanding of the fracture process. For obvious reasons, interest concentrated initially on metals, which still dominate the subject, but developments in this area have been reflected in polymer engineering: there is now an extensive literature devoted to fracture mechanics studies in both plastics and rubbers, and fracture mechanics procedures have become standard in many industrial laboratories, although the impact of this work has yet to be felt in the more practical areas of data specification and engineering design. The main benefits have been a clearer recognition of the role of stress concentrators and stored strain energy in controlling crack initiation and propagation, and the development of more soundly based procedures for ranking polymers in order of fracture resistance in service. The basic principles of fracture mechanics are expounded in a book by Knott,[1] and in a number of other texts. An outline of the subject is presented below.

9.1   LINEAR ELASTIC FRACTURE MECHANICS

The foundations of fracture mechanics were laid by Griffith,[2] who concluded that the strengths of crystalline or glassy solids were much lower than the theoretical values because these materials contained small defects or flaws. Theoretical strengths can be calculated from the force–displacement curves of interatomic bonds, since the strength of a solid is limited ultimately by the strength of its bonds: typical values are of the order of $E/10$, where E is the Young's modulus of the solid. Strengths of this magnitude can be achieved only if all bonds are stressed equally, a condition that is approached in the recently developed technology of strong fibres and whiskers[3] by eliminating flaws and defects from crystalline materials. In general, however, materials exhibit strengths that are two or three orders of magnitude below the theoretical value, clearly indicating that the distribution of stresses upon individual bonds is highly non-uniform, even in an apparently homogeneous material under uniform stress. Zhurkov's infra-red measurements, which were described in Section 6.2.3, provide experimental evidence for a distribution of bond stresses in polymers, and show that a small fraction of highly stressed bonds breaks under load. The resulting defects form nuclei for cracks. Almost all samples, whether polymers, metals, ceramics or other materials, contain some defect that can act as a stress concentrator and initiate a crack.

9.1.1 Griffith criterion

The treatment proposed by Griffith is based upon a thermodynamic principle: that a crack will propagate only if the energy available is sufficient to support its growth. The energy is provided by the release of potential energy from the strained material in the neighbourhood of the crack, and is absorbed in the formation of fresh crack surfaces. The first problem is therefore to calculate the strain energy U stored in the body, as a function of the crack length.

Griffith considered the case illustrated in Fig. 9.1, which represents an infinite plate of unit thickness, containing a sharp planar crack of length $2a$, and subjected to an applied tensile stress σ at a distance. The stress distribution in this case had already been calculated by Inglis.[4] By integrating the product of stress and strain in each element of the plate, Griffith obtained the following expression for the strain energy release rate $-(\partial U/\partial a)$:

$$-\frac{\partial U}{\partial a} = \frac{\sigma^2 \pi a}{E} \qquad \textit{in plane stress} \qquad (9.1)$$

The calculation assumes that the material is *linearly elastic* (*i.e.* that stress is everywhere proportional to strain), and that the plate is thin, so that the material is in a state of *plane stress, i.e.* that there is no component of stress in the through-thickness direction. In the case of a thick plate, which is in a

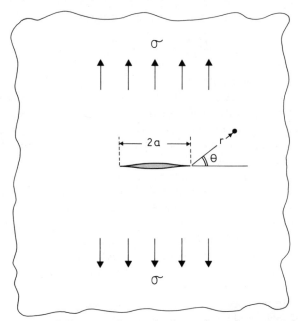

Fig. 9.1   *Central planar crack in wide plate subjected to uniaxial tension.*

state of *plane strain*, the strain energy release rate is given by a slightly different expression, as follows:

$$-\frac{\partial U}{\partial a} = \frac{\sigma^2 \pi a}{E}(1 - v^2) \qquad \textit{in plane strain} \qquad (9.2)$$

Since Poisson's ratio v is approximately 0·4 for glassy polymers, the difference in $(\partial U/\partial a)$ predicted by the two expressions is only 16%.

In an ideally brittle material, the energy absorbed in crack propagation is equal to the work done against intermolecular attractions in forming the fracture surface. As there are two surfaces, the rate of absorption of energy $(\partial U'/\partial a)$ is given by

$$\frac{\partial U'}{\partial a} = 2\gamma \tag{9.3}$$

where γ is the fracture surface energy per unit area.

The criterion for crack propagation is

$$-\frac{\partial U}{\partial a} \geq \frac{\partial U'}{\partial a}$$

Thus, for a plate in plane strain, fracture occurs when

$$\frac{\sigma^2 \pi a}{E}(1 - v^2) \geq 2\gamma \tag{9.4}$$

This energy-balance criterion, based essentially upon the first law of thermodynamics, avoids the problem of calculating the stresses at a crack tip, which are critically dependent on the tip radius. The treatment requires that the crack be sharp enough to produce stresses in the tip region that are in excess of the interatomic or intermolecular forces, but does not specifically consider the tip radius.

Experiments on plates containing introduced cracks show that the critical stress σ_c at the point of crack initiation is proportional to $a^{-1/2}$, as predicted by eqn. (9.4), in a wide range of metals and plastics. However, the measured values of fracture surface energy are several orders of magnitude greater than those calculated from interatomic bonding energies.[5, 6] The discrepancy led Orowan and Irwin[7] to suggest that the energy of crack propagation consisted largely of energy dissipated in plastic work at the crack tip. Provided both the crack length and the plate width W are large compared with the plastic zone, the elastic stress distributions in the plate are very little affected by a limited amount of yielding, and eqns. (9.1) and (9.2) remain valid. Equation (9.4) can therefore be adapted to represent brittle fracture in metals and plastics simply by replacing γ with γ_p, the plastic work per unit area of fracture surface. Following Irwin, many authors prefer to define the plastic work of crack formation in terms of the *critical strain energy release rate* $\mathcal{G}_c$, which is equal to $2\gamma_p$. The Griffith equation (9.4) then becomes

$$\sigma_c^2 = \frac{E\mathcal{G}_c}{\pi a(1 - v^2)} \qquad \textit{in plane strain} \tag{9.5}$$

This equation is valid only when the plastic zone is small and $\mathcal{G}_c$ is constant, a point that must be checked experimentally for each material. It is a fundamental principle of linear elastic fracture mechanics that the fracture behaviour of a brittle material can be defined in terms of materials properties such as γ_p and $\mathcal{G}_c$, which are independent of specimen geometry. Section 9.4 discusses the factors affecting $\mathcal{G}_c$.

9.1.2 Stress intensity criterion

There is an alternative criterion for brittle fracture, namely that a crack will propagate when the stress distribution around the crack tip reaches a critical state. As in the Griffith approach, the stresses very close to the crack tip are not considered, but the tip radius is small compared with the crack length, so that it has no influence upon the general stress distribution around the crack. The stress analysis, due to Westergaard,[8] shows that for a central crack in a large plate loaded perpendicular to the crack plane, as illustrated in Figs. 9.1 and 9.2, the stresses are given by

$$\sigma_{11} = \frac{\mathcal{K}}{(2\pi r)^{1/2}} \cos\frac{\theta}{2}\left(1 + \sin\frac{\theta}{2}\sin\frac{3\theta}{2}\right) + \cdots$$

$$\sigma_{22} = \frac{\mathcal{K}}{(2\pi r)^{1/2}} \cos\frac{\theta}{2}\left(1 - \sin\frac{\theta}{2}\sin\frac{3\theta}{2}\right) + \cdots$$

$$\sigma_{12} = \frac{\mathcal{K}}{(2\pi r)^{1/2}} \sin\frac{\theta}{2}\cos\frac{\theta}{2}\cos\frac{3\theta}{2} + \cdots$$

$$\sigma_{13} = \sigma_{23} = 0$$

$$\sigma_{33} = \frac{2v\mathcal{K}\cos\frac{\theta}{2}}{(2\pi r)^{1/2}} \quad \textit{in plane strain}$$

$$\sigma_{33} = 0 \quad \textit{in plane stress} \tag{9.6}$$

where $\mathcal{K} = \sigma(\pi a)^{1/2}$.

With the appropriate choice of the *stress intensity factor* $\mathcal{K}$, the same equations describe the stresses near a crack tip under all loading configurations. The advantage of the stress intensity factor is that it combines in a single parameter the three test variables: applied stress, crack length and specimen geometry. In the general case

$$\mathcal{K} = Y\sigma(a)^{1/2} \tag{9.7}$$

where Y is a geometrical factor. Formulae for stress intensity factors for important test configurations are to be found in standard texts.[1] For example, in the case of a plate of width W under tensile stress σ and having a central crack of length $2a$

$$\mathscr{K} = \sigma W^{1/2}\left[\tan\left(\frac{\pi a}{W}\right) + 0\cdot 1 \sin\left(\frac{2\pi a}{W}\right)\right]^{1/2} \qquad (9.8)$$

The symbol $\mathscr{K}_1$ is used to denote *opening mode* (Mode I) fracture, in which the crack is opening normally under tensile stress in the manner

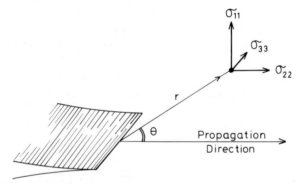

Fig. 9.2   Coordinate system for specification of stresses at a crack tip.

discussed above. Mode II refers to shear stresses acting parallel to the direction of crack propagation, and Mode III refers to shear stresses acting parallel to the crack tip. As neither $\mathscr{K}_{II}$ nor $\mathscr{K}_{III}$ has received significant attention in the polymer literature, these modes of stressing will not be considered further.

In discussing conditions at the crack tip, it is convenient to direct attention to the plane of the crack, where $\theta = 0$, and the stresses are therefore at a maximum given by

$$\sigma_{11} = \sigma_{22} = \frac{\mathscr{K}}{(2\pi r)^{1/2}}$$

$$\sigma_{12} = 0$$

$$\sigma_{33} = \frac{2v\mathscr{K}}{(2\pi r)^{1/2}} \qquad \textit{in plane strain} \qquad (9.9)$$

One problem is immediately apparent from eqns. (9.6) and (9.9), namely that the stresses σ_{11} and σ_{22} tend to infinity as the distance r from the crack tip tends to zero. A continuum stress analysis is of course meaningless when one is discussing distances of the order of atomic dimensions, but even at rather larger values of r the equations predict extremely high stresses. The usefulness of Westergaard's analysis lies in the fact that the general form of the stress distribution is substantially unaltered by blunting of the crack tip, and consequent reduction of stresses in the immediate vicinity of the tip. Figure 9.3 compares stresses calculated using eqns. (9.9) for an ideally sharp crack with those calculated for a blunt notch with a tip radius of 100 μm. In the latter case, σ_{11} has a finite value at the notch tip, equal to $20 \cdot 58\sigma_{appl}$, where σ_{appl} is the applied tensile stress. The stress σ_{22}, acting parallel to the direction of crack propagation, is zero at the tip of the notch because a free surface cannot support a normal stress, and rises to a maximum at some distance from the tip. Consequently, the stress becomes more triaxial in character. Despite the comparatively large radius of the notch, the stress distribution illustrated in Fig. 9.3b shows an obvious similarity to the solution for an ideally sharp crack. As the tip radius is decreased, the solutions for a rounded notch become virtually identical with those for a sharp crack, except in the small region very close to the crack tip.

The stress intensity criterion simply states that a crack will propagate when the stress intensity factor $\mathcal{K}$ reaches a critical value. For opening mode fracture, this condition can be expressed as follows:

$$\mathcal{K}_I \geq \mathcal{K}_{IC} \tag{9.10}$$

The quantity $\mathcal{K}_{IC}$ is known as the plane strain opening mode fracture toughness, or more simply as *fracture toughness*. Like $\mathcal{G}_c$, $\mathcal{K}_{IC}$ is a materials property, and must not be confused with $\mathcal{K}_I$, which is a test variable. Both $\mathcal{K}_I$ and $\mathcal{K}_{IC}$ are measured in MN/m$^{3/2}$.

Although the stress intensity criterion lacks the physical basis that is brought out so clearly in the energy-balance analysis, the two treatments are equivalent, and lead to the same results. For a large centrally cracked plate

$$\mathcal{K}_{IC} = \sigma_c(\pi a)^{1/2}$$

Hence from eqn. (9.5)

$$\mathcal{K}_{IC}^2 = \frac{E\mathcal{G}_c}{(1 - v^2)} \tag{9.11}$$

9.1.3 Non-linearity

The assumption that stress and strain are linearly related is not strictly necessary to fracture mechanics. The energy balance criterion works

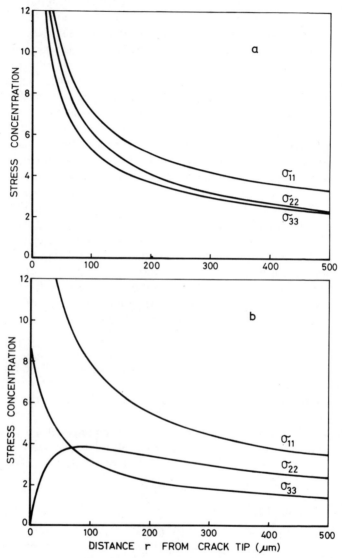

Fig. 9.3 *Calculated stress distributions in the crack plane around the tip of a central 2 cm crack in a wide plate: (a) for ideally sharp crack, root radius zero; (b) for blunt crack, root radius* 100 *μm. Stresses are expressed as the ratio of local stress to applied tensile stress, and* σ_{33} *is calculated for a Poisson's ratio of* 0·40.

equally well for non-linear elastic materials. Indeed, in the field of polymer science, the work of Rivlin and Thomas on the tearing of rubbers,[9] which are highly non-linear materials, considerably predates fracture mechanics studies of approximately linear materials such as PMMA. The experiments are complicated by the need to measure strain energy release rates experimentally, but the basic approach is unchanged.

9.2   CRACK-OPENING DISPLACEMENT CRITERION

Linear elastic fracture mechanics is obviously not ideally suited to materials such as rubber-toughened plastics, which form extensive yield zones ahead of a propagating crack. However, with suitable modifications, fracture mechanics can be extended to these more ductile materials. The problem is to define the conditions under which the crack and its associated yield zone will grow in response to the applied stress.

The dimensions of a wedge-shaped plastic zone ahead of a crack tip in a sheet under plane stress were calculated by Dugdale.[10] The principles upon which his calculations were based are illustrated in Fig. 9.4: the plate contains a central crack of length $2a$ lying normal to the applied tensile

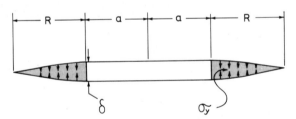

Fig. 9.4   Dugdale's model for yielding at a crack tip.[10] The stress has a constant value σ_y within the yield zone.

stress, and two yield zones, each of length R, in the crack plane. Yielding will occur in this plane only if the sheet is thin enough to allow through-thickness necking. The stress within the plastic zones is assumed to be uniform, and equal to the yield stress σ_y of the material. The elastic stress distributions in the plate are then calculated for a crack of length $2(a + R)$ in which closing forces σ_y are acting over a distance R at each end. Using this model, Dugdale showed that the length of the yielded zone is given by

$$\frac{a}{(a + R)} = \cos\left(\frac{\pi\sigma}{2\sigma_y}\right) \qquad (9.12)$$

When the applied stress σ is less than $0\cdot 3\sigma_y$, eqn. (9.12) reduces to

$$\frac{a}{(a + R)} = 1 - \frac{\pi^2 \sigma^2}{8\sigma_y^2} \qquad (9.13)$$

A further approximation, again assuming that σ/σ_y is small, then yields

$$R = \frac{\pi^2 \sigma^2 a}{8\sigma_y^2} = \frac{\mathscr{K}^2}{8\sigma_y^2} \qquad (9.14)$$

The crack opening displacement, δ, which characterises the width of the wedge-shaped plastic zone, is given by[11]

$$\delta = \frac{8a\sigma_y}{\pi E} \ln \sec \left(\frac{\pi \sigma}{2\sigma_y} \right) \qquad (9.15)$$

By expanding in series, and neglecting higher order terms, it can be shown that eqn. (9.15) reduces at low stresses to

$$\delta = \frac{\sigma^2 \pi a}{\sigma_y E} = \frac{\mathscr{K}^2}{\sigma_y E} \qquad (9.16)$$

These equations form the basis for a new fracture criterion related to the crack opening displacement (COD) δ: when the COD reaches the critical value δ_c, both the crack and the yield zone will extend.

The critical COD criterion may be tested by direct experimental measurement of δ, or by plotting R/a against σ^2: according to eqn. (9.14), such a plot should give a straight line of slope $\pi^2/8\sigma_y^2$. Experiments show that the criterion successfully predicts crack propagation in a number of metals and alloys.[12, 13] Although some reservations remain concerning the general validity of the approach,[1] the work on metals has stimulated interest in applying Dugdale's stress analysis and the COD criterion to plastics.[14, 15]

Much of this interest has been concentrated upon crazed specimens.[16-18] As required by the model, crazes are plastic zones lying in the crack plane. To a good approximation, they are also formed at a well defined craze yield stress. The Dugdale model should therefore be applicable to individual crazes formed at a crack tip, and to multiple-craze zones. The experimental evidence supports this view in general.[18, 19] However, as discussed in Section 6.2.2, there are some conditions under which the plastic zone analysis does not accurately represent the stress distributions around the tip of a non-linearly viscoelastic craze.[20] When the plastic zone or craze length R is small compared with the crack length a, the COD criterion becomes equivalent to the Griffith energy balance criterion, since from eqns. (9.5) and (9.16)

$$\delta_c = \frac{\sigma_c^2 \pi a}{\sigma_y E} = \frac{\mathscr{G}_c}{\sigma_y} \qquad \textit{in plane stress} \qquad (9.17)$$

This equation relates the critical strain energy release rate $\mathscr{G}_c$ to the work done when a stress σ_y acts through a distance δ_c. On reaching the critical strain represented by δ_c, the craze or plastic zone begins to tear, and the crack propagates.

9.3   TEST METHODS

The first step in any fracture mechanics study is to establish the validity of the approach, by showing that the critical conditions for crack propagation are independent of specific specimen geometry. Tests should preferably be carried out on as wide a variety of test configurations as possible. For this reason, a range of different methods has been developed. Figure 9.5 shows some of the specimens that have been used in experiments on rigid polymers.[21]

In principle, the simplest test specimen is the single edge notch (SEN) specimen (a), which behaves as a semi-infinite plate with an edge crack of length a. The specimen is derived from the infinite plate with an internal crack which forms the basis of the preceding discussion. Each experiment requires a large number of specimens containing cracks of different lengths. Load is applied at constant strain rate until the critical stress for crack propagation σ_c is reached, and the crack propagates, usually in a catastrophic manner. A linear relationship between σ_c^2 and $1/a$ indicates that a linear elastic fracture mechanics approach is valid. To a first approximation, $\mathscr{G}_{IC}$ can be calculated from eqn. (9.5), but a more accurate value is obtained from eqn. (9.18), which takes account of the difference in stress distribution between an infinite and semi-infinite plate:[21]

$$\sigma_c^2 = 1 \cdot 12 \frac{E \mathscr{G}_{IC}}{\pi a (1 - v^2)} \qquad (9.18)$$

The double edge notch (DEN) specimen (b) is a simple variant of the SEN specimen, and is used in exactly the same way.

The main problems in SEN and DEN testing arise from the method of introducing the edge cracks. Published estimates of $\mathscr{G}_{IC}$ for polystyrene vary by more than a factor of two, depending upon the notching technique,[22]

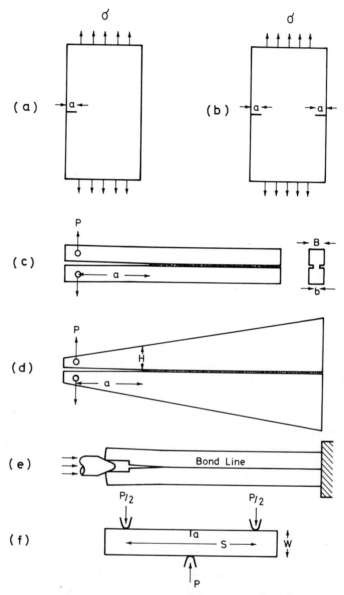

Fig. 9.5 *Test specimens for fracture mechanics measurements:* (a) *single edge notch (SEN);* (b) *double edge notch (DEN);* (c) *double cantilever beam (DCB);* (d) *tapered DCB;* (e) *wedge opening load;* (f) *three-point bending.*

and the temperature at which the notch was introduced.[23] The most satisfactory solution to the problem is to use a standard procedure that produces a consistent, sharp notch. Suitable techniques include slow, controlled pressure applied to a razor blade,[21] and fatigue crack growth.

The double cantilever beam (DCB) specimen (c) is also widely used in fracture mechanics measurements on rigid polymers. Its main advantage is

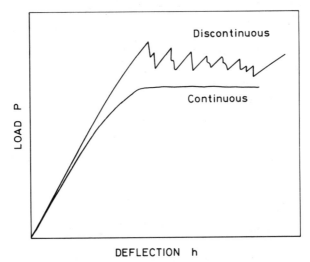

DEFLECTION h

Fig. 9.6   Load–deflection curves for DCB specimens fracturing by continuous and discontinuous crack propagation.

that it produces more stable crack growth than SEN or DEN specimens, so that several measurements can be made with the same specimen. The main disadvantage is that specimen preparation involves extensive machining, especially in cutting the fine side grooves that guide the crack along the centre of the specimen.

Measurements are usually made by increasing the grip separation at a constant rate, and recording applied load P and crack length a, with the aid of a scale marked on the specimen. In some materials, crack propagation is continuous, and the crack speed $\dot{a}$ is measured; in others, propagation is discontinuous (stick–slip mode), and two sets of results are obtained, one for crack initiation, and one for crack arrest. The alternative procedure, using a fixed load and measuring displacement, is much less common because it produces less stable crack propagation. Figure 9.6 shows curves of load against displacement h for continuous and stick–slip propagation.

The strain energy U per unit thickness, in a specimen of crack width B, is given by

$$U = \frac{Ph}{2B}$$

The strain energy release rate $\mathcal{G}_I$ is obtained by differentiating with respect to crack length:

$$\mathcal{G}_I \equiv \frac{dU}{da} = \frac{\left(P\dfrac{dh}{da} + h\dfrac{dP}{da}\right)}{2B}$$

In order to evaluate this expression, calculations are made of specimen compliance C, defined as the ratio of h to P:

$$C = \frac{h}{P}$$

from which

$$\frac{dC}{da} = \frac{1}{P}\frac{dh}{da} - \frac{h}{P^2}\frac{dP}{da}$$

Combining the above expressions:

$$\mathcal{G}_I = \frac{\left(P^2\dfrac{dC}{da}\right)}{2B} \quad \text{at constant load } P$$

The fracture energy $\mathcal{G}_{IC}$ and the fracture toughness $\mathcal{K}_{IC}$ are then given by

$$\mathcal{G}_{IC} = \frac{P_c^2}{2B}\frac{dC}{da} \tag{9.19}$$

and

$$\mathcal{K}_{IC} = \frac{P_c^2 E}{2B}\frac{1}{(1 - v^2)}\frac{dC}{da} \tag{9.20}$$

which are known as the Irwin–Kies relationships.[24, 25]

To some extent, the DCB specimen overcomes the problem of introducing a sharp crack, since the tip is formed by crack propagation, except at the beginning of a test. However, the problem is not entirely eliminated, as the tip radius is affected by the conditions of testing, including the rate of grip separation and the width of the side grooves.

The tapered DCB specimen (d) was developed by Mostovoy et al.[26] The specimen is so designed that the compliance increases linearly with crack length, and dC/da is therefore constant. The stress intensity factor is then dependent only upon the applied load. According to beam theory, the compliance dependence is given by

$$\frac{dC}{da} = \frac{6}{Eb}\left(\frac{4a^2}{H^3} + \frac{1 + v}{H}\right) \tag{9.21}$$

where b is the total thickness of the beam, and H is the height of a single arm of the specimen at a distance a from the loading point. When dC/da is set equal to a constant, the equation becomes

$$\frac{4a^2}{H^3} + \frac{1 + v}{H} = M \tag{9.22}$$

where M is a constant. The required specimen shape can conveniently be calculated from eqn. (9.22) by choosing a value for M, inserting values for H, and calculating a. The fracture energy $\mathscr{G}_{IC}$ is calculated by combining eqns. (9.19), (9.21) and (9.22) to give

$$\mathscr{G}_{IC} = \frac{3MP^2}{EbB} \tag{9.23}$$

The tapered DCB specimen is particularly useful in studying crack propagation, as opposed to initiation. Under constant applied load, the stress intensity factor at the crack tip is independent of crack length. Consequently, the crack speed is usually constant throughout a test, and the relationship between stress intensity and crack speed can be studied.[27, 28] In comparison with the SEN and DEN specimens, the improved stability of crack propagation in DCB test pieces is a considerable advantage in testing the more brittle polymers.[29]

The disadvantages of the DCB specimen include the requirement for relatively large quantities of material, and the increased complexity of specimen preparation. The latter problem can be reduced by approximating the curved outline of the DCB by straight sides.[27] The main drawback is the tendency of the arms to twist and distort excessively when the material under test is tough and ductile, and the loads are therefore relatively high; this tendency is not found in brittle materials. In order to reduce this distortion, Kobayashi and Broutman developed a laminated specimen, in which the test material was sandwiched between two plates of a more rigid material.[30] After bonding the layers together with an adhesive, side grooves were cut through the outer plates into the test material, and the specimen was machined to shape. Due allowance was made for the stiffness of the reinforcing plates in the calculation.

The wedge-opening method (e) is another variant of the DCB specimen. One advantage of the method is that the compressive stress produced in the longitudinal direction by the wedge tends to stabilise the crack, preventing it from deviating into the side arms. The procedure for calculating $\mathscr{G}_{IC}$ is given by Kanninen.[31]

The three-point bending specimen (f) is of interest principally as a Charpy impact test piece, and will be discussed further in Chapter 10. Under this type of loading, the stress in the 'outer fibre', which contains the crack, is given by elastic bending theory as

$$\sigma = \frac{3PS}{2BW^2} \qquad (9.24)$$

The stress intensity factor is calculated by inserting this value of σ into eqn. (9.7), together with the appropriate value of the geometrical parameter Y, which depends upon the span-to-width ratio S/W. For $S/W = 4$

$$Y = 1 \cdot 93 - 3 \cdot 07\left(\frac{a}{W}\right) + 14 \cdot 53\left(\frac{a}{W}\right)^2 - 25 \cdot 11\left(\frac{a}{W}\right)^3 + 28 \cdot 80\left(\frac{a}{W}\right)^4 \dots$$

An expression for $\mathscr{G}_{IC}$ is obtained by combining eqns. (9.7), (9.11) and (9.24), as follows:

$$\mathscr{G}_{IC} = \left[\frac{9S^2aY^2(1 - v^2)}{4B^2W^4E}\right]P_f^2 \qquad (9.25)$$

This equation forms the basis for the fracture mechanics analysis of impact behaviour in Chapter 10.

9.4   FACTORS AFFECTING FRACTURE RESISTANCE

A glance at the literature of fracture mechanics is sufficient to show that Griffith's simple concept of fracture surface energy as a single-valued materials constant is not valid. In polymers, the measured $\mathscr{G}_{IC}$ and $\mathscr{K}_{IC}$ values not only vary with temperature, molecular weight and orientation, as might be expected, but are also quoted as functions of crack speed. In a more general context, Clausing[32] has shown that $\mathscr{G}_{IC}$ varies markedly with specimen size, specimen geometry and loading system. This is a serious objection to the whole approach, which means that fracture mechanics data must be treated with considerable caution. The most reliable comparisons are those made by a single technique on similar materials, and especially on relatively brittle materials. The problems involved in testing relatively ductile materials, and applying yielding fracture mechanics, are reviewed by Turner.[33] Nevertheless, provided its limitations are recognised, fracture mechanics forms a convenient basis for the discussion of fracture resistance in rubber-toughened plastics: alternative methods are no more satisfactory.

9.4.1 Crack speed

In presenting fracture mechanics data for plastics, it is common practice to plot $\mathscr{K}_{IC}$ or $\mathscr{G}_{IC}$ against crack velocity $\dot{a}$, as shown in Fig. 9.7. The justification for this practice is not entirely clear. If the crack is already propagating, the measured fracture resistance is in no sense a critical value. It would be more logical to regard the stress intensity factor $\mathscr{K}_I$ or the strain energy release rate $\mathscr{G}_I$ as the independent variable, and to plot $\dot{a}$, as the dependent variable, against $\mathscr{K}_I$ or $\mathscr{G}_I$. The limiting value of $\mathscr{G}_I$ as crack speed tended to zero would then define $\mathscr{G}_{IC}$.

Figure 9.7 presents typical data for fracture energy $\mathscr{G}_{IC}$ as a function of crack speed in PMMA at 22 °C. The results fall on three separate curves, each rising with crack speed. The intervals, between 6·5 cm/s and 2·5 m/s, and between 20 and 53 m/s, indicate regions in which fracture resistance

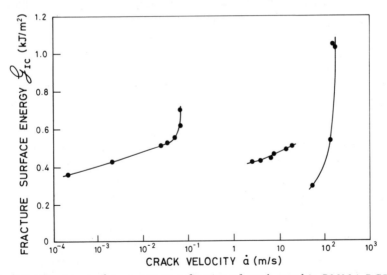

Fig. 9.7 *Fracture surface energy as a function of crack speed in PMMA DCB specimens at 22 °C (from L. J. Broutman and T. Kobayashi, ref. 34, reproduced with permission).*

appears to fall with increasing crack speed, so that the crack accelerates to the next region of stable crack propagation.[34]

Broutman and Kobayashi[34] suggest that the peaks in fracture resistance at 7 cm/s and 20 m/s are due to viscoelastic processes. By plotting $\ln \dot{a}$ against $1/T$ at constant $\mathscr{G}_{IC}$, they obtain estimated activation energies for the two processes, which correspond with activation energies obtained from dynamic mechanical tests. On the basis of this evidence, they identify the peak at 7 cm/s with the glass transition of PMMA, and the peak at 20 m/s with the secondary β-relaxation due to methyl rotation. The idea is not a new one: the relationship between viscoelastic processes and fracture resistance is well established in rubbers.[35, 36]

The maximum attainable velocity of crack propagation in a solid is just over one-third of the velocity of sound in the solid. In PMMA, the limiting velocity is 670 m/s, above which the crack forks.[37] Forking is the ultimate result of the change in stress distribution that takes place around the crack as its velocity increases. In effect, the crack accelerates ahead of its own stress concentrations, and at high velocities the peak stress concentrations are no longer in the crack plane, but above and below that plane.[38] Splitting of the stress field occurs when the crack is travelling at about two-thirds of the velocity of a transverse wave in the material. This change in the stress field is probably a major factor in the rise in $\mathscr{G}_{IC}$ at speeds above 100 m/s. Berry has pointed out that the effects of crack speed on stress distribution have largely been ignored in fracture mechanics analyses.[39]

9.4.2 Rubber content

Fracture energy is also a function of crack speed in rubber-toughened plastics, as illustrated in Fig. 9.8.[28] The maxima in the curves are probably viscoelastic in origin, as in PMMA, but have not been identified positively.

The main feature of Fig. 9.8, however, is the rubber-toughening effect. As little as 4% rubber raises $\mathscr{G}_{IC}$ by more than an order of magnitude. Stress-whitening on the fracture surface confirms that the toughening is due to multiple craze formation. The relationship between $\mathscr{G}_{IC}$ and rubber content is more explicitly illustrated in Fig. 9.9, in which fracture resistance is plotted against composition at three different crack speeds. It is clear that the relationship is not a simple one: the trend of the curves is generally upwards, but there are peaks in two of them, and a suggestion of a peak in the third. The reasons for the unexpected decrease in fracture resistance with increasing rubber content, which is particularly marked at 50 μm/s, are not well understood. In view of the ductility of rubber-toughened plastics, especially at high rubber content ϕ_2, an explanation should perhaps be sought in terms of yielding fracture mechanics, by noting that both COD and yield stress are functions of rubber content and crack speed, so that eqn. (9.17) can be written

$$\mathscr{G}_{IC} = \delta_c(\phi_2, \dot{a}, T)\sigma_y(\phi_2, \dot{a}, T) \tag{9.26}$$

As the rubber content increases, the COD also increases, but the crazing yield stress decreases. The overall effect on $\mathscr{G}_{IC}$ depends upon the precise

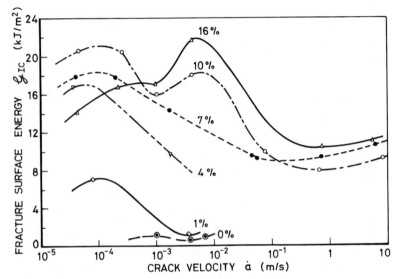

Fig. 9.8   *Fracture surface energy as a function of crack speed in AMBS polymers, poly(styrene-co-acrylonitrile-co-methyl methacrylate), containing 0–16% rubber (from T. Kobayashi and L. J. Broutman, ref. 28, reproduced with permission).*

form of both functions: under some conditions, the fall in σ_y appears to be more significant than the rise in δ_c. A similar argument can be applied to the effects of crack speed: in general, σ_y increases and δ_c decreases with increasing crack speed.

Fracture mechanics tests have been used to demonstrate the toughening due to addition of rubber in a number of polymers, including HIPS,[18, 21] ABS[40] and acrylic copolymers.[28, 30, 34] Rubber-toughened epoxy resins have been studied extensively, especially using the DCB method.[29, 41–47] In many of the studies, the analysis was based on a simple linear elastic fracture mechanics approach, which appears to be relatively successful provided the crazed zone is not too large. Williams *et al.* have shown that yielding fracture

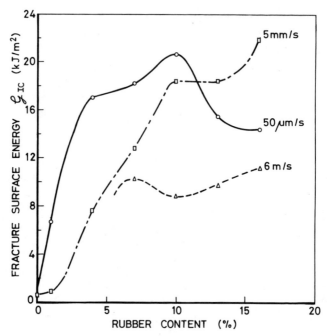

Fig. 9.9   Fracture surface energy as a function of rubber content in AMBS polymers (from T. Kobayashi and L. J. Broutman, ref. 28, reproduced with permission).

mechanics provides a better basis for the formulation of fracture criteria in the tougher rubber-modified polymers, which tend to form large crazed zones.[18, 21]

9.4.3 Temperature

Fracture resistance, like most polymer properties, varies with temperature.[6, 48 - 52] A decrease in $\mathscr{G}_{IC}$ with increasing temperature, as shown in Fig. 9.10 for PMMA, appears to be typical of brittle glassy polymers. On the other hand, $\mathscr{G}_{IC}$ reaches a maximum at a temperature $85°K$ below T_g in polysulphone and polyphenylene oxide, which are relatively ductile glassy polymers.[52]

As in the case of variations with crack speed, the reasons for this temperature-dependence are not well understood. Viscoelastic energy dissipation around the crack tip is probably a contributory factor, but the main cause is probably the effect of temperature on the kinetics of crazing. Increasing the temperature increases the rate of crazing at a given stress, or, conversely, decreases the stress required to form a craze within a given time interval. In other words, the critical stress for craze formation falls on heating. Referring to eqn. (9.26), raising the temperature causes a reduction in σ_y. In brittle glassy polymers, the critical COD δ_c is the thickness of a single craze at fracture, which appears to vary relatively little with temperature, so that the magnitude of $\mathcal{G}_{IC}$ is governed by σ_y.

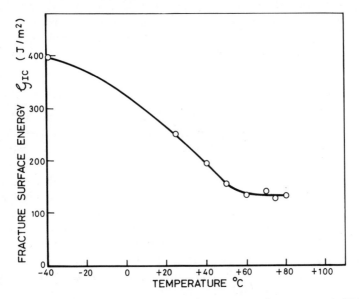

Fig. 9.10 *Relationship between fracture surface energy and temperature in PMMA (from L. J. Broutman and F. J. McGarry, ref. 49, reproduced with permission).*

No strictly comparable study of $\mathcal{G}_{IC}$ as a function of temperature has been published for a rubber-toughened polymer. However, evidence from impact and other tests indicates that $\mathcal{G}_{IC}$ *increases* with temperature in most rubber-modified polymers. Referring again to eqn. (9.26), these observations can be explained by an increase in δ_c on heating, which outweighs the decrease in σ_y. In a rubber-toughened polymer, the density of crazes in the crack tip region increases rapidly with temperature, and the COD is determined not by the behaviour of a single craze but by the number of crazes formed before the crack moves forward. The elongation at break in the crack tip region increases with temperature in the same way as the elongation at break in a tensile test.

At low temperatures or high strain rates, the relaxation behaviour of the rubber becomes a controlling factor. A rubber below its T_g is effectively just a glassy filler, and has little or no effect upon the parent polymer. The composite behaves simply as a brittle glass. At higher temperatures, the transition from tough to brittle fracture depends critically upon strain rate. Strain rates are extremely high in the region around the tip of a rapidly propagating crack, and it is in this region that the relaxation properties of the rubber become critical. Even at room temperature, many rubber-toughened plastics show a sharp drop in $\mathcal{G}_{IC}$ at high crack speeds because the time required for the rubber particles to relax through their glass transition is long compared with the time that the crack takes to propagate through the stress field.

9.4.4 Molecular weight

Polymers of very low molecular weight are extremely brittle, and it is difficult to make measurements of fracture resistance on such materials. However, Robertson has developed an elegant experimental technique, based on the wedge-splitting method, that overcomes this problem so successfully that measurements can be made on polymers with a degree of polymerisation of only 30.[53] Samples of low molecular weight polystyrene were dissolved in methylene chloride and sandwiched between layers of high molecular weight PMMA. After drying, the sandwich specimens were subjected to the wedge-splitting test, and $\mathcal{G}_{IC}$ was calculated using the standard equations for wedge-splitting, but with the modulus of PMMA rather than polystyrene in the relevant expressions, since it is the PMMA that supplies the energy for crack propagation.

Robertson's data are combined in Fig. 9.11 with data obtained by Kusy and Turner for PMMA of high molecular weight.[54] The result is a sigmoidal curve rising steeply at a molecular weight of about 10^5, but levelling out at high and low molecular weights. The limiting value of $\mathcal{G}_{IC}$ at low molecular weights is approximately $0.04\,\mathrm{J/m^2}$, which is the surface energy of molten polystyrene extrapolated to $20\,^\circ\mathrm{C}$.[54, 55]

As explained in Chapter 6, the relatively high fracture resistance of thermoplastics is due essentially to the stability of the entanglements formed by the macromolecules. In the absence of such entanglements, stable crazes and yield zones do not form, and the material is very brittle. This is exactly what happens when the molecular weight is reduced. As the chains become shorter, it becomes more difficult to establish stable craze fibrils, and $\mathcal{G}_{IC}$ falls rapidly, especially in the range of molecular weight between 100 000 and 30 000. Wellinghoff and Baer have demonstrated this change in craze stability directly, by preparing thin films of various molecular weights, and stretching them in the electron microscope.[56]

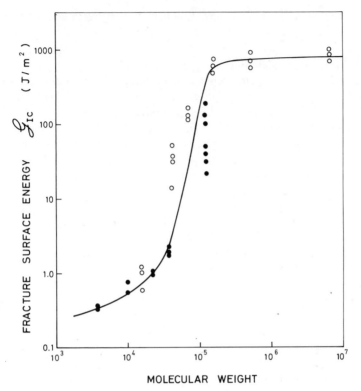

Fig. 9.11 *Relationship between fracture surface energy and molecular weight:* (●) *in PS;* (○) *in PMMA (after R. E. Robertson, ref. 53, and R. P. Kusy and D. T. Turner, ref. 54, reproduced with permission).*

9.4.5 Fibre reinforcement

The success of CTBN rubbers in toughening epoxy resins raised hopes that it might be possible to prepare fibre-reinforced thermosetting resins of greatly increased fracture resistance. Unfortunately, these hopes have not been fully realised. Fracture mechanics studies show that the improvements resulting from adding rubber are much smaller in fibre-reinforced polymers than in the unreinforced polymer. In experiments on epoxy resins, Scott and Phillips showed that addition of 9 % CTBN rubber raised $\mathscr{G}_{IC}$ from 0·3 to 3·0 kJ/m². However, in the presence of 60 vol. % of carbon fibres, the increase on adding the same rubber was only from 0·24 to 0·49 kJ/m², for a crack running parallel to the aligned fibres. These observations suggest that rubber toughening is far less effective in a constrained system. Experiments on thin adhesive layers of rubber-toughened epoxy resin lend support to this view.[57]

9.4.6 Molecular orientation

The qualitative effects of molecular orientation on the strength of polymers are well known. Parallel to the orientation, polymers exhibit high strengths, which are exploited commercially in textile fibre-spinning and in the manufacture of oriented film and sheet. The low strength in the perpendicular direction is a problem in the design of injection mouldings.

Because of the marked anisotropy of strength in oriented polymers, it is difficult to make meaningful measurements of fracture energy other than in planes parallel to the orientation direction. Irrespective of the method of loading, cracks tend to turn into these planes, and most of the published data therefore refer to fracture parallel to these weaker planes. Cleavage test results presented in Fig. 9.12 illustrate the marked reduction in $\mathscr{G}_{IC}$ observed as a result of orientation in polystyrene.[58] Uniaxial hot-drawing at 115 °C

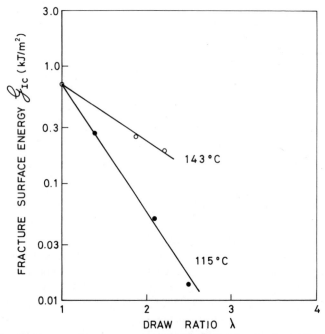

Fig. 9.12   *Fracture surface energy measured parallel to the draw direction in polystyrene drawn uniaxially at two different temperatures (from L. J. Broutman and F. J. McGarry, ref. 58, reproduced with permission).*

reduces $\mathscr{G}_{IC}$ by almost two orders of magnitude at a draw ratio of 2·5. Drawing at 143 °C has a smaller effect upon properties, because substantial relaxation takes place during drawing, so that the residual orientation in the sheet is much lower than that indicated by the draw ratio.

Hot drawing has a similar effect on the fracture behaviour of rubber-toughened plastics. As explained in Chapter 8, the residual orientation can be determined directly by measuring the shape of the deformed rubber particles, which is conveniently expressed in terms of the ratio of major to minor axes of the elliptical sections seen under the microscope. Figure 9.13 presents cleavage test results for HIPS specimens drawn uniaxially at temperatures between 105 and 140 °C and tested with the crack running

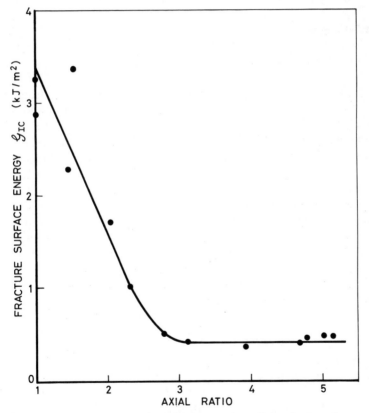

Fig. 9.13   Fracture surface energy at 20 °C measured parallel to the draw direction in uniaxially hot-drawn HIPS plotted against axial ratio of elliptical rubber particles. Crack speed 0·8 mm/s (J. C. Reid and C. B. Bucknall, to be published).

parallel to the draw direction.[59] As the orientation, represented by the ratio of the major to minor axes of the elliptical particles, increases, there is a sharp drop in $\mathscr{G}_{IC}$, which levels out at a value $0.4\,\mathrm{kJ/m^2}$ when the axial ratio reaches 3.

The isotropic material exhibits dense stress-whitening on the fracture surface, indicating extensive crazing. The stress-whitening decreases in intensity as a result of drawing, and disappears when the axial ratio of the elliptical particles is greater than 2·5. These changes show that the depth of the plastic zone, and therefore the critical COD δ_c, decreases with orientation. Tensile tests on drawn samples show that the yield stress σ_y also decreases. Both factors combine to produce a fall in $\mathscr{G}_{IC}$, as predicted by eqn. (9.26).

9.4.7 Fatigue

Cyclic loading introduces a number of additional factors into the problem of fracture. It is necessary to take into consideration frequency, waveform, upper and lower strain or stress limits, in addition to the usual factors such as temperature, humidity, fluid environment and method of stressing which affect all tests. Adequate test programmes are therefore difficult to devise, especially as fatigue tests are individually rather long by their very nature.

The effects of test parameters and materials structure upon the fatigue of polymers generally are discussed in an excellent review by Manson and Hertzberg.[60] Polymers present particular difficulties because they combine high mechanical hysteresis with low thermal conductivity. The temperature of a polymeric specimen therefore tends to rise significantly under cyclic loading. Failures due to thermal softening are observed in PMMA at frequencies as low as 2 Hz. The total temperature rise depends, of course, on the rate of heat loss, and therefore on the thickness of the specimen. Even if there is no direct failure due to thermal softening, temperature rises caused by mechanical hysteresis can affect fracture behaviour. In the absence of heat losses, the rate of heating is given by

$$\frac{\mathrm{d}T}{\mathrm{d}t} = \frac{\pi f J''(f, T)\sigma_{max}^2}{\rho C_p} \tag{9.27}$$

where f = frequency, J'' = loss compliance, σ_{max} = peak stress, ρ = density and C_p = specific heat.

Fatigue in many polymers, including both rubbers and plastics, may be represented by the following equation:

$$\frac{\mathrm{d}a}{\mathrm{d}n} = A\,\Delta\mathscr{K}^N \tag{9.28}$$

where n = number of cycles, A and N are materials constants, and $\Delta\mathscr{K}$ is the range of stress intensity factor during a single cycle. This equation was first proposed by Paris and Erdogan for metals.[61, 62] The value of N for most polymers is between 4 and 5.[63–65]

The mean stress intensity factor and the frequency also affect fatigue

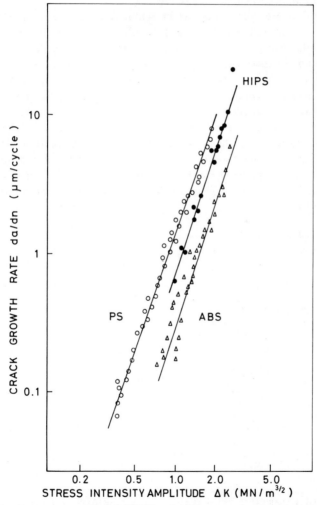

Fig. 9.14   *Fatigue curves for PS, HIPS and ABS at room temperature (from R. W. Hertzberg et al., ref. 66, reproduced with permission).*

crack propagation, but their influence appears in general to be considerably less than that of $\Delta\mathcal{K}$. If the mean stress is not zero, the test is further complicated by creep of the specimen under the mean applied load. As in all failure experiments, temperature, fluid environment and materials structure are also important. Some increase in temperature is to be expected in the crack tip region, where the stress cycle is a maximum, but little attention has been paid to this point.

The work of Hertzberg et al., which is illustrated in Fig. 9.14, shows that fatigue crack growth in PS, HIPS and ABS follows eqn. (9.28).[60, 66] The value of the exponent N, determined from the slope of the log–log relationship, is between 4 and 5 for all three polymers. The effect of rubber-toughening is to reduce the constant A, causing a reduction in the rate of crack propagation per cycle at a given level of $\Delta\mathcal{K}$. DiBenedetto has shown that fatigue in toughened epoxy resins also follows eqn. (9.28).[67]

Scanning electron microscopy of fatigue crack surfaces in HIPS shows that crazes are formed ahead of the main crack front, at an angle to the fracture surface.[68] This observation demonstrates that multiple crazing in the tip region is responsible for the fracture resistance of rubber-toughened plastics under cyclic loading, in a similar manner to that observed under static loading.

REFERENCES

1. J. F. Knott, *Fundamentals of Fracture Mechanics*, Butterworth, London, 1973.
2. A. A. Griffith, *Phil. Trans. Roy. Soc.* **A221** (1921) 163.
3. A. A. Kelly, *Strong Solids*, 2nd edn., Clarendon Press, Oxford, 1973.
4. C. E. Inglis, *Trans. Inst. Naval Arch.* **60** (1913) 219.
5. E. Orowan, *Trans. Inst. Engrs Shipbldrs Scot.* **89** (1945) 165.
6. J. P. Berry, *J. Poly. Sci.* **50** (1961) 107, 313.
7. G. R. Irwin, J. A. Kies and H. L. Smith, *ASTM Proc.* 58 (1958) 640.
8. H. M. Westergaard, *J. Appl. Mech.* **61** (1939) A49.
9. R. S. Rivlin and A. G. Thomas, *J. Poly. Sci.* **10** (1953) 291.
10. D. S. Dugdale, *J. Mech. Phys. Solids* **8** (1960) 100.
11. F. M. Burdekin and D. E. W. Stone, *J. Strain Anal.* **1** (1966) 145.
12. A. A. Cottrell, *Iron Steel Inst. Spec. Rept.* **69** (1961) 281.
13. A. A. Wells, *Crack Propagation Symp. Proc.*, College of Aeronautics, Cranfield **1** (1961) 210.
14. J. G. Williams, *Int. J. Fract. Mechs.* **8** (1972) 393.
15. P. S. Theocaris, *J. Strain Anal.* **9** (1974) 197.
16. G. P. Marshall, L. E. Culver and J. G. Williams, *Proc. Roy. Soc.* **A319** (1970) 165.
17. M. Kitagawa and K. Motomura, *J. Poly. Sci. (Phys.)* **12** (1974) 1979.
18. J. G. Williams and G. P. Marshall, *Proc. Roy. Soc.* **A342** (1975) 55.
19. I. Narisawa and T. Kondo, *J. Poly. Sci. (Phys.)* **11** (1973) 223.
20. T. L. Peterson, D. G. Ast and E. J. Kramer, *J. Appl. Phys.* **45** (1974) 4220.

21. R. J. Ferguson, G. P. Marshall and J. G. Williams, *Polymer* **14** (1973) 451.
22. G. P. Marshall, L. E. Culver and J. G. Williams, *Int. J. Fract. Mechs* **9** (1973) 295.
23. L. C. Cessna and S. S. Sternstein, *J. Poly. Sci.* B3 (1965) 825.
24. G. R. Irwin and J. A. Kies, *Weld. J. Res. Suppl.* **33** (1954) 1935.
25. G. R. Irwin, *Appl. Mat. Res.* **3**(2) (1964) 65.
26. S. Mostovoy, P. B. Crosley and E. J. Ripling, *J. Materials* (ASTM) **2** (1967) 661.
27. G. P. Marshall, L. E. Culver and J. G. Williams, *Plast. Polym.* **37** (1969) 75.
28. T. Kobayashi and L. J. Broutman, *J. Appl. Polymer Sci.* **17** (1973) 2053.
29. A. C. Meeks, *Polymer* **15** (1974) 675.
30. T. Kobayashi and L. J. Broutman, *J. Appl. Polymer Sci.* **17** (1973) 1909.
31. M. F. Kanninen, *Int. J. Fracture* **9** (1973) 83.
32. D. P. Clausing, *Int. J. Fract. Mechs.* **5** (1969) 211.
33. C. E. Turner, *J. Strain Anal.* **10** (1975) 207.
34. L. J. Broutman and T. Kobayashi, in *Dynamic Crack Propagation*, G. C. Sih (ed.), Noordhoff, Leyden, 1973, p. 215.
35. H. W. Greensmith and A. G. Thomas, *J. Poly. Sci.* **18** (1955) 189.
36. H. W. Greensmith, *J. Poly. Sci.* **21** (1956) 175.
37. B. Cotterell, *Appl. Mat. Res.* **4** (1965) 227.
38. E. H. Yoffe, *Phil. Mag.* **42** (1951) 739.
39. J. P. Berry, in *Fracture Processes in Polymeric Solids*, B. Rosen (ed.), Interscience, New York, 1964, p. 157.
40. H. R. Brown, *J. Mater. Sci.* **8** (1973) 941.
41. A. C. Meeks, *Polymer* **15** (1974) 675.
42. A. C. Meeks, *Brit. Poly. J.* **7** (1975) 1.
43. F. J. McGarry, *Proc. Roy. Soc.* **A319** (1970) 59.
44. J. N. Sultan and F. J. McGarry, *Poly. Engng Sci.* **13** (1973) 29.
45. A. D. S. Diggwa, *Polymer* **15** (1974) 101.
46. E. H. Rowe, A. R. Siebert and R. S. Drake, *Mod. Plast.* **47** (Aug. 1970) 110.
47. D. Peretz and A. T. DiBenedetto, *Eng. Fract. Mech.* **4** (1972) 979.
48. J. J. Benbow, *Proc. Phys. Soc.* **78** (1961) 970.
49. L. J. Broutman and F. J. McGarry, *J. Appl. Polymer Sci.* **9** (1965) 585.
50. J. P. Berry, *J. Poly. Sci.* A1 (1963) 993.
51. N. L. Svenson, *Proc. Phys. Soc.* **77** (1961) 876.
52. A. T. DiBenedetto and K. L. Trachte, *J. Appl. Polymer Sci.* **14** (1970) 2249.
53. R. E. Robertson, *ACS Div. Org. Coat. Plast. Prepr.* **34**(2) (1974) 229.
54. R. P. Kusy and D. T. Turner, *Polymer* **15** (1974) 394.
55. S. Wu, *J. Adhesion* **5** (1973) 39.
56. S. T. Wellinghoff and E. Baer, *J. Macromol. Sci.* (*Phys.*) **B11** (1975) 367.
57. J. M. Scott and D. C. Phillips, *J. Mater. Sci.* **10** (1975) 551.
58. L. J. Broutman and F. J. McGarry, *J. Appl. Polymer Sci.* **9** (1965) 609.
59. J. C. Reid and C. B. Bucknall, unpublished work.
60. J. A. Manson and R. W. Hertzberg, *CRC Crit. Revs Macromol. Sci.* **1** (1973) 433.
61. P. C. Paris and F. Erdogan, *Trans. ASME* **D85** (1963) 528.
62. H. H. Johnson and P. C. Paris, *Eng. Fract. Mech.* **1** (1968) 3.
63. R. W. Hertzberg, H. Nordberg and J. A. Manson, *J. Mater. Sci.* **5** (1970) 521.
64. N. H. Watts and D. J. Burns, *Poly. Engng Sci.* **7** (1967) 90.
65. B. Mukherjee, L. E. Culver and D. J. Burns, *Exp. Mech.* **9** (1969) 90.
66. R. W. Hertzberg, J. A. Manson and W. C. Wu, ASTM STP536 (1973) 391.
67. A. T. DiBenedetto, *J. Macromol. Sci.* **B7** (1973) 657.
68. J. A. Manson and R. W. Hertzberg, *J. Poly. Sci.* (*Phys.*) **11** (1973) 2483.

CHAPTER 10

IMPACT STRENGTH

Impact strength is often the deciding factor in materials selection. Many otherwise satisfactory plastics are rejected for engineering and other applications simply because they have a tendency towards brittle fracture under impact loading. The problem particularly concerns materials that would normally be considered ductile on the evidence of tensile tests at low or moderate strain rates. Brittle fractures are especially likely to occur in this type of material if the component contains a stress concentrator, either as a design feature or in the form of an accidental scratch or other defect. The designer should always be aware of the possibility of brittle fracture in impact.

Numerous empirical impact tests have been devised. Lever and Rhys cite no fewer than 15 separate methods in a list that is by no means exhaustive.[1] This plethora of tests serves to emphasise the difficulty of correlating standard test results with service performance. Each test measures the energy absorbed in breaking a standard specimen using a standard striker. The methods fall broadly into two categories: (i) limiting energy methods, in which the striker energy is adjusted until only a set fraction of the specimens breaks; and (ii) excess energy methods, in which the kinetic energy of the striker is considerably greater than the energy required to break the specimens. The falling weight test falls into the first category, and the Charpy, Izod and tensile impact tests fall into the second. These four methods are included in official ASTM, BS, DIN and ISO standards, and are therefore used more frequently than other impact tests.[2, 3]

The standard impact tests have the advantages of being easily and rapidly performed, and very widely quoted. The chief disadvantage is that there is no obvious way in which the impact energies that they measure can be correlated with service performance. For this reason, many manufacturers prefer to rely upon component testing, despite the high cost of this approach. Standard impact tests should perhaps be regarded not as sources of quantitative data on fracture resistance but as methods for ranking materials in order of impact toughness. Even in this respect there are major anomalies, since tests on specimens of differing geometry often rank materials in different orders of merit.

Most of these problems arise because impact strength (or, more correctly, *impact energy*) is a composite quantity, consisting of energy terms arising from various stages of the failure process. The energy is absorbed in elastic and viscoelastic deformation of the specimen during the initial impact, local plastic yielding before fracture and, finally, deformation of the material around the tip of the propagating crack. Variations in specimen geometry alter the relative importance of these separate energy terms, so that different aspects of the material's fracture resistance are emphasised. Tests on sharply notched specimens tend to measure resistance to crack propagation, whereas tests on unnotched specimens place a greater emphasis upon ductility prior to crack initiation. The two types of tests will not necessarily rank a series of materials in the same order.

The problem of specimen geometry can be approached in a number of ways. Valuable information can be obtained by making systematic variations in notch depth and radius, and by instrumenting the striker, to give force–deflection curves during impact. Measurements over a range of temperatures will reveal the existence of ductile–brittle transitions, and can often be used as an alternative to varying notch sharpness. Visual inspection of the fracture surface is a much-neglected aid in interpreting impact energy measurements. Finally, an analysis of the results in terms of fracture mechanics resolves many of the problems in relating energy absorbed to specimen geometry. The application of these procedures to the study of impact energy absorption in rubber-toughened plastics is the subject of this chapter.

10.1 CHARPY AND IZOD TESTS

In the Charpy test, illustrated in Fig. 10.1, a bar specimen rests on horizontal supports against two upright pillars, and is struck centrally by a pendulum. The subsequent swing of the pendulum measures the kinetic energy abstracted from the pendulum during impact. The method is essentially a high-speed three-point bending test, with a striker speed of 2·44 m/s in the British standard procedure.[2] The bar usually contains a notch, as shown in the diagram, which is placed on the tensile face, away from the striker. The depth and radius of the notch are specified in the standard.

The notched Izod test, also illustrated, is similar to the Charpy test in using a centrally notched bar. The difference between the two is that the Izod bar is clamped at the lower end, and is subjected to cantilever loading during impact. Clamping generates complex stresses in the specimen, so

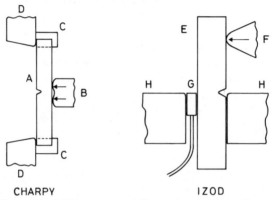

CHARPY   IZOD

Fig. 10.1   The notched Charpy test: (A) specimen; (B) striker; (C) specimen supports; (D) main pillars of instrument. An instrumented Izod test: (E) specimen; (F) striker; (G) pressure transducer; (H) grips.

that it is more difficult to analyse the stresses around the notch root in the Izod specimen than in the Charpy specimen. Variations in clamping force can cause significant differences in impact energy absorption: use of a torque wrench overcomes this problem by standardising the clamping force.

10.1.1 Fracture mechanics

Fracture mechanics provides the most informative analysis of the Charpy impact test. The standard specimen containing a rounded notch is not ideally suited to such an analysis, however, and it is therefore convenient to base the initial discussion upon a sharply notched specimen, as illustrated in Fig. 9.5f. In the case of brittle fracture, a crack initiates in such a specimen when the applied load P reaches a critical value P_f, which is related to specimen geometry and to the fracture surface energy of the material by eqn. (9.25) as follows:

$$\mathscr{G}_{IC} = \left(\frac{9S^2 a Y^2 (1 - v^2)}{4B^2 W^4 E} \right) P_f^2 \tag{9.25}$$

In instrumented tests, the peak force P_f is measured directly. In the standard type of test, only the impact energy I is measured, and it is necessary to calculate $\mathcal{G}_{IC}$ from I. Since the specimen deforms elastically in a brittle fracture, I is related to the critical elastic deflection δ_c at the point of impact by

$$I = \frac{P_f \delta_c}{2} = \frac{P_f^2 C}{2} \tag{10.1}$$

where the specimen compliance C is given by

$$C = \frac{\delta}{P}$$

From the Irwin–Kies relationship, eqn. (9.19)

$$\mathcal{G}_{IC} = \frac{P^2}{2B} \frac{dC}{da} \tag{9.19}$$

Combining eqns. (10.1) and (9.19) yields

$$I = \frac{CB\mathcal{G}_{IC}}{\left(\dfrac{dC}{da}\right)} = BW\left[\frac{C}{\dfrac{dC}{d(a/W)}}\right]\mathcal{G}_{IC} \tag{10.2}$$

or

$$I = BWZ\mathcal{G}_{IC} \tag{10.3}$$

where the factor Z can be obtained experimentally or can be calculated from[4]

$$Z = \frac{\dfrac{SW}{18} + \displaystyle\int Y^2 a \, da}{Y^2 Wa} \tag{10.4}$$

For short notches in a three-point bend test, $Y^2 \approx \pi$, so that in Charpy specimens containing sharp notches[5]

$$Z = \frac{1}{2}\left(\frac{a}{W}\right) + \frac{1}{18\pi}\left(\frac{S}{W}\right)\left(\frac{W}{a}\right) \tag{10.5}$$

Measured values of Z are in good agreement with values calculated from eqns. (10.4) and (10.5).[5]

Equation (10.3) predicts that a graph of I against BWZ should give a straight line of slope $\mathcal{G}_{IC}$, passing through the origin. Tests on PMMA and other polymers show that the relationship is linear, but that the extrapolated impact energy I has a positive value when BWZ is zero.[4] Marshall et al. attribute the discrepancy to a kinetic energy term I_{ke}, which must be subtracted from the total impact energy in order to obtain the potential energy stored in the specimen when the crack is initiated.[4] Equation (10.3) then becomes

$$\mathscr{G}_{IC} = \frac{I - I_{ke}}{BWZ} \tag{10.6}$$

Another possibility is that some of the energy of impact is dissipated by viscoelastic processes. Whatever the cause of the descrepancy, the main practical result is that $\mathscr{G}_{IC}$ cannot be calculated directly from a single measurement of impact energy I. Either I_{ke} must be determined in a separate calibration, or I must be plotted against BWZ, and $\mathscr{G}_{IC}$ obtained from the slope.

With the reservations noted earlier concerning the stress distributions around the notch, the Izod test can be analysed in a similar way.[5] In order to produce an equivalent deflection of the specimen, the Izod pendulum must travel twice as far as the Charpy pendulum, but the force exerted is reduced to one half. The work done is therefore the same in both cases. For the purposes of fracture mechanics analysis, the two tests can be regarded as approximately equivalent when comparable specimen dimensions and pendulum velocities are employed. Ideally, the Izod pendulum should strike the specimen at twice the velocity of the Charpy specimen. Plati and Williams suggest the following expression for the factor Z in the Izod test:[5]

$$Z = \frac{1}{2}\left(\frac{a}{W}\right) + \frac{1}{36}\left(\frac{S}{W}\right)\left(\frac{W}{a}\right) \tag{10.7}$$

Measured values of Z for Izod specimens differ from values calculated from eqn. (10.7), owing to rotation of the specimen in the clamp, which is not ideally rigid.[5]

The main conclusion to be drawn from a fracture mechanics analysis of the notched Charpy and Izod tests is that the impact energy I increases linearly with the fracture surface energy $\mathscr{G}_{IC}$ in the case of brittle fracture initiated from a sharp notch. This result is applicable to rubber-toughened plastics at low temperatures, or after ageing. More generally, the root of the notch is rounded, and some yielding occurs prior to fracture, so that the simple linear elastic fracture mechanics treatment presented above must be modified.

10.1.2 Blunt notches

The stress concentrations at the root of a notch of depth a and of known root radius r_0 can be calculated by treating the notch as approximately elliptical in shape. The results are shown diagrammatically in Fig. 9.3. The maximum stress concentration in the direction of the applied stress is given by[6]

$$\sigma_{11} = \sigma \left[1 + 2 \left(\frac{a}{r_0} \right)^{1/2} \right] \tag{10.8}$$

which reduces for relatively sharp notches ($a \gg r_0$) to

$$\sigma_{11} = 2\sigma \left(\frac{a}{r_0} \right)^{1/2} \tag{10.9}$$

One of the basic principles of linear elastic fracture mechanics is that the fracture stress for a notched or cracked specimen is relatively insensitive to variations in root radius, provided the radius is small. The difference between the stress field around a sharp crack and the corresponding stress field for a rounded notch is significant only over a small region extending a distance $r_0/4$ from the notch tip, as illustrated in Fig. 10.2. The Irwin–Westergaard approach to linear elastic fracture mechanics

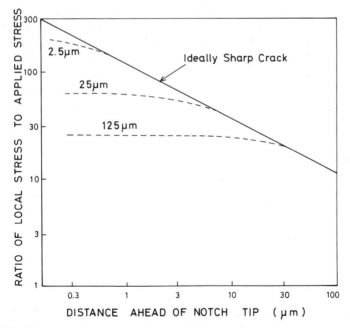

Fig. 10.2   *Stress concentrations* $\sigma_{11}/\sigma_{appl}$ *near the tip of a central 5 cm crack in a wide plate for various values of notch root radius.*

specifically states that this difference can be neglected for small r_0, as discussed in Chapter 9.

Experiments show that measured values of fracture toughness $\mathscr{K}_{IC}$ do vary with r_0. For example, in high-strength steel $\mathscr{K}_{IC}$ is independent of r_0 up to a notch root radius of 6 μm, and thereafter increases linearly with $r_0^{1/2}$.[7] Similar observations have been made in PMMA[8] and in polystyrene.[9] In the case of polystyrene, the increase in apparent fracture toughness is due largely to an increase in the number of crazes formed at the base of the notch. In more ductile materials, the effect is due to plastic yielding, which is the subject of the following section.

In the absence of a notch, or in the presence of a very blunt notch, brittle fracture initiates at random flaws in the specimen. The Griffith equation can be applied if the average size of the inherent flaws is known, using eqn. (9.24) to calculate the 'outer fibre stress' for the beam in three-point bending. The outer fibre stress is the local stress acting on surface flaws in an unnotched specimen. In the case of fracture from a blunt notch, the stress concentration due to the notch must also be taken into account, using eqn. (10.8) or some more complete expression. Alternatively, brittle fracture can be treated simply by employing a critical stress criterion. The two approaches are equivalent, since the critical stress for crack initiation is related to the average intrinsic flaw size by the Griffith equation (9.5).

The foregoing analysis is valid only for the case of brittle fracture. In practical terms, the expression 'brittle fracture' means that the extent of yielding in the specimen is very limited, so that the radius R_y of the plastic zone at the crack tip is small in relation to the crack length a. The condition for a valid measurement of $\mathscr{G}_{IC}$ is

$$R_y \leq 0 \cdot 02a \qquad (10.10)$$

Brittle fracture, as defined by eqn. (10.10), is observed in rubber-toughened plastics at low temperatures, or after prolonged exposure to sunlight, but under normal conditions there is extensive yielding around the crack tip, which has a major influence upon impact properties.

10.1.3 Fracture with yielding

The effects of notch radius upon fracture resistance are even more striking in ductile materials than in brittle materials. Charpy and Izod tests show that most ductile polymers are notch sensitive. Impact energies fall rapidly with decreasing notch radius because the presence of a sharp notch drastically reduces the extent of plastic yielding. This effect is illustrated in

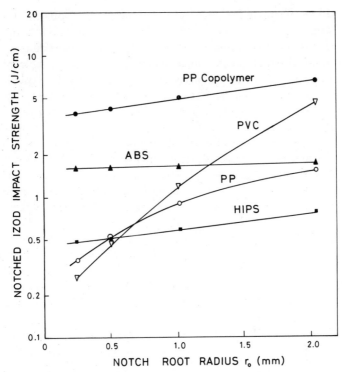

Fig. 10.3 Relationship between Izod impact strength and notch root radius for toughened and untoughened polymers at 23 °C (from A. C. Morris, ref. 9, reproduced with permission).

Fig. 10.3, notably in the case of rigid PVC. The rubber-toughened polymers included in the diagram are less notch sensitive.

In uniaxial tension, unnotched specimens of PVC, polycarbonate and polysulphone are ductile. In other words, the yield stress is well below the stress required for crazing and fracture. Once necking has occurred, there is a local increase in stress, but the stress concentration is in a region of high molecular orientation in the neck, and crazing is therefore suppressed. In thin plate specimens containing notches, these polymers also behave in a ductile manner. The stress in the through-thickness direction of the thin plate is zero, and the response is therefore known as *plane-stress* deformation. Yielding occurs at constant volume, involving through-thickness contraction.

In a thick plate containing a sharp crack, the situation is very different.

The stresses at the tip of the crack are triaxial, as given in equation. (9.6) for plane strain. The reason for the state of triaxial tension is that the material in the tip of the crack is trying to contract in the plane perpendicular to the applied stress, in order to maintain constant volume, but is unable to do so because it is constrained by the surrounding material. The resulting change in the form of the stress tensor from uniaxial or biaxial tension to triaxial tension is sufficient to raise the yield stress above the crazing stress, with consequent reduction in fracture resistance. The material reaches the critical tensile strain for craze initiation before the shear strain energy density is sufficient to cause shear yielding. Referring to Chapter 6, the envelope for crazing lies inside the shear yielding envelope in the neighbourhood of balanced triaxial tension. The reduction in impact strength resulting from this limitation on yielding is known as the plane-stress to plane-strain transition.

In impact tests, notches not only introduce triaxiality into the stress field, but also alter the strain rate in the highly stressed region around the root of the notch. In an elastic material containing a sharp crack of length a, the stress and strain rates increase linearly with $a^{1/2}$. This increase in strain rate also has some effect upon fracture behaviour, since it tends to raise the yield stress relative to the fracture stress.

These effects do not mean that yielding cannot occur at the tip of a sharp crack, merely that the extent of yielding is likely to be smaller. Fracture

TABLE 10.1
FRACTURE PARAMETERS FOR A RANGE OF POLYMERS
(after Plati and Williams[5])

Polymer	$\mathscr{G}_{IC}(kJ/m^2)$	$\mathscr{G}_{B}(kJ/m^2)$ $(r_0 = 1\,mm)$	$U_y(MJ/m^3)$
PVC	1·4	7	10
PC	4·8	62·0	94
PMMA	1·3	8·5	12·5
HDPE	3·4	6·4	7·5
MDPE	8·1	62	91
HIPS	15 $(\mathscr{J}_{IC})$	15	2·9
ABS	49 $(\mathscr{J}_{IC})$	79	9·5

mechanics measurements using sharply notched specimens show that $\mathscr{G}_{IC}$ is indeed low for many apparently tough and ductile materials, including many ductile polymers. This point is brought out clearly in Table 10.1 and in Fig. 10.4, which compare $\mathscr{G}_{IC}$ with $\mathscr{G}_{B}$, the apparent critical strain energy release rate in specimens containing rounded notches of specified tip radius.[5] Also shown in the table is the elastic energy per unit volume stored in the material at yield, U_y, defined by

$$U_y = \frac{\sigma_y^2}{2E} \tag{10.11}$$

The results demonstrate that the high impact strengths of polycarbonate (PC) and of medium-density polyethylene (MDPE) in a standard test using

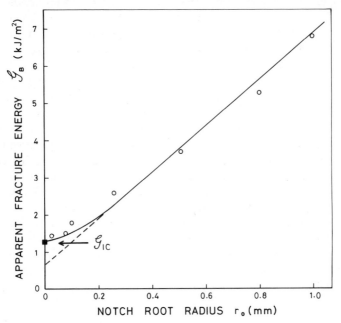

Fig. 10.4 *Apparent fracture surface energy for blunt-notched specimens as a function of notch root radius in rigid PVC Charpy specimens (from E. Plati and J. G. Williams, ref. 5, reproduced with permission).*

a relatively blunt notch are attributable to a high strain energy density at yield. The difference in a *sharp-notch* test between polycarbonate and PMMA, which would generally be considered typical of brittle glassy polymers, is relatively small.

Two mechanisms of yielding are possible under plane-strain conditions: shear band formation and crazing. As explained in Chapter 6, shear bands form in planes of maximum resolved shear stress when the polymer is constrained in two dimensions. Crazes form in planes perpendicular to the direction of maximum tensile strain. Both mechanisms are possible in a sharply-notched specimen, but the extent of shear band formation will be reduced, for the reasons already given. The change from plane-stress to plane-strain conditions would not be expected to affect HIPS to the same extent as polycarbonate or other ductile polymers, since HIPS deforms largely by multiple craze formation, which is essentially a plane-strain process. The figures given in Table 10.1 support this view: increasing the notch radius to 1 mm has no significant effect upon the measured value of strain energy release rate for HIPS. Measurements on ABS, on the other hand, show an appreciable notch sensitivity, which can be related to the larger contribution of shear processes to the deformation of ABS. Increasing the triaxiality of the stress field reduces the extent of shear deformation in ABS, and consequently reduces the critical strain energy release rate. Nevertheless, the fracture resistance of the ABS polymer in sharp-notch tests is far higher than any of the other polymers included in the table.

In sharply notched specimens of PVC, PC, PMMA, HDPE and MDPE, the plastic zone size is sufficiently small to permit a valid measurement of $\mathscr{G}_{IC}$.[5] However, in HIPS and ABS, yielding by multiple crazing is too extensive to be included within a linear elastic fracture mechanics analysis, and it is therefore necessary to employ yielding fracture mechanics. The alternative is simply to exclude HIPS and ABS from the discussion, recognising that they are not brittle under the conditions of the test.

The status of yielding fracture mechanics is not clearly established. Two fracture criteria are in current use for metals, and have received some attention in the polymer literature. The critical crack opening displacement, δ_c, was discussed in Chapter 9. An alternative parameter is $\mathscr{J}_{IC}$, the plastic work parameter, defined by [5, 7]

$$\mathscr{J}_{IC} = \sigma_y \delta_c \qquad (10.12)$$

In the elastic case, $\mathscr{J}_{IC}$ is equal to $\mathscr{G}_{IC}$ (cf. eqn. (9.17)). The plastic work parameter has been used to correlate ductile failures in metals, and is quoted in Table 10.1 for HIPS and ABS, since the conditions for a valid measurement of $\mathscr{G}_{IC}$ are not met in these polymers. Further work will be required, however, before it can be established whether $\mathscr{J}_{IC}$ is a satisfactory parameter for correlating fracture data for rubber-toughened plastics.

Figure 10.5 is a schematic diagram showing elastic–plastic stress distributions in a loaded Charpy specimen which has yielded near the root of the notch.[10] Within the plastic zone, σ_{11} increases initially owing to an increase in the transverse stresses σ_{22} and σ_{33}, and then levels out at the yield stress σ_y. The elastic stress falls rapidly with distance from the end of the plastic zone, and becomes compressive over the region of the specimen opposite the notch.

Very ductile materials are able to yield in the compression region. With increasing ductility, the extent of the yield zones in both tension and compression increases, until eventually the zones meet, and the specimen reaches a state of *general yield*. General yield is unlikely to occur in sharply

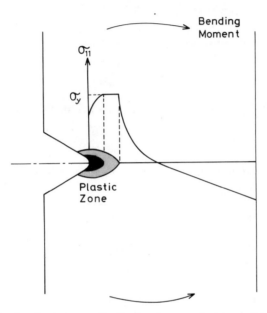

Fig. 10.5 *Elastic–plastic stress distribution in a notched bar subjected to plane-strain bending (from T. R. Wilshaw et al., ref. 10, reproduced with permission).*

notched specimens, but is a distinct possibility in unnotched Charpy and Izod specimens of rubber-toughened polymers. The problem is therefore discussed under that heading.

10.1.4 Unnotched specimens—brittle fracture and general yield

Brittle fracture means that the specimen deforms elastically up to the point of crack initiation, so that the measured impact strength is the energy of elastic bending in the case of a Charpy or Izod specimen. This problem can be approached by applying classical beam theory,[11] and assuming that brittle fracture occurs when the maximum stress in the 'outer fibre' reaches a critical value σ_c. The maximum stress σ_{max} in a centrally loaded and simply supported beam, as illustrated in Fig. 10.6a, is given by

$$\sigma_{max} = \frac{3PS}{2BW^2} \tag{10.13}$$

Hence the load P_c at failure, when $\sigma_{max} = \sigma_c$, is

$$P_c = \frac{2BW^2\sigma_c}{3S} \tag{10.14}$$

The deflection X at the centre of the beam is

$$X = \frac{PS^3}{4EBW^3} \tag{10.15}$$

The elastic energy U is then equal to the work done on the beam:

$$U = \tfrac{1}{2}PX = \frac{P^2S^3}{8EBW^3} \tag{10.16}$$

Setting $P = P_c$ and combining eqns. (10.14) and (10.16) then yields

$$I = U_c = \frac{\sigma_c^2}{18E} BSW \tag{10.17}$$

Thus the impact energy in brittle fracture increases linearly with the dimensions of the bar. Using the Griffith equation (9.5), the fracture condition can alternatively be stated in terms of $\mathscr{G}_{IC}$ and the intrinsic flaw size a'.

Ductile failure is illustrated in Fig. 10.6b and c, for the case of a material obeying the Tresca yield criterion. Yielding occurs simultaneously at the upper and lower surfaces of the bar, and spreads through both the tensile and compressive zones to meet in the middle (Fig. 10.6c). The bar is then unable to support a further increase in load, and is said to be in a state of general yield. The load P_{gy} at the general yield point is[11]

$$P_{gy} = \frac{BW^2}{S}\sigma_y \tag{10.18}$$

Beyond the general yield point, the work w_p done in deforming the bar plastically is given by

$$w_p = P_{gy}X_p = \left(\frac{X_p}{S}\right)BW^2\sigma_y \tag{10.19}$$

where X_p is the deflection subsequent to general yield. At large deflections,

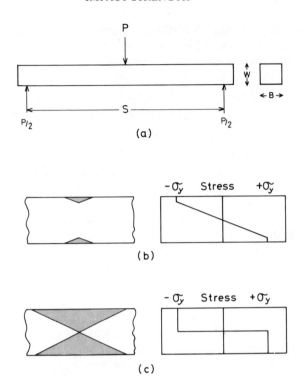

Fig. 10.6 *Simply supported rectangular beam subjected to central loading:* (a) *specimen geometry;* (b) *first yield;* (c) *general yield. Yield zones are shaded.*

the elastic contribution to deformation can be neglected, and X can be substituted for X_p, to give an approximate expression for the total work done on the bar. If the final rupture of the bar occurs at a critical deflection-to-span ratio X/S given by X_c, the impact strength I is

$$I = X_c B W^2 \sigma_y^2 \qquad (10.20)$$

which would predict an increase in impact strength with the square of specimen width W, as compared with a linear increase for brittle fracture (eqn. (10.17)).

This treatment of ductile failure in bending is far from rigorous. No account is taken of the difference between tensile and compressive yield stress in polymers, nor of strain-hardening due to molecular orientation. Furthermore, the criteria for rupture within the plastic yield zone require clarification. Nevertheless, this approximate analysis is useful in indicating how the impact energy might be expected to depend upon specimen geometry and materials properties in the absence of a notch, and serves as an introduction to the subject of falling weight tests on plate specimens, which is discussed in the following section.

10.2   FALLING WEIGHT TEST

In the falling weight test, plate specimens are simply supported by an annular ring, and subjected to impact by a falling dart of variable mass. The dart has a hemispherical head, and is dropped from a fixed height above the specimen. In the standard procedure, each specimen is subjected to only one impact, after which it is removed from the apparatus and examined.[3] If no failure has occurred, the mass of the dart is raised by one increment for the next specimen. If the specimen has fractured, or failed by ductile tearing, the mass is reduced by one increment. The first three readings are regarded as 'ranging shots', and are discarded. The mean of at least 20 subsequent trials is taken as the falling weight impact strength.

The falling weight test, as described above, is essentially a limiting-energy test for plastics sheet. As such, it is used extensively to test extruded HIPS, ABS and other rubber-toughened plastics. The aim is to measure the energy that is just sufficient to break 50 % of the specimens. In order to achieve this ideal, a very large number of specimens would be required. The standard procedure is a compromise aimed at obtaining meaningful results without excessive expenditure of effort. Experiments on HIPS and ABS sheet show that a very much larger number of trials would be required in order to eliminate observer bias and other sources of error. The recorded impact strength can be increased or decreased by as much as 20 % by altering the impact level at which the first specimen is tested.[12]

An alternative and more informative method of conducting falling weight tests is to subject a fixed number of specimens to impact at each of a series of energy levels, and to record the fraction of fractures at each impact energy level. Using this 'Probit' method, Morris has shown that the distribution of failures is Gaussian in ABS and a number of other polymers.[9] As shown in Fig. 10.7, a linear relationship is obtained when the fraction broken is plotted on a normal probability scale against impact energy. The relationship is sigmoidal when both quantities are plotted on a linear scale.[12] The 'Probit' method requires a large number of specimens, but overcomes many of the problems associated with the standard 'staircase' method, and is very much more informative. In particular, the 'Probit'

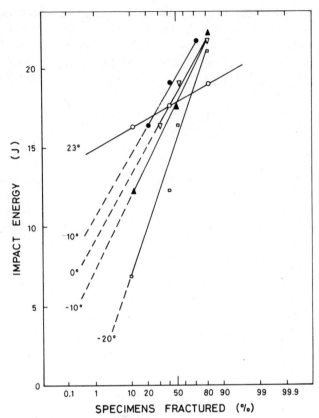

Fig. 10.7 *'Probit' curves from falling-weight tests on ABS over a range of temperatures. Plate specimens 2 mm thick, injection-moulded at 240°C (from A. C. Morris, ref. 9, reproduced with permission).*

method provides information about the impact energies required to cause failure in a small fraction of the specimens, *e.g.* 1%. This incidence of failures is usually of more direct interest than the 50% failure rate to users of plastics sheet. In this context, it is interesting to note the effect of temperature on impact behaviour in Fig. 10.7. The energy required to produce 50% failures in ABS changes little with temperature over the range −20° to 23 °C. By contrast, the energy level for 1% fractures is very sensitive to temperature within this range. The Probit method therefore reveals an important aspect of falling weight impact behaviour that is obscured by the standard staircase technique.

10.2.1 Elastic deformation of circular plate—energy for brittle fracture

A simply supported circular plate subjected to point loading P at the centre can be treated in a similar manner to the centrally loaded rectangular bar. An elastic stress analysis by Timoshenko and Woinowsky-Krieger[13] shows that the deflection X at the centre of the plate, and the maximum stress σ_{max} in a plate of thickness W and radius r, are given by

$$X = \left[\frac{3(3 + v)(1 - v^2)}{4\pi(1 + v)E}\right]\left[\frac{Pr^2}{W^3}\right] \qquad (10.21)$$

and

$$\sigma_{max} = \left[(1 + v)\left(0\cdot485\log_e\frac{r}{W} + 0\cdot52\right) + 0\cdot48\right]\frac{P}{W^2} \qquad (10.22)$$

where v and E are Poisson's ratio and Young's modulus respectively. Assuming that brittle fracture occurs when the maximum stress reaches a critical value ($\sigma_{max} = \sigma_c$) as before, the load P_c at fracture can be expressed in terms of σ_c using eqn. (10.22). The impact strength I is given by

$$I = U_c = \tfrac{1}{2}P_cX$$

Combining these equations yields

$$I = AWr^2\sigma_c^2 \qquad (10.23)$$

where

$$A = \left(\frac{3(3 + v)(1 - v^2)}{8(1 + v)E}\right)\left[(1 + v)\left(0\cdot485\log_e\frac{r}{W} \times 0\cdot52\right) + 0\cdot48\right]^{-2}$$

There are obvious similarities between eqn. (10.23) for brittle fracture of a plate and eqn. (10.17) for fracture of a bar. The impact energy in both cases is proportional to σ_c^2, and hence to $\mathscr{G}_{IC}$, and also in both cases is proportional to the volume of the specimen. A linear relationship is predicted between falling weight impact strength and plate thickness.

10.2.2 Falling weight impact strength *vs* plate thickness—experimental data

Figure 10.8 presents experimental data on falling weight impact strength as a function of plate thickness in HIPS, ABS, polypropylene and toughened polypropylene. The results clearly show the difference in fracture resistance between ABS and HIPS, and the toughening effect of rubber upon polypropylene. In none of the materials is the impact strength exactly proportional to specimen thickness W. The curves for HIPS and PP are almost linear, but the curves for the two tougher polymers, ABS and polypropylene copolymer, show an increase in slope with W. By comparison with the rectangular beam (eqn. 10.20)) a relationship of this type is to be expected in the more ductile polymers such as ABS. Equation (10.23) should apply to brittle materials, in which the deformation is purely elastic, but at room temperature most rubber-toughened plastics are sufficiently tough and ductile to cause some deviation from the result predicted by elasticity theory. No really satisfactory quantitative treatment of ductile failure in impact tests on unnotched specimens is available.

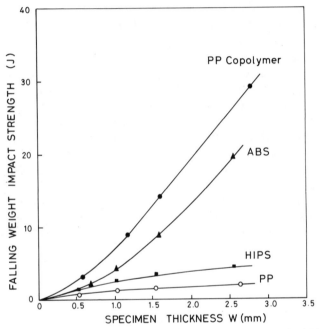

Fig. 10.8   *Effect of thickness upon the falling-weight impact strength of ABS, HIPS and polypropylene materials at 23 °C (from A. C. Morris, ref. 9, reproduced with permission).*

10.2.3 Foil-coated sheet

An interesting example of brittle fracture in falling weight impact tests occurs in foil-coated HIPS sheet. Polystyrene foil approximately 25 μm in thickness is laminated to one side of extruded HIPS sheet as a means of improving the gloss on the surface.[14, 15] Impact tests with the foil coat upwards show little effect of the foil coating. The polystyrene is in compression, and does not have a significant influence upon the ductile failure of the sheet. Much lower impact strengths are recorded when the foil is on the lower, tensile surface of the sheet.[14] The thin layer of polystyrene is sufficient to cause brittle fracture of the whole disc specimen. The reason is that crazes form in polystyrene at low strains, below 1 %, and grow to form cracks equal in length to the foil thickness. These cracks propagate from the tensile face of the sheet through the remainder of the strained specimen. Foil lamination is therefore equivalent to the introduction of sharp cracks into the surface of the sheet.

10.3   EFFECT OF TEMPERATURE ON IMPACT STRENGTH

Temperature has a profound influence upon impact behaviour in all plastics, not least in rubber-toughened polymers. At very low temperatures, the rubber phase is hard and glassy, and the rubber-toughened polymer is brittle. At higher temperatures, the multiple-crazing mechanism becomes active, and impact strength rises. In some toughened polymers, *e.g.* toughened PVC, there is a third stage, in which the polymer yields without crazing at sufficiently high temperatures. These changes are not only of intrinsic interest, but also form a convenient basis for discussion of the principles set out in the earlier part of this chapter.

Figures 10.9 and 10.10 illustrate the relationship between impact behaviour and temperature in a typical rubber-toughened polymer.[16] The polymer in question is MBS, a transparent material based on

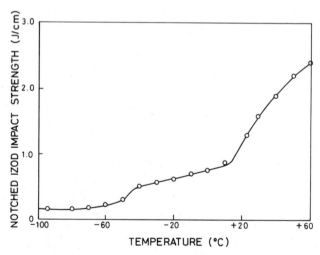

Fig. 10.9   Notched Izod impact strength of MBS over a range of temperatures showing three distinct regions of fracture behaviour.

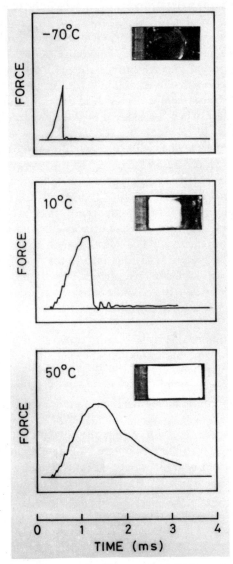

Fig. 10.10 *Fracture behaviour of MBS in notched Izod impact tests. Oscilloscope traces of force applied to the specimen as a function of time, and photographs of fracture surfaces, with notch to the left. Three contrasting types of fracture are observed over the range of temperature.*

styrene–methyl methacrylate copolymer, in which the matrix is chosen to match the refractive index of the polybutadiene rubber. Stress-whitening can be observed more clearly in transparent polymers of this type than in the ordinary HIPS and ABS polymers. Figure 10.10 shows how the extent of stress-whitening on the fracture surface increases with temperature, and compares load–deflection curves, obtained by instrumenting the Izod test, at three widely spaced temperatures. The supplementary information contained in Fig. 10.10 is of considerable value in interpreting the impact energy data in Fig. 10.9.

Below $-65\,°C$, the polymer is brittle. Fracture surfaces are rough and broken, with no sign of stress-whitening. The force–displacement curve is approximately linear up to the point of fracture, falling sharply as the crack propagates catastrophically across the specimen. Brittle behaviour of this type is observed in most rubber-toughened plastics below the glass transition of the rubber. The oscilloscope traces show that the critical strain energy density is reached within $0.3\,ms$ of impact. The rubber phase has little effect upon the fracture behaviour unless it can relax from the glassy to the rubbery state within this interval. In other words, the T_g of the rubber *measured at approximately* $10\,kHz$ must be below the test temperature if it is to function as intended. At test temperatures below $-65\,°C$, this condition is not satisfied.

Under the conditions of brittle fracture observed at low temperatures, linear elastic fracture mechanics is applicable, and the Izod impact strength is predicted by eqns. (10.6) and (10.7). Similarly, eqn. (10.23) predicts falling weight impact strength. The rubber-toughened polymer has a fracture surface energy $\mathscr{G}_{IC}$ comparable with that of an unmodified homogeneous glassy polymer such as polystyrene or PMMA.

Immediately above the glass transition region of the rubber particles, the impact strength begins to rise. Examination of the fracture surfaces reveals a stress-whitened zone near the root of the notch, which increases in extent with increasing temperature. The fracture surface is relatively smooth in this stress-whitened zone, but over the remaining area is rough and broken, as at low temperatures. This observation suggests slow, controlled fracture in the whitened zone, followed by rapid crack propagation beyond this zone. Stress-whitening to a depth of 2 mm can be seen in the transparent polymer near the root of the notch.

The rise in impact strength between $-65°$ and $12\,°C$ in MBS is due to an increase in the elastic energy in the specimen at the point of crack initiation. Yielding occurs at the base of the notch, and a higher peak force P_f is required for crack initiation. With increasing temperature, the extent of the whitened zone increases, and the oscilloscope records higher values of P_f. These trends are illustrated in Fig. 10.10. Linear elastic fracture mechanics is no longer applicable to the crack initiation, as there is extensive yielding prior to fracture.

In this intermediate temperature region, crack *propagation* is essentially brittle, and does not require additional energy from the pendulum. At a velocity of 500 m/s, the time taken by the crack to travel across an Izod specimen is 20 μs. Between $-65°$ and $12°C$, the rubber particles in MBS play no part in the crack propagation stage of the fracture, because they are unable to relax from the glassy to the rubbery state as the rapidly moving crack overtakes them. The glass transition of interest in this situation is T_g measured at perhaps 10 GHz. Time–temperature superposition means that the effective T_g of the rubber particles shifts upwards $-60°$ to $10°C$ as far as the crack is concerned.

Above $12°C$, MBS exhibits a third type of fracture behaviour. The surface is fully stress-whitened and relatively smooth, indicating that there is no brittle crack propagation. The force–deflection curves show a gradual fall from the peak stress, in contrast to the sharp drop observed at lower temperatures, and the crack speed falls. This is the type of failure studied by Plati and Williams[5] and by Brown[17] using sharply notched Charpy specimens. Brown showed that the impact energy of a typical ABS polymer at room temperature is proportional to the fracture surface energy:

$$I = B(W - a)\mathcal{J}_{IC} \tag{10.24}$$

where $\mathcal{J}_{IC}$ is the average energy absorbed per unit area of fracture surface in propagating the crack. The $\mathcal{J}_{IC}$ figures quoted in Table 10.1 are calculated on the basis of a different model, assuming general yield before crack initiation, and must be divided by a factor of 2 for comparison with Brown's data. Both studies show that large amounts of impact energy are absorbed in multiple craze formation over the entire ligament area. The outstanding question is whether this energy is absorbed by general yielding before crack formation, as suggested by Plati and Williams, or during crack propagation, as proposed by Brown.

The available evidence supports the view of Bucknall and Street[16] and of Brown[17] that impact energy is controlled by crack propagation behaviour, in tests on sharply notched tough specimens. General yielding does undoubtedly precede crack initiation in tests on unnotched specimens of tough polymers, notably in falling weight tests on sheet, but there is no direct evidence that it is the controlling factor in notched Charpy or Izod specimens of rubber-toughened plastics.

In order to influence crack propagation, the rubber must be able to relax through the glass–rubber transition extremely rapidly, so that it can initiate crazing ahead of the rapidly propagating crack. If the energy absorbed in multiple crazing is large compared with the stored elastic energy in the specimen, additional energy will be required from the pendulum to maintain the propagation, and the impact strength will consequently increase. This mechanism explains the additional increase in the impact strength of MBS at temperatures above 12 °C. The rubber is sufficiently far above its T_g to be able to behave in a rubbery manner even in the rapidly rising stress field ahead of a fast crack.

Falling weight impact strength
Temperature has a similar effect upon falling weight impact strength. Impact energies are low at temperatures below the glass transition of the rubber, and the fracture is of the typically brittle type. At higher temperatures, yielding occurs around the point of impact, but brittle crack propagation follows as the crack extends beyond the stress-whitened zone. At high temperatures, the polymer stress-whitens at the point of impact, and begins to tear, but the crack is arrested before it can propagate across the specimen. This pattern of behaviour has been demonstrated in HIPS by Morris[9] and in toughened polypropylene (polypropylene 'copolymer' with 25 % ethylene) by Fujioka.[18]

This analysis explains the effect of temperature upon the Probit curves for ABS shown in Fig. 10.7. Between $-20°$ and 10 °C, crack propagation is brittle, and the falling weight impact strength is determined by the extent of multiple craze formation before a crack is initiated. Fracture can occur at very low impact energies if the tensile surface of the plate specimen contains a sufficiently large flaw. In other words, the material is notch-sensitive. A small percentage of specimens fail at low impact energies, whilst the majority exhibit a high apparent impact strength. At 23 °C, by contrast, brittle crack propagation at low energies is no longer possible. Because of the rapid response of the rubber, the ABS is able to control crack growth from adventitious flaws, and the specimen fails by general yield in biaxial tension as the falling weight extends the sheet at the point of contact. At this temperature, the minimum energy for failure is high, and the Probit curve is almost flat.

Classification of impact behaviour
It follows from the preceding discussion that there are two separate factors controlling impact properties:

(i)   energy absorbed in initiating a crack;
(ii) energy absorbed in propagating a crack.

With increasing temperature, both energy contributions show a transition from a characteristic brittle response to a tough response. In rubber-toughened plastics, the temperature of the transition is governed by the relaxation behaviour of the rubber. The following three types of fracture can be identified:

I   Low Temperatures: Rubber particles glassy at times up to 10^{-4} s. Brittle crack initiation and propagation. Impact strength low. Energy absorbed in elastic bending of specimen to point of fracture.

II   Moderate temperatures: Rubber particles exhibit glass transition between 10^{-8} and 10^{-4} s. Yielding can occur by multiple crazing before crack initiation, but crack is able to accelerate from a notch or flaw and propagate in a brittle manner. Impact strength depends upon geometry. Material is notch sensitive. Generally, energy is absorbed in elastic bending of specimen. High impact strengths correspond to a high value of $\mathscr{G}_B$. $\mathscr{G}_{IC}$ for fast crack propagation is low. No stress-whitening occurs during propagation.

III   High temperatures: Rubber particles rubbery at 10^{-8} s. $\mathscr{J}_{IC}$ for fast crack propagation relatively high. Crack does not accelerate from a flaw and propagate in a brittle manner. Unnotched specimens tend to exhibit general yield. Notched specimens absorb significant amounts of energy during crack propagation. Entire fracture surface is stress-whitened, except in very ductile polymers, which may show extensive shear yielding.

These three types of response, exemplified in Figs. 10.9 and 10.10, are observed in a wide variety of rubber-toughened plastics. An understanding of the classification helps in discussing the effects of composition and of ageing upon impact strength.

10.3.1 Glass transition of rubber

The relationship between impact strength and the glass transition temperature of the rubber particles is well known. Matsuo *et al.* measured loss tangents for a series of ABS polymers containing various styrene–butadiene copolymers, and showed that the Charpy impact strength rose sharply at the temperature of the secondary loss peak of each ABS polymer.[19] A similar correlation was observed by the same authors in PVC blends containing butadiene–acrylonitrile copolymer rubbers.

Figure 10.11 presents results obtained by Bucknall and Street in notched Izod tests on two HIPS polymers, one based on polybutadiene, and the other on poly(butadiene-*co*-styrene).[16] Dynamic mechanical measurements show T_g at $-98°$ and $-16°$C respectively for the two rubbers. The low-temperature impact data correlate well with these T_g values: the impact strength begins to rise at about $-90°$C in the HIPS containing

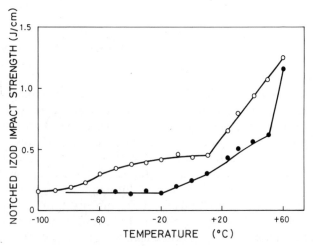

Fig. 10.11   *Notched Izod impact strength over a range of temperatures for HIPS polymers:* (○) *toughened with polybutadiene;* (●) *toughened with poly(butadiene-co-styrene).*

polybutadiene, and at about $-20°$C in the HIPS containing copolymer rubber. The first signs of stress-whitening at the root of the notch are observed at these temperatures. At higher temperatures, both show a further sharp rise in impact strength. Examination of the fracture surfaces shows that the rise corresponds with the temperature at which full stress-whitening is observed on the fracture surface. Significantly, this second transition, like the first, occurs at a lower temperature in the HIPS containing polybutadiene. The difference between the toughening action of polybutadiene and the copolymer rubber in the higher temperature range is the strongest evidence that a significant amount of energy is absorbed in crack propagation. The simplest explanation of the observations is that the polybutadiene is able to control rapid crack growth at temperatures above $10°$C, whilst poly(butadiene-*co*-styrene) is ineffective in this respect at temperatures below $50°$C, owing to its higher T_g. The energy absorbed in crack propagation depends partly upon the relaxation behaviour of the rubber. Other significant factors include the rate of crazing of the matrix, which also increases rapidly with temperature (*see* Chapter 6), and the volume fraction of rubber particles in the polymer.

10.4 EFFECT OF POLYMER STRUCTURE ON IMPACT STRENGTH

In many respects, the relationships between polymer structure and fracture behaviour are the same in impact as in other tests. The distinguishing feature of impact tests is that the load is applied to the specimen relatively rapidly. An increase in strain rate is important if the shear yield envelope is close to the crazing envelope, so that shear deformation gives way to brittle fracture. This type of transition occurs not only in impact but also in those tests at low strain rate in which rapid crack propagation plays an important part. Fracture mechanics studies on sharply notched specimens of PVC, for example, show that this polymer exhibits brittle fracture from a notch even when the load is applied slowly.

The factors affecting toughness in rubber-modified plastics were discussed in Chapters 7–9. The size, composition and volume fraction of the rubber particles, the adhesion between rubber and matrix, and the mechanisms of deformation of the matrix are all important. A further factor, the relaxation behaviour of the rubber, was discussed in the preceding section. Each of these factors affects impact strength, but as they have already been considered in some detail, the present discussion will be limited to only two fundamental features of structure: rubber content and matrix composition.

10.4.1 Rubber content

The effect of rubber content on impact strength over a range of temperatures is illustrated in Fig. 10.12, which compares SAN with a series of ABS polymers containing respectively 6%, 10%, 14% and 20% of polybutadiene.[16] Below −75°C, all of the materials are brittle, and the rubber has little effect upon impact strength, behaving simply as a hard filler. Just above the T_g of polybutadiene, the impact strength of the ABS polymers begins to rise, and the fracture surfaces show stress-whitening near the notch. The extent of the rise increases with rubber content, as might be expected. In this moderate temperature range, the impact energy is controlled by the amount of yielding around the root of the notch. The apparent fracture surface energy for blunt notch specimens, $\mathscr{G}_B$, increases with rubber content and with temperature, because both factors affect the elongation-to-break.

At about $-10\,°C$, the ABS containing 20 % rubber shows a further sharp increase in impact strength, accompanied by stress-whitening of the entire fracture surface. A similar change occurs at $-5\,°C$ in the ABS with 14 %

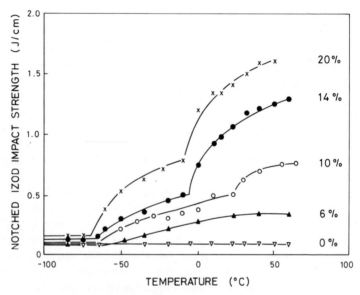

Fig. 10.12 *Notched Izod impact strength over a range of temperatures for SAN copolymer* (0 % *rubber*) *and for a series of ABS polymers containing* 6–20 % *polybutadiene.*

rubber, and at 20 °C in the ABS containing 10 % rubber. The rubber content therefore affects not only the amount of energy absorbed in the impact test, but also the manner in which it is absorbed. At high rubber contents, the energy absorbed in crack propagation at room temperature is higher than the energy stored elastically in the specimen when the crack initiated, so that additional energy is abstracted from the pendulum during the propagation stage. At lower rubber contents, the energy abstracted from the crack during propagation is smaller, and the available elastic energy is sufficient to complete the fracture of the specimen. In fracture mechanics terms, $\mathscr{J}_{IC}$ is a function of rubber content and of temperature. There is a positive feedback effect in crack propagation: as elastic energy is abstracted from the specimen, the crack speed falls and $\mathscr{J}_{IC}$ consequently increases.

Similar relationships between impact strength and rubber content are observed in other polymers. In toughened PVC, the transition from Type II (moderate temperature) to Type III (high temperature) fracture is accompanied by a very dramatic rise in impact strength.[16] Considerable shear deformation is apparent in the high-temperature range, and in some materials the degree of shear yielding is sufficient to suppress crazing almost entirely.

10.4.2 Matrix composition

The role of the rubber particles is to modify the deformation behaviour of the matrix polymer. The properties and composition of the matrix are therefore of paramount importance in determining the impact strength of a rubber-toughened polymer. The first rubber-toughened plastics were based on polystyrene, which is a relatively brittle polymer, but later commercial developments include toughened grades of PVC, polycarbonate and polypropylene (*see* Chapter 1). These latter polymers are capable of exhibiting very high impact strengths.

The reason for adding rubber to apparently tough polymers such as polypropylene is that these materials are notch sensitive. Table 10.1 and Figs. 10.3 and 10.4 illustrate the problem. Toughened polypropylene shows a clear advantage over polypropylene in notched impact tests, and in corresponding service conditions, especially at reduced temperatures. Triaxial tension due to the presence of a notch, a high strain rate and a low temperature all combine to raise the shear yield stress relative to the crazing and fracture stress. The energy absorbed in crack growth consequently falls. However, in the presence of rubber particles the number of crazes formed under a given stress is greatly increased, and toughening results. In relatively ductile materials, the yield envelope is not far above the crazing envelope even under the adverse conditions of the impact test, so that crazing is accompanied by shear yielding, and the increase in toughness on adding rubber is greater than would have been achieved by the crazing mechanism alone. For this reason, the toughest rubber-modified plastics are those based on relatively ductile matrix polymers. In the creep test, these very tough rubber-modified plastics exhibit little if any crazing. The multiple-crazing mechanism operates essentially at high strain rates and in triaxial stress fields, as an alternative to brittle fracture. This combination of properties is quite attractive: in ordinary service, under low static or dynamic loads, the material deforms by shear, avoiding the deleterious effects resulting from craze formation; under impact, when the shear mechanism is unable to operate effectively, multiple crazing comes into play, and catastrophic fracture is avoided.

The effects of matrix ductility upon impact strength are illustrated in Fig. 10.13, which presents data obtained by Deanin and Moshar in notched Izod tests on PVC/ABS blends.[20] Adding PVC to ABS increases the ductility of

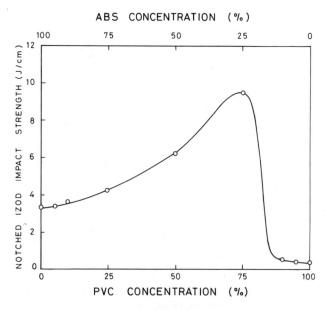

Fig. 10.13 *Notched Izod impact strength of PVC/ABS blends at* 23°C *showing the maximum in toughness achieved by balancing matrix ductility against rubber content (from R. D. Deanin and C. Moshar, ref. 20, reproduced with permission).*

the matrix, but at the same time reduces the rubber content. Over the greater part of the composition range, the dominant factor is the increase in matrix ductility. The impact strength rises with the PVC/ABS ratio, reaching a maximum in the blends containing only 25 % ABS. Further reduction of the rubber content results in a sharp drop in impact strength, reflecting the notch sensitivity of unmodified PVC. The peak in impact strength marks the optimum balance between an adequate rubber content and the preferred combination of multiple crazing with interacting shear deformation.

10.5   AGEING

Ageing is a perennial problem in rubber-toughened plastics, especially those based on butadiene polymers and copolymers. On exposure to sunlight, HIPS, ABS, toughened PVC and related polymers suffer a drastic drop in impact strength. The cause is well known: the ultraviolet component of sunlight initiates photo-oxidation of the rubber, rendering it ineffective as a toughening agent.

The chemical aspects of ageing were discussed briefly in Chapter 3. An oxidised layer is produced in the exposed surface of the polymer, the thickness of the layer increasing linearly with log(exposure time) (see Fig. 3.11).[21] Trans-1,4-diene groups disappear, and hydroxyl and carbonyl groups are formed.[22, 23] Cross-linking of the rubber causes the glass transition to shift from $-80°$ to $20°C$ (see Fig. 3.12). Figure 5.6 illustrates a similar shift due to cross-linking with sulphur.

The degradation affects only the surface layer. The remainder of the material is unaffected, so that the final result is a component with a brittle surface. As discussed in Section 10.2.3, a brittle surface layer only $20\,\mu m$ thick can cause brittle fracture of the whole specimen if it is placed in tension during an impact test. Cracks are generated in the surface layer, where the T_g of the rubber is at or above room temperature, and propagate into the interior of the component.

Bucknall and Street demonstrated that outdoor ageing could be simulated by laminating a brittle polymer layer onto the broad side-face of a standard notched Izod bar.[24] Their results are illustrated in Figs. 10.14 and 10.15. The brittle layer of SAN laminated to ABS not only affects the measured impact energies in the same way as outdoor exposure, but also gives the same type of fracture surface and the same force–deflection traces as the naturally aged samples. The oscilloscope traces show that $\mathscr{G}_{IC}$ is reduced in the surface layer even at very low temperatures. However, the impact strengths of aged and unaged samples are approximately the same up to $-75°C$, presumably because the aged samples absorb some energy after the crack has jumped forward through the embrittled surface.

Between $-75°$ and $0°C$, the brittle surface greatly reduces the Izod impact strength. The effect is similar to a reduction in notch radius. As explained earlier, $\mathscr{G}_{IC}$ is really quite low in this moderate temperature region, so that the toughened polymer is notch sensitive. Ageing has the maximum effect upon fracture resistance under these conditions. The oscilloscope traces again show that the stored elastic energy in the specimen at the point of crack initiation is very low—lower indeed than in the unaged specimen at

−80 °C, and consequently insufficient to cause the crack to propagate across the specimen. A continuous input of energy is required from the pendulum, and the force–deflection curve is long and low. The fracture surfaces show a very flat clear region near the notch, characteristic of slow crack growth at low energy, and a stress-whitened area over the remainder of the aged specimen, indicating energy absorption during crack propagation.

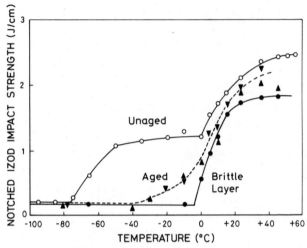

Fig. 10.14   *Effects of surface embrittlement on the notched Izod impact strength of an ABS polymer:* (○) *unaged;* (▼) *after* 635 h *exposure to xenon lamp;* (▲) *after* $3\frac{1}{2}$ *months exposure to English weather;* (●) *unaged ABS laminated with a* 0·5 *mm layer of SAN (from C. B. Bucknall and D. G. Street, ref.* 24, *reproduced with permission).*

Above 0 °C, the aged specimens behave in a very similar manner to unaged specimens. Impact strengths are only slightly reduced, and the oscilloscope traces differ only in being somewhat lower. Fracture surfaces are fully stress-whitened in all cases. The effects of ageing are less apparent over this temperature range because the ABS absorbs energy through multiple-craze yielding over the entire fracture surface, as represented by eqn. (10.24). The conditions for crack initiation are therefore much less significant in determining impact strength than they are between − 75 ° and 0 °C.

The small drop in impact strength at higher temperatures is sometimes misinterpreted. Tests on the ABS polymer at 20 °C could lead to the conclusion that prolonged exposure to sunlight had little or no effect upon

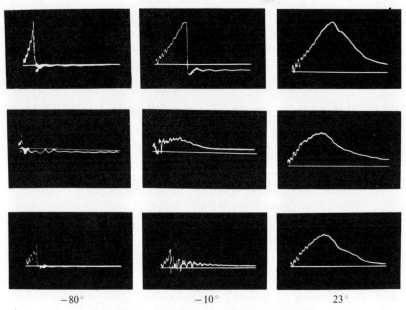

-80°   -10°   23°

Fig. 10.15   Oscilloscope traces of force as a function of time during notched Izod impact tests on ABS showing the effects of ageing and of laminating with SAN on fracture behaviour at three different temperatures. (Top) Unaged; (centre) aged 2 years; (bottom) brittle layer. Base lines are of 3 ms duration (from C. B. Bucknall and D. G. Street, ref. 24, reproduced with permission).

impact strength. If the material is to be used at 20 °C or above, this conclusion causes few problems. If, however, the component is expected to operate at temperatures below 0 °C, the tests at 20 °C serve only to mask the ageing that has taken place, and the problem will become apparent during the service life of the component. In general, it is not safe to infer that a polymer has good ageing resistance simply on the basis of tests at room temperature.

Ageing problems can be alleviated in a number of ways. Carbon black offers considerable protection by screening the polymer from ultraviolet light, but results in a lowering of initial impact strength,[25] and is of course not a universal solution. Other pigments offer a smaller level of protection. Antioxidants are useful over a limited period, but tend to become depleted relatively rapidly. A more radical solution is to replace the diene rubber with a saturated elastomer such as butyl acrylate. Acrylate rubbers show considerably improved resistance to ageing,[26-29] but are not completely resistant, and do not have the low T_g of polybutadiene.

REFERENCES

1. A. E. Lever and J. Rhys, *The Properties and Testing of Plastic Materials*, Temple Press, London, 1957.
2. BS 2782, British Standards Institution, London.
3. ASTM Standards Part 27, American Society for Testing and Materials, Philadelphia, 1971.
4. G. P. Marshall, J. G. Williams and C. E. Turner, *J. Mater. Sci.* **8** (1973) 949.
5. E. Plati and J. G. Williams, *Poly. Engng Sci.* **15** (1975) 470.
6. C. E. Inglis, *Trans. Inst. Naval Arch.* **60** (1913) 219.
7. J. F. Knott, *Fundamentals of Fracture Mechanics*, Butterworth, London, 1973.
8. P. E. Reed and H. V. Squires, *J. Mater. Sci.* **9** (1974) 129.
9. A. C. Morris, *Plast. Polym.* **36** (1968) 433.
10. T. R. Wilshaw, C. A. Rau and A. S. Tetelman, *Eng. Fract. Mech.* **1** (1968) 191.
11. J. G. Williams, *Stress Analysis of Polymers*, Longmans, London, 1973.
12. A. A. Bibeau, *Poly. Engng Sci.* **15** (1975) 294.
13. S. Timoshenko and S. Woinowsky-Krieger, *Theory of Plates and Shells*, 2nd edn., McGraw-Hill, New York, 1959.
14. R. O. Carhart, D. A. Davies and R. Giuffria, *SPE J.* **18** (1962) 440.
15. R. Giuffria, R. O. Carhart and D. A. Davies, *J. Appl. Polymer Sci.* **7** (1963) 1731.
16. C. B. Bucknall and D. G. Street, SCI Monograph No. 26 (1967) 272.
17. H. R. Brown, *J. Mater. Sci.* **8** (1973) 941.
18. K. Fujioka, *J. Appl. Polymer Sci.* **13** (1969) 1421.
19. M. Matsuo, A. Ueda and Y. Kondo, *Poly. Engng Sci.* **10** (1970) 253.
20. R. D. Deanin and C. Moshar, *ACS Poly. Prepr.* **15**(1) (1974) 403.
21. E. Priebe and J. Stabenow, *Kunststoffe* **64** (1974) 497.
22. A. Ghaffar, A. Scott and G. Scott, *Eur. Poly. J.* **11** (1975) 271.
23. A. Casale, O. Salvatore and G. Pizzigone, *Poly. Engng Sci.* **15** (1975) 286.
24. C. B. Bucknall and D. G. Street, *J. Appl. Polymer Sci.* **12** (1968) 1311.
25. P. G. Kelleher and B. D. Gesner, *J. Appl. Polymer Sci.* **11** (1967) 1731.
26. E. Zahn, *Appl. Polymer Symp.* **11** (1969) 209.
27. J. Zelinger and E. Wolfova, *Kunststoffe* **63** (1973) 319.
28. D. W. Langer, *Gummi Asbest Kunst.* **28**·(1975) 88.
29. G. Menzel, *Plastverarbeiter* **26** (1975) 259.

CHAPTER 11

MELT RHEOLOGY AND PROCESSING

Processing characteristics are amongst the most important properties of any polymer, and not least of rubber-toughened plastics. In terms of tonnage, injection moulding is the principal method for processing HIPS, ABS, modified PPO and toughened polypropylene. Extrusion ranks second in importance. In both processes the melt rheology of the polymers is a fundamental factor, which controls the quality of the product. Substantial amounts of extruded HIPS and ABS are subsequently thermoformed. The forming process focuses attention upon a rather different aspect of polymer rheology, namely the extensional viscosity and elasticity of the melt. The elastico-viscous behaviour of the polymer above its glass temperature exerts a strong influence upon the choice of processing conditions and the quality of the product in all three of these processes. Measurements of shear viscosity and elongational viscosity together with data on melt elasticity provide the basic information required in any study of moulding, extrusion or thermoforming.

Users of toughened plastics are interested not simply in the behaviour of the polymer in the processing machinery, but also in the quality of the product, especially the performance of the article in service. Amongst the properties affected by processing conditions and melt flow characteristics are impact strength, flexural strength, dimensional stability and surface gloss. These properties are strongly influenced by molecular orientation in the matrix, and by deformation, aggregation, comminution or migration of the rubber particles. The study of structural changes in two-phase liquids during flow is a branch of *microrheology*, a subject that is discussed in Section 11.3.

11.1   ELASTICO-VISCOUS PROPERTIES

Polymer melts differ from ordinary liquids in a number of ways. The term 'viscoelastic' is often applied to polymers above T_g, but a more correct description is *elastico-viscous*.[1] Elastico-viscosity, or melt elasticity, means that the polymer melt is capable of elastic recovery from an imposed strain. The recovery is usually only partial, but at temperatures just above T_g polymers often exhibit rubber-like behaviour, recovering virtually completely from large imposed strains.

Stress relaxation is a standard technique for measuring elastic properties above the glass transition temperature. A fixed tensile strain ε is applied to the specimen, and the time-dependent stress $\sigma(t)$ is monitored as the specimen relaxes. The stress-relaxation modulus $E(t)$ is then given by

$$E(t) = \frac{\sigma(t)}{\varepsilon} \tag{11.1}$$

Figure 11.1 shows typical stress relaxation data for an ABS polymer over the temperature range $-40°$ to $+120\,°C$.[2] The modulus $E(t)$ is multiplied by T_0/T_e, where T_0 is the reference temperature ($100\,°C$), and T_e is the 'effective temperature', which takes account of free volume effects, and is defined by

$$T_e = \left(T + \frac{\xi(T)}{\Delta\beta_f} \right) \qquad \text{at} \qquad T \geq T_g$$

$$T_e = \left(T_g + \frac{\xi(T)}{\Delta\beta_f} \right) \qquad \text{at} \qquad T \leq T_g$$

where $\xi(T)$ is the frozen free volume at temperature T, and $\Delta\beta_f$ is the temperature coefficient of equilibrium free volume.

The data presented in Fig. 11.1 can be superimposed by shifting each curve horizontally along the logarithmic time axis by a shift factor a_T, to produce a master curve of $\log[E(t)T_0/T_e]$ against $\log(t/a_T)$, as shown in Fig. 11.2. Thus the time–temperature superposition principle is applicable to this set of measurements on ABS, as it is to homogeneous polymers. The superposition principle does not apply generally to two-phase polymer blends,[3] and the observed agreement in the case of ABS therefore deserves comment. Time–temperature superposition fails when two polymers contribute simultaneously to the relaxation of a composite because the shift factors are generally different for the two components of the composite. In ABS, however, the glass transition temperature of the polybutadiene rubber is very low, so that the rubber probably contributes little to the measured

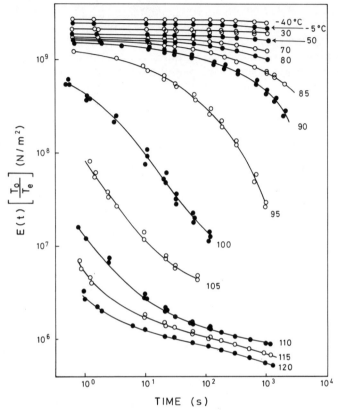

Fig. 11.1 *Stress relaxation curves for ABS;* $T_0 = 100\,^\circ C$ *(from K. C. Rusch, ref. 2, reproduced with permission).*

stress relaxation over the temperature range -40° to $120\,^\circ C$, and simply acts as an elastic filler in the viscoelastic SAN matrix. In other words, the superposition principle works for ABS because it behaves like SAN.

It is convenient to divide the relaxation master curve into four sections as follows:

(i)   the glassy region, in which the modulus of the ABS is high;
(ii) the glass transition region of the SAN matrix;
(iii) the entanglement plateau, where the ABS behaves like an uncross-linked rubber, with temporary cross-links due to entanglements;
(iv) the flow region, in which the melt becomes more like a viscous liquid.

Polymers are heated to temperature region (iii) for thermoforming and to region (iv) for injection moulding and extrusion.

Experimental comparisons of ABS with an equivalent SAN copolymer are lacking. In papers by Scalco et al.[4] and by Rusch,[2] ABS is compared with polystyrene. As shown in Fig. 11.2, the two materials differ in a number of ways in their relaxation behaviour. In the glassy region, the modulus of

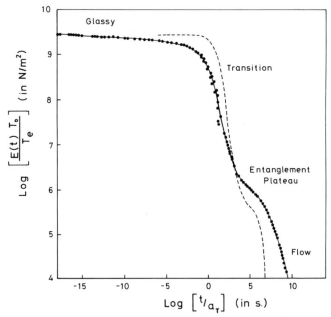

Fig. 11.2   Master curve of reduced modulus against time for ABS at a reference temperature of 100°C. Dashed line shows master curve for polystyrene (from K. C. Rusch, ref. 2, reproduced with permission).

ABS falls more rapidly as the polymer approaches T_g. The transition region is slightly wider in ABS, and at higher temperatures the entanglement plateau is higher than, and approximately twice as broad as, the entanglement plateau in polystyrene. In the flow region, ABS shows a more gradual fall in modulus with temperature. These differences might be due to the presence of the rubber particles,[4] but the published evidence is inconclusive: differences in molecular weight distribution might also be responsible. A more systematic study is required.

One of the consequences of melt elasticity is die swell. The relationship between the two phenomena is imperfectly understood theoretically, but is nevertheless well established.[5] On emergence from the die of an extruder, polymer melts expand laterally, and the extrudate expansion ratio B is an empirical measure of melt elasticity. However, a direct relationship between the two has not been demonstrated for two-phase polymer melts.

Experiments on rubber-toughened plastics show that the expansion ratio B decreases on adding rubber to polystyrene or SAN,[6, 7] indicating that the die swell effect is due largely to the matrix polymer, as in the stress relaxation experiments. Figure 11.3 presents data obtained by Casale *et al.* from ABS

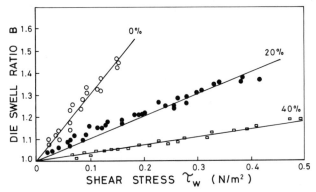

Fig. 11.3   *Die swell of ABS polymers made by blending SAN with 0–40 % SAN-grafted rubber. Polymers extruded by capillary rheometer between 180° and 240°C. SAN $\overline{M}_w = 145\,000$ (from A. Casale et al., ref. 7, reproduced with permission).*

polymers made by blending SAN mechanically with 0–40 % of grafted rubber.[7] The expansion ratio B increases linearly with τ_w, the shear stress at the wall of the extruder die, and decreases with increasing rubber content, as already stated. At a given τ_w, die swell is independent of temperature over the range 180–240 °C. Increasing polydispersity increases B, but, at a given polydispersity, B is effectively independent of the weight-average molecular weight.[7] Since even water exhibits a die swell of 13·5 %, it is not clear why B tends to zero at low shear rates in Fig. 11.3.

At high shear rates, polymer melts exhibit flow instabilities, which are in many cases associated with melt elasticity. The elastic strain energy stored in the deformed melt provides the driving force for the fracture. Saito *et al.* studied flow instabilities in PS, HIPS, SAN and ABS at temperatures between 180° and 240°C, and showed that the critical shear stress at the onset of melt fracture increased on adding rubber.[8] The recoverable shear strain at instability decreased with increasing rubber content.

11.2   VISCOSITY

The term 'viscosity', used without qualification, usually refers to the shear viscosity of a liquid, defined as the ratio of shear stress to shear rate:

$$\eta = \frac{\tau}{\dot{\gamma}} \tag{11.2}$$

A liquid is described as *Newtonian* if its viscosity is independent of shear rate at constant temperature. Many ordinary liquids approximate to Newtonian behaviour. By contrast, most polymer melts exhibit marked

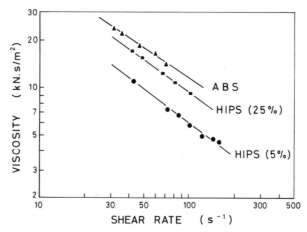

Fig. 11.4   *Non-Newtonian flow curves for ABS and for HIPS polymers of differing rubber contents at* 200°C (*from C. D. Han, ref. 6, reproduced with permission*).

non-Newtonian behaviour at moderate or high strain rates, in the form of a decrease in viscosity with increasing shear rate. This type of non-Newtonian behaviour is described as *pseudoplastic*. Figure 11.4 illustrates pseudoplastic behaviour over a narrow range of shear rates in HIPS and ABS polymers.[6]

Non-Newtonian behaviour considerably complicates the problems of measuring and applying viscosity data. Only in a few specific cases, such as in the cone-and-plate viscometer, is the shear rate uniform throughout the flowing melt. In cases of more general interest in polymer processing, the shear rate, and therefore the viscosity, varies within the melt. In injection moulding and extrusion, for example, the polymer flows through tubes and channels under the influence of a pressure drop. In this type of flow, the shear rate is a maximum at the wall of the tube, and zero in the centre. The behaviour of polymeric melts under these conditions is discussed in a number of standard texts.[9, 10]

The simplest case of flow under the influence of a pressure drop is Poiseuille flow, which is the term used to indicate isothermal flow through a long circular capillary, as in a capillary rheometer. This type of rheometer is widely used for measuring viscosities of polymers, because it is simple in operation and capable of working at high strain rates. The polymer is forced through a capillary of radius r_w and length L under an applied pressure p, and the volume flow rate Q is measured. Several corrections are necessary in order to obtain accurate results. For a Newtonian liquid, the viscosity is given by the Poiseuille equation:

$$\eta = \frac{\pi \, \Delta p r_w^4}{8LQ} \tag{11.3}$$

where Δp is the pressure drop along the length of the capillary. This equation neglects the kinetic energy of the flowing liquid, which also contributes to the flow rate. More accurate expressions include a term to take account of the kinetic energy.

In presenting data on non-Newtonian liquids, the standard practice is to plot the shear stress against shear rate. Some authors ignore the variations in viscosity with radial distance in the capillary, and simply apply the equations developed for Newtonian liquids to their measurements. Usually, however, the results are corrected for non-Newtonian behaviour by means of the Rabinowitsch equation, which relates the shear rate $\dot{\gamma}_w$ at the wall to the pressure drop Δp and the flow rate Q, as follows:[9 − 11]

$$-\dot{\gamma}_w = \frac{1}{\pi r_w^3}\left(3Q + p\frac{dQ}{d\,\Delta p}\right) \tag{11.4}$$

The negative sign indicates that the shear rate is positive when the flow is in the negative directions, and *vice versa*. The shear stress τ_w at the wall of the capillary is given by

$$\tau_w = \frac{r_w \, \Delta p}{2L} \tag{11.5}$$

Both equations are of quite general application to homogeneous fluids, independent of the nature of the fluid.

Over a limited range of shear rates, the viscosity can be represented by a power law of the form

$$\tau = C\dot{\gamma}^n$$

or

$$\eta = \eta_0 \left| \frac{\dot{\gamma}}{\dot{\gamma}_0} \right|^{n-1} \tag{11.6}$$

where C is a constant, and η_0 and $\dot{\gamma}_0$ are the viscosity and shear rate of the liquid in some arbitrary standard state. Han has applied the power-law relationship to melt flow behaviour in rubber-toughened plastics, with $n = 0.29$ for HIPS and $n = 0.38$ for ABS.[6] Figure 11.5 compares the velocity

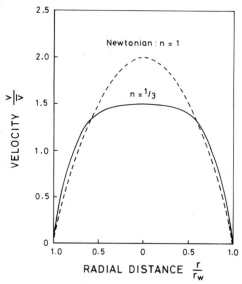

Fig. 11.5   *Velocity profiles in isothermal flow through a tube. Comparison between pseudoplastic and Newtonian behaviour (from J. M. McKelvey, ref. 9, reproduced with permission).*

profile of a non-Newtonian liquid having a low value of n (0.33) with the corresponding profile for a Newtonian liquid ($n = 1$), and clearly illustrates the increase in shear rate near the wall of the capillary in pseudoplastic liquids such as HIPS and ABS melts. In this diagram, the velocity v at a radial distance r from the centre of the tube is compared with the average velocity $\bar{v}$, defined as $Q/\pi r_w^2$.

11.2.1 Factors affecting viscosity
The viscosity of a polymer melt depends upon the conditions under which it is flowing. The temperature, pressure and shear rate all have a strong

influence upon viscosity. Important structural factors include molecular weight, molecular weight distribution and branching of the polymer chain.

The addition of rubber particles produces a sharp increase in the viscosity of a polymer melt.[7, 11] Many manufacturers offset this increase by reducing the molecular weight of the matrix polymer, accepting a small reduction in fracture resistance (below what might otherwise have been achieved) in return for better moulding behaviour in the rubber-toughened product. Thus the molecular weight of the polystyrene matrix in a typical HIPS is usually lower than the molecular weight of a comparable general-purpose polystyrene. For this reason, a simple comparison of HIPS with PS, or ABS with SAN, is often misleading. In assessing the effect of rubber particle content on viscosity, it is necessary to ensure that all other parameters are equal in the series of polymers under investigation. Of the few published papers relating to the melt rheology of rubber-toughened polymers, there appear to be only two that satisfy this requirement.[7, 12]

The work of Casale et $al.$ has already been mentioned in connection with die swell.[7] By blending SAN mechanically with 0–40 % of grafted rubber, these authors ensured that the molecular weight of the SAN component was at least approximately the same in each member of the series, including the material containing no rubber. They prepared three SAN copolymers, with weight average molecular weights between 68 000 and 150 000, which they then used to make three series of blended ABS polymers. Viscosities were measured with a 1·25 mm diameter capillary rheometer, using the Rabinowitsch correction to calculate shear rates at the wall.

Figure 11.6 shows data obtained in experiments on a series of ABS blends, all based upon a single SAN copolymer ($\overline{M}_w = 68\,000$). Results obtained over a range of temperatures between 180° and 240 °C have been correlated in this figure, using the method of reduced variables.[13] The flow curves relating $\log \tau_w$ to $\log \dot{\gamma}$ have been shifted vertically by a factor $\log(\rho_0 T_0/\rho T)$, and horizontally by a factor $\log a_T$, where ρ is density, ρ_0 is the density at an arbitrary reference temperature T_0, and a_T is the ordinary shift factor. The result is a set of master curves, one for each rubber content, at a reference temperature of 210 °C. The increase in viscosity with increasing rubber content is immediately apparent. The effect of temperature upon viscosity follows the Arrhenius equation: $\log a_T$ is linear with $1/T$.

All four of the flow curves presented in Fig. 11.6 are of the same shape, and can be superimposed by shifting them horizontally by a factor $\log a_\phi$, where a_ϕ is a new shift factor, relating flow behaviour to rubber content ϕ_2. This second shift produces a master curve at a reference temperature of

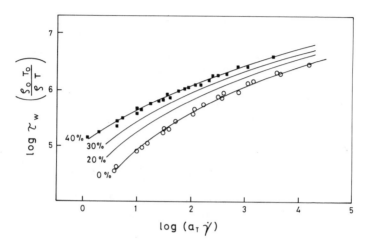

Fig. 11.6　*Master curves at a reference temperature of* 210 °C *for blends of SAN* ($\overline{M}_w = 68\,000$) *with* 0–40% *SAN-grafted rubber* (*from A. Casale* et al., *ref. 7, reproduced with permission*).

210 °C and a reference rubber content of 0%. The shift $\log a_\phi$ is proportional to the rubber content ϕ_2.

Having reduced the data to 210 °C and 0% rubber, Casale *et al.* were able to correlate results for blends containing SAN of various molecular weights by means of another shift factor a_M. Figure 11.7 shows the final result: a single master curve of $\log \tau_w(\rho_0 T_0/\rho T)$ against $\log (a_T a_\phi a_M \dot{\gamma})$ correlates 500 separate measurements made over a range of temperatures, rubber contents and matrix molecular weights. This correlation applies to SAN polymers of similar molecular weight distribution. The data cannot be superposed if the SAN copolymers differ in polydispersity. On the other hand, acrylonitrile contents within the range 20–33% have little effect upon the shape of the flow curve.

These experiments illustrate the problem of achieving a satisfactory balance of properties. Adding 40% graft rubber to SAN of $\overline{M}_w = 68\,000$ has approximately the same effect upon flow behaviour as doubling the molecular weight to 140 000. Increases in viscosity of this kind are not usually of critical importance in making polymers for extrusion, but in the case of injection-moulding grades, the manufacturer must select the molecular weight of the matrix polymer with great care in order to strike the optimum balance between flow characteristics and fracture toughness.

The melt viscosity of ABS and other toughened polymers is determined by the *effective volume concentration* of rubber particles, which includes not just the volume fraction of polybutadiene, nor even the total volume of rubber plus hard sub-inclusions, but also involves the layer of grafted SAN copolymer (or other grafted polymer) attached to the outer surface of the rubber particle. This skin or shell surrounding the particle may be up to 30 nm thick.[12] Its contribution to the effective rubber volume in HIPS is negligible, because the rubber particles are large compared with the skin thickness. In ABS polymers, however, the grafted layer can have a

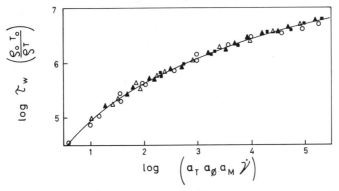

Fig. 11.7   Master flow curve for blends of SAN with 0–40% SAN-grafted rubber. Reference conditions: melt temperature 210°C; rubber content 0%; $\overline{M}_w$ of SAN 68 000 (from A. Casale et al., ref. 7, reproduced with permission).

measurable effect upon viscosity. Figure 11.8 shows results obtained by Huguet and Paxton in experiments on ABS polymers containing 20% polybutadiene.[12] During emulsion polymerisation of the ABS, the monomer–rubber ratio was varied, thus producing grafted layers of various thicknesses. The polybutadiene content of the material was then adjusted to 20% by latex blending. Polymers containing 88 nm particles have higher viscosities than those containing 280 nm particles simply because the smaller polybutadiene particles have a larger surface area to volume ratio, and therefore a larger proportion of grafted SAN copolymer attached to the surface. The viscosities of both ABS polymers increase with graft layer thickness over the range 10–30 nm, as the effective volume of the particles increases. However, decreasing the graft layer thickness below 10 nm also causes the viscosity to increase. The reason is that incompletely grafted particles tend to form aggregates, again increasing the effective volume of the particles. The aggregates trap ungrafted SAN copolymer. Whether

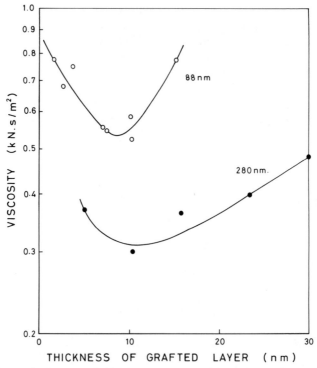

Fig. 11.8   *Melt viscosity of ABS emulsion polymers at* 260 °C *and* 100 s⁻¹. *All specimens contain* 20% *polybutadiene particles of diameter 88 or 280 nm. Viscosity varies with thickness of grafted SAN layer surrounding the particles (from M. G. Huguet and T. R. Paxton, ref. 12, reproduced with permission).*

chemically grafted or not, SAN copolymer that is effectively bound to the polybutadiene particles causes an increase in the melt viscosity of ABS.

11.3   MICRORHEOLOGY

Microrheology is the study of flow behaviour in relation to microstructure. In the melt, rubber-toughened plastics act as suspensions or dispersions of deformable spheres in a non-Newtonian elastico-viscous liquid. In a shear field, the suspended particles can respond in a number of different ways. In addition to deforming and rotating in response to the stresses generated in the surrounding polymer melt, the particles might break down into smaller

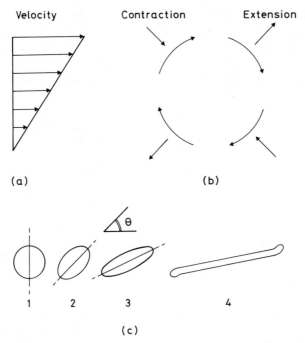

Fig. 11.9 *Shear flow in two-phase fluids: (a) uniform velocity gradient; (b) principal axes of deformation, and direction of rotation; (c) changes in shape during deformation and breakup of suspended droplets (from H. L. Goldsmith and S. G. Mason, ref. 13, reproduced with permission).*

particles, aggregate to form large groups of particles, or migrate away from the region of maximum shear rate. All of these responses have been observed experimentally, and most of them can be explained in terms of the general microrheology of dispersions.[13]

A rigid spherical particle suspended in a liquid subjected to a shear rate of $\dot{\gamma}$ rotates with a period of $4\pi/\dot{\gamma}$. A fluid particle, on the other hand, both rotates and deforms in response to the shear forces within the suspending medium. As described in Chapter 6, a shear deformation consists of an extension in one direction combined with a contraction at right angles to the extension, the direction of maximum extension being at 45° to the shear plane. The fluid particle or drop therefore deforms, against the resistance of viscous and surface tension forces, to an ellipsoid with its longest axis at 45° to the flow direction, as shown in Fig. 11.9. At low shear rates, the longest axis remains in this orientation as the liquid circulates around the drop. At higher shear rates, however, the orientation θ of the principal axis of the drop decreases from $45°$ towards $0°$ (along the flow direction) because the rate of accommodation of the drop shape to the applied stresses is insufficient to maintain the angle of $45°$. At yet higher shear rates, the disruptive forces acting on the drop become greater than the stabilising forces of surface tension, and the drop breaks up. Each of these stages is shown in Fig. 11.9.

Uncross-linked rubber particles fall into the category of fluid drops, and deform, rotate and disintegrate as described above. Baer has demonstrated the reduction in particle size produced by an extruder in HIPS containing uncross-linked, or undercross-linked, grafted natural rubber.[14] Similar reductions in particle size occur in other uncross-linked rubbers. Indeed, the process of mechanically blending rubbers with plastics depends upon the dispersion of the uncross-linked rubber under the shearing action of an extruder or other equipment.

A more typical rubber-toughened polymer contains cross-linked grafted rubber, often incorporating sub-inclusions of the same composition as the matrix. These particles are usually capable of large deformations in the sheared melt, but show little sign of breaking down, even at very high shear rates. Hemispherical particles are sometimes seen in HIPS, but are uncommon, and are almost certainly formed by fracture of the polymer below its T_g, perhaps during pelletising. Most of the rubber particles in HIPS are obviously unaffected by shear during processing, retaining their original identity throughout the extrusion and moulding processes to which they are subsequently subjected.

Deformation of the rubber particles in a flowing melt is of particular interest in connection with processing, because the deformed shapes are partially retained in the moulded or extruded product, especially if the polymer is cooled rapidly. Both the rubber particles and the matrix tend to recover towards the isotropic state whilst the matrix is well above T_g, but the rate of recovery is very low just above T_g, and rapid quenching therefore preserves the material in a very similar state of orientation to that in the melt. Because of the low thermal conductivity of polymers, quenching rates are highest near the surface, and low in the interior. The techniques of optical and electron microscopy described in Chapter 3 are especially suitable for studying rubber particle shapes in items moulded from rubber-toughened plastics. Both optical and scanning electron microscopy enable the investigator to examine a wide area of the specimen relatively quickly, and to assess the pattern of residual orientation. Examples are included in Sections 11.4 to 11.6.

11.3.1 Rubber particle migration

Under certain melt flow conditions, especially during extrusion, rubber particles migrate away from the wall of the channel through which they are flowing, leaving a depleted surface layer, which can be observed in the material on cooling.[15] This type of migration, across the planes of shear in a flowing liquid, is by no means peculiar to polymers, but has also been observed in water drops suspended in silicone oil[16] and in rigid PMMA spheres in a neutrally buoyant liquid.[17]

Two types of radial migration can be distinguished in flowing suspensions. One mechanism applies only to deformable spheres at low values of 'particle Reynolds number' Re_p, defined as follows:

$$Re_p = \frac{2R|v_f - v_p|\rho}{\eta} \tag{11.7}$$

where R is the radius of the particle, v_f and v_p are the velocities of the fluid and of the particle respectively parallel to the shear direction, and η and ρ are the viscosity and density of the suspension respectively. The deformation of the particle produces a perturbation of the flow pattern in the surrounding liquid, resulting in an interaction with the wall which produces an inward radial migration given by[13]

$$-\dot{r} = \frac{\left(\frac{d\dot{\gamma}}{dr}\right)^2 R^4\eta}{\mathscr{T}} \left(\frac{r^2}{r_w^2 - r^2}\right)\left(\frac{99(9\eta_{21}^2 + 17\eta_{21} + 9)(19\eta_{21} + 16)}{2240(\eta_{21} + 1)^3}\right) \tag{11.8}$$

where $\dot{r}$ is the radial migration rate at radius r in a tube of radius r_w in Poiseuille flow, $\dot{\gamma}$ is shear rate, $\mathscr{T}$ is interfacial tension and η_{21} is the viscosity ratio of suspended phase to suspending phase. The migration rate thus increases rapidly with shear gradient, with particle radius and with radial distance from the centre of the tube. Figure 11.10 illustrates this type of migration in a suspension of water droplets in silicone oil.[13]

Another mechanism of particle migration operates at higher particle Reynolds numbers. The effect has its origins in the inertial terms of the flow equation, which are negligible at very low Reynolds numbers. Inertial effects are observed in suspensions of both deformable and rigid particles, but the results are different in the two cases. In Poiseuille flow, deformable particles migrate away from the walls towards the centre of the tube, whereas rigid spheres migrate both inwards from the walls and outwards from the centre, to a radial distance of 0·6r.[13]

Migration effects have never been investigated systematically in rubber-toughened plastics, although they are known to occur. The most probable explanation of the observed migration is the first one outlined above, involving viscosity rather than inertial terms, and relating to low particle Reynolds numbers. Selecting arbitrary values for the parameters of eqn. (11.7), let $R = 0.5\,\mu m$, $v_f - v_p = 1$ mm/s, $\rho = 1$ g/cm^3 and $\eta = 1$ N.s/m^2. The value of Re_p obtained from these figures is 10^{-6}, within the range in which migration is observed due to wall effects in ordinary liquids.

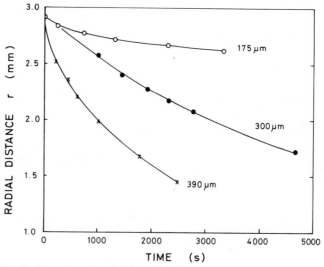

Fig. 11.10 *Radial migration of deformable droplets during Poiseuille flow: results for water droplets of various diameters suspended in silicone oil (from H. L. Goldsmith and S. G. Mason, ref. 13, reproduced with permission).*

One important practical result of particle migration is that viscosities of suspensions, as measured by a capillary rheometer, tend to be spuriously high, especially at high shear rates. There is no available information concerning the magnitude of this effect in rubber-toughened plastics. In most cases it is probably negligible, but the point deserves further investigation.

11.3.2 Necklace formation
The term *necklace formation* is aptly employed to describe the association of particles into long strings in a flowing suspension. The 'necklaces' are aligned parallel to the flow direction. The effect has been observed in

blood[13] and has also been reported in ABS[18] and in HIPS.[19] Necklace formation appears to be related to particle migration. The particles move across the flow lines until a number of them lie together in the same plane. The association is temporary rather than permanent in a flowing suspension, but in rubber-toughened plastics is preserved on cooling the flowing melt. Examples can be seen in Fig. 11.16.

11.3.3 Extensional flow

The rubber particles do not rotate when an isotropic ABS or HIPS melt is subjected to tensile deformation. As the melt deforms, the major principal axes of the ellipsoids remain in the direction of maximum tensile strain rate, so that the particles undergo continuous elongation without rotation. Under these conditions, highly deformed rubber particles are often produced. At the same time, the molecules of the surrounding matrix also extend in the tensile direction, so that the rubber-toughened polymer becomes highly anisotropic in structure.

The level of orientation produced depends upon temperature and strain rate. At high temperatures, the polymer relaxes as it is deformed; but at low temperatures, especially at high strain rates, relaxation effects are minimised, and the material becomes highly anisotropic. The orientation produces strain hardening: the modulus increases with elastic strain in the melt, in a similar manner to the strain hardening observed in ductile polymers during tensile tests below T_g. This change in melt elasticity is usually accompanied by a change in extensional viscosity. Cogswell has observed a decrease in extensional viscosity with applied stress in transparent ABS at each of five temperatures in the range 130–200 °C.[20]

The extensional flow of HIPS at moderate temperatures has been studied by Lai and Holt,[21] who showed that the relationship between true stress σ and true strain ε at an elapsed time t under uniaxial tension could be expressed as follows:

$$\sigma = K t^m \varepsilon^n \tag{11.9}$$

where K is a constant. At a temperature of 122 °C, the measured values of the exponents were $m = -0.33$ and $n = 1.1$ for HIPS.

11.4   THERMOFORMING

Thermoforming is an important commercial process for the production of hollow articles from HIPS, ABS and other toughened plastics. Vacuum-formed refrigerator liners and coffee cups are examples of thermoformed products. The process consists of heating a sheet of the polymer to the temperature of the entanglement plateau (*see* Fig. 11.2), which is usually between $(T_g + 10\,^\circ\text{C})$ and $(T_g + 40\,^\circ\text{C})$, and forming the material by application of vacuum or pressure to one side of the sheet. Forming plugs are often used to assist in the operation.

At the relatively low temperatures employed in thermoforming, the melt behaves in a rubber-like manner, and the thickness of the formed sheet can be predicted with reasonable accuracy using large-strain rubber elasticity theory.[22] In general, the material in the sheet is subjected to biaxial tensile deformation. For example, in blowing a hemispherical bubble from a flat sheet, the material at the centre of the sheet is in balanced biaxial extension. The material at the rim of the hemisphere is in a state of plane strain, since the circumference of the rim does not change during forming. The remainder of the sheet is in an intermediate state between balanced biaxial extension and plane strain. The rubber particles conform to this strain pattern, although the level of residual strain in the sheet may be reduced by relaxation, especially at high forming temperatures.

Differences in thermal history often produce variations in the state of orientation through the thickness of the wall of a thermoformed component. The surface of the sheet cools more rapidly than the interior, especially when it is in contact with a cold mould, so that the surface retains a higher level of orientation than the centre of the section, as illustrated in Fig. 11.11. An additional cause of through-thickness variations is the existence of a temperature gradient between the upper and lower surfaces of the sheet before the polymer is formed. Thick sheets are heated to a uniform temperature in an oven, but thinner sheets are brought to the forming temperature by radiant heaters placed above the upper surface. Consequently, the lower part of the sheet is formed at a rather lower temperature than the upper part. These orientation gradients can cause the formed sheet to warp.[23]

The effects of orientation upon mechanical properties were discussed in Chapters 7–10. In thermoformed components, the most important effect is the reduction in fracture resistance in planes parallel to the orientation direction. This reduction in strength is particularly noticeable in the walls of deep-drawn containers, and other items of similar geometry. Areas of balanced biaxial deformation are stronger.

In any thermoforming process, the sheet will become thinner at some points than at others. Successful forming requires that these areas of local thinning become stabilised, and do not become progressively thinner and then tear. Two factors help to stabilise the sheet in the thinned regions:

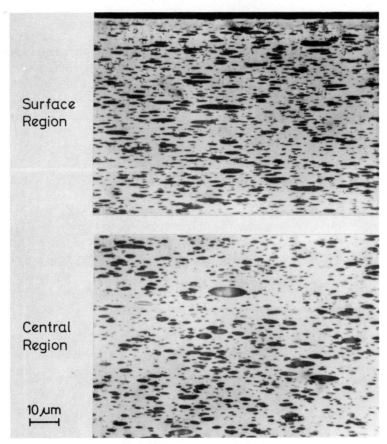

Fig. 11.11 *Vacuum-formed HIPS sheet, showing high orientation in rapidly cooled surface, and lower orientation in centre. Specimen polished by microtoming, etched with chromic acid and observed in reflected light.*

firstly, cooling is accelerated and, secondly, the material begins in some cases to strain harden. Both factors increase the resistance of the material to further thinning, so that an acceptable distribution of thicknesses is achieved.

The principal effect of the rubber phase upon vacuum forming characteristics is to give the surface a matt appearance. Broadening of the entanglement plateau, as discussed in Section 11.1, might also be significant, but has not been studied systematically. The value of rubber particles as strain indicators is demonstrated in Fig. 11.11.

11.5 EXTRUSION

Extrusion is an important process for the production of sheet, pipe and rod. Special extrusion grades of HIPS and ABS are offered by manufacturers for these applications. Since flow rates are relatively low, extrusion grades can have higher viscosities than injection-moulding grades, with the result that rather better fracture toughness can be achieved, by increasing molecular weight or rubber content or both.

Orientation tends to be relatively low in extruded products, although the melt is subjected to both shear and extensional flow through the die of the extruder. The reasons are that melt temperatures are relatively high (approximately 100 °C above T_g), shear rates are low and cooling rates in contact with air on emergence from the die are fairly low.

When orientation is observed in extruded products, and especially in sheet, it is usually due to stretching during the take-off stage of the process, as the material leaves the die slit. At this stage, the temperature of the sheet is falling, so that any imposed strain is more likely to produce residual orientation. Hot-drawing of the sheet during take-off is one method of compensating for die-swell, and may also be used to make sheet of non-standard thicknesses, for which dies are not available. Giuffria demonstrated orientation produced in this way in HIPS sheet extruded at various take-off speeds.[24] Measurements of particle shape and of shrinkage on reheating indicated residual strains of up to 43 % in the extruded sheet.

Stretching during take-off from a sheet extruder approximates to a plane strain deformation, since the hot strip of sheet emerging from the die is constrained by the die behind it and the cooled, rigid sheet ahead of it. Since the deformation is rubber-like, the volume remains constant, so that

$$\lambda_1 \lambda_2 \lambda_3 = 1 \tag{11.10}$$

where λ is draw ratio. Taking 1, 2, and 3 as the width, thickness and length directions of the sheet, eqn. (11.10) reduces to

$$\lambda_2 \lambda_3 = 1 \quad \textit{in plane strain} \tag{11.11}$$

since $\lambda_1 = 1$ for the reasons already given. If there is no relaxation, the ratio a/b of major to minor of the ellipsoidal particles is then given by

$$\frac{a}{b} = \frac{\lambda_3}{\lambda_2} = \lambda_3^2 = (1 + e_3)^2 \tag{11.12}$$

where e_3 is the engineering strain in the extrusion direction. In addition to this plane strain deformation on emergence from the extruder, the polymer also exhibits die swell, which further affects the shape of the rubber particles.

Not all of the orientation in extruded sheet is due to hot drawing after emergence from the die. Blyumental' and co-workers[25] extruded HIPS sheet without applying tension during the take-off stage, and used a thermal shrinkage test to demonstrate residual orientation. Shrinkages of up to 8 % were observed on heating the material at 150 °C for 30 min. Quantitatively, this test is of little value, since the orientation is inhomogeneous, being restricted largely to the surface layers of the sheet. A more satisfactory procedure is to measure shrinkage in microtomed sections cut parallel to

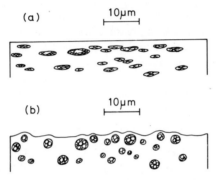

Fig. 11.12   *Mechanism for formation of a matt surface on HIPS: (a) polymer in contact with die wall or mould surface; (b) the same region after relaxation.*

the plane of the sheet,[26] rather than measuring the shrinkage of the sheet as a whole. Nevertheless, the bulk shrinkage test is a useful qualitative method in the present case.

The appearance of the extruded surface is an important consideration in many applications of rubber-toughened plastics. A glossy surface is usually preferred, but the presence of rubber particles gives rise to a distinctly matt finish, especially in HIPS, which has relatively large particles. On emergence from the extruder, the deformed rubber particles in the surface layer of the sheet are able to relax, causing undulations in the surface on a scale comparable with the wavelength of light, as illustrated in Fig. 11.12. The problem is aggravated in thermoforming, when the extruded sheet is reheated, and further relaxation becomes possible.

Several methods have been used to improve surface gloss. The simplest is to laminate a thin foil of polystyrene onto the HIPS sheet as it emerges from the extruder. The disadvantage of this method is that it embrittles the sheet, as discussed in Chapter 10.[27] Perhaps the most interesting technique is to extrude the HIPS at a high shear rate. Under these conditions, the rubber particles migrate inwards from the surface, leaving a rubber-depleted layer up to 10 μm thick.[15] The tendency to migrate is most apparent in the larger particles. Possible reasons for this effect were discussed in Section 11.3.1. Like foil lamination, extrusion at high shear rates embrittles the sheet by producing a brittle layer on the surface, in which cracks can nucleate.

11.6 INJECTION MOULDING

Injection moulding is undoubtedly the main processing technique for rubber-toughened plastics. Large quantities of HIPS, ABS and toughened polypropylene are injection moulded each year. As illustrated in Fig. 11.13, the pattern of orientation in a typical moulded component is quite complex. Fortunately, the local variations in orientation can be studied relatively easily in rubber-toughened plastics, by employing the etching technique described in Chapter 3.[28] Optical microscopy reveals at least three distinct regions in the moulding, which can clearly be distinguished in Fig. 11.13:

Surface region: The particles are aligned with their major axes parallel to the flow direction. The ellipticity of the particles is greatest at the surface, and decreases with depth below the surface.

Shear region: The particles are also elliptical in section, but the major axes are tilted towards the centre of the flow channel.

Central region: The rubber particles are approximately spherical, and there is little evidence of orientation.

The development of this pattern is explained in a paper by Tadmor.[29] Orientation develops in two stages. In the first stage, material in the flow front is subjected essentially to extensional deformation as it advances into the mould. On contact with the cold mould, it solidifies, constricting the flow channel and freezing in the orientation. In the second stage, the remaining material flows between the solidified layers. As there is a temperature gradient between the centre of the channel and the solidified layer, the Poiseuille equations do not apply.

The development of orientation in the advancing flow front is illustrated in Fig. 11.14. There is a 'fountain effect', in which particles from the centre of the channel decelerate as they approach the slower moving air interface at the tip of the flow front, and elongate as they move outwards towards the wall of the channel.[30] This orientation is retained through rapid cooling on contact with the mould wall, especially in the surface; some relaxation

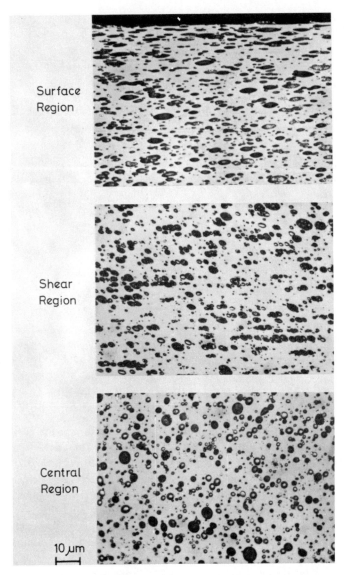

Surface
Region

Shear
Region

Central
Region

10 μm

Fig. 11.13   *Orientation patterns in injection-moulded HIPS. Flow direction from left to right. Rubber particles are elongated parallel to flow axis in the surface region, but at an angle of* 45° *in the shear region; the central region is essentially isotropic. Microtomed block etched with chromic acid and observed in reflected light.*

COLD WALL

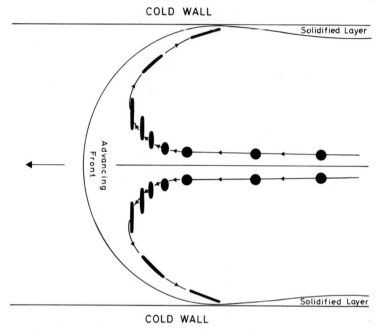

COLD WALL

Fig. 11.14   *Schematic representation of flow pattern in injection moulding showing deformation of initially spherical particles (from Z. Tadmor, ref. 29, reproduced with permission).*

occurs in the sub-surface layers. At high melt temperatures or low injection rates, relaxation also takes place in the advancing flow front, so that the residual orientation in the quenched surface of the moulding is reduced. During subsequent filling of the mould, the thickness of the solidified layer slowly increases, as the cube root of time.[31] The velocity profile across the channel consequently changes with time, as illustrated in Fig. 11.15.[32]

The velocity gradients shown in Fig. 11.15 explain the orientation patterns in the shear region and the central region of the moulding (*see* Fig. 11.13). The shear rate is a maximum not at the wall of the flow channel, as in isothermal flow, but just below the solidified skin. In the shear region, the material is beginning to cool towards T_g through conduction, and the orientation of the rubber particles in the flowing melt is consequently preserved. As expected in shear flow, the axes of the particles lie at or about 45° to the flow direction. In the central region, on the other hand, shear rates are low or zero, and the orientation of the rubber particles in the flowing melt is correspondingly low. Any orientation that might be present is lost by relaxation during cooling, especially if the melt temperature is high.

Molecular orientation in the surface is a major cause of fracture problems in injection-moulded items. The orientation is usually unidirectional and parallel to the flow direction in the mould. Its effects are particularly evident in thin mouldings, in which the surface layers represent

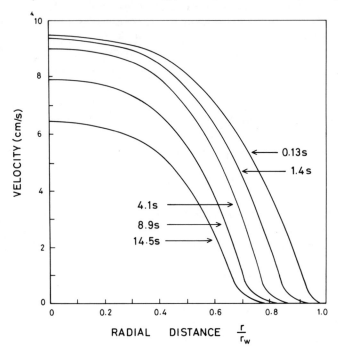

Fig. 11.15   *Velocity profiles for radial flow into a disc-shape cooled mould at a fixed location and at various times (from C. G. Gogos, ref. 32, reproduced with permission).*

a substantial fraction of the total thickness. In standard mechanical tests, including tensile, tensile impact, Izod, Charpy and flexural tests, surface orientation tends to raise the measured strength of the polymer because the specimens are stressed along the orientation direction. However, tests perpendicular to the flow direction reveal much lower strengths in injection-moulded HIPS and ABS components, as in other polymers.[33, 34] For flat, plate-like mouldings, the most satisfactory method of studying fracture resistance is perhaps the falling dart test, which produces biaxial tension in the surface of the plate, causing fracture in the weak direction, wherever it may lie. Falling dart results correlate well with tests specifically designed to stress the component across the pre-determined orientation direction. As already indicated, the anisotropy of strength is greatly reduced by raising the temperature of the melt.

Another cause of weakness in injection mouldings is the formation of 'weld lines' or 'knit lines' where two flow fronts meet. The effect is easily demonstrated by injection moulding a tensile bar with double end-gating, so that the flow fronts meet to form a weld line in the middle of the bar. Comparisons with bars moulded in the ordinary way show that the weld considerably reduces the strength and elongation-at-break of an ABS moulding.[35] A contributory cause of this weakness is the presence of a surface notch at the weld line, which acts as a stress concentrator. Some increase in strength is obtained on machining away the notch, but weakness is still apparent.[35] Possible causes are transverse orientation at the weld line, and incomplete adhesion between the two flow fronts.

Weakness parallel to the surface is sometimes found in injection-moulded HIPS and ABS, especially near the gate. The cause of this type of weakness is illustrated in Fig. 11.16. In the region of maximum shear rate, the particles have formed long strings or 'necklaces' parallel to the flow direction. This structure allows the moulding to delaminate, and was termed the 'delamination layer' by Kato.[18] Necklace formation is observed most clearly at high melt temperatures, and appears to be related to similar phenomena in blood and other non-polymeric suspensions (see Section 11.3.2). Although the 'necklaces' lie parallel to the flow direction, the axes of the rubber particles of which they are comprised lie at about 45° to the flow, as expected in shear flow of deformable particles.

The surfaces of HIPS and ABS injection mouldings sometimes show blemishes that are described variously as scars, blushes and surface fractures. These defects, which occur especially near the gate to the mould cavity, appear to be due to tearing of the cooled layer in contact with the mould: failure occurs when the shearing stress exerted by the melt exceeds the cohesive strength of the surface layer plus the polymer-to-metal frictional forces at the mould surface. Microscopy shows that the fracture is brittle, indicating failure below T_g in the outer surface of the moulding. The defect is not usually observed in homogeneous polymers such as polystyrene and SAN. Ballman et al. found that the tendency to fracture during moulding is high when the melt viscosity and shear rate are high, and when the melt elasticity and strength of the polymer are low.[36] A high coefficient of friction between polymer and metal reduces the tendency to fracture. The increased incidence of surface fracture in rubber-toughened

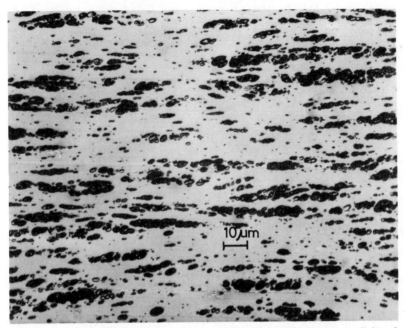

Fig. 11.16 *Association of rubber particles to form 'necklaces' lying parallel to the flow direction in injection-moulded HIPS. Microtomed block etched with chromic acid and observed by reflected light.*

plastics is explained by the effect of rubber upon the properties listed. Rubber particles raise melt viscosity and lower melt strength and elasticity.

As in extrusion, the surface quality of injection-moulded HIPS and ABS is affected by the presence of the rubber particles. In general, ABS mouldings have a glossier appearance than HIPS mouldings because the particle size is smaller in ABS. Gloss is determined by the extent to which the particles disturb the surface of the moulding by relaxing from their deformed state, and can be improved significantly by increasing the pressure in the mould cavity,[37] which also reduces the extent of pitting due to release of volatile constituents of the ABS.

REFERENCES

1. M. Reiner and G. W. Scott Blair, in *Rheology*, Vol. 4, F. R. Eirich (ed.), Academic Press, New York, 1967, p. 461.
2. K. C. Rusch, *J. Macromol. Sci. Phys.* **B2** (1968) 421.

3. D. Kaplan and N. W. Tschoegl, in *Recent Advances in Polymer Blends, Grafts and Blocks*, L. H. Sperling (ed.), Plenum Press, New York, 1974, p. 415.
4. E. Scalco, T. W. Huseby and L. L. Blyler, *J. Appl. Polymer Sci.* **12** (1968) 1343.
5. J. R. A. Pearson, in *Polymer Science*, A. D. Jenkins (ed.) North Holland, Amsterdam, 1972, p. 443.
6. C. D. Han, *J. Appl. Polymer Sci.* **15** (1971) 2591.
7. A. Casale, A. Moroni and C. Spreafico, *ACS Poly. Prepr.* **15**(1) (1974) 334.
8. Y. Saito, T. Sakita and T. Yuki, *Kobunshi Ronbunshu* (Engl. edn.) **3** (1974) 1705.
9. J. M. McKelvey, *Polymer Processing*, Wiley, New York, 1962.
10. S. Middleman, *The Flow of High Polymers*, Interscience, London, 1968.
11. H. Kubota, *J. Appl. Polymer Sci.* **19** (1975) 2299.
12. M. G. Huguet and T. R. Paxton, in *Colloidal and Morphological Behaviour of Block and Graft Copolymers*, G. E. Molau (ed.), Plenum, New York, 1971, p. 183.
13. H. L. Goldsmith and S. G. Mason, in *Rheology*, Vol. 4, F. R. Eirich (ed.), Academic Press, New York, 1967, p. 86.
14. M. Baer, *J. Appl. Polymer Sci.* **16** (1972) 1125.
15. R. Giuffria, R. O. Carhart and D. A. Davies, *J. Appl. Polymer Sci.* **7** (1963) 1731, S51.
16. H. L. Goldsmith and S. G. Mason, *Nature* **190** (1961) 1095.
17. G. Segré and A. Silberberg, *Nature* **189** (1961) 209.
18. K. Kato, *Polymer* **9** (1968) 225.
19. C. B. Bucknall, I. C. Drinkwater and W. E. Keast, *Polymer* **13** (1972) 115.
20. F. N. Cogswell, ACS Symposium on Fibre and Yarn Processing, Philadelphia, April 1975.
21. M. O. Lai and D. L. Holt, *J. Appl. Polymer Sci.* **19** (1975) 1805.
22. J. G. Williams, *J. Strain Anal.* **5** (1970) 49.
23. J. C. Reid, C. B. Bucknall and I. C. Drinkwater, unpublished observations.
24. R. Giuffria, *J. Appl. Polymer Sci.* **7** (1963) 333.
25. M. G. Blyumental', V. P. Volodin, V. V. Lapshin and M. S. Akutin, *Soviet Plastics* **8**(8) (1966) 26.
26. G. Menges and G. Wubken, SPE 31st ANTEC (1973) 519.
27. R. O. Carhart, D. A. Davies and R. Giuffria, *SPE J.* **18** (1962) 440.
28. C. B. Bucknall and I. C. Drinkwater, *Polymer* **15** (1974) 254.
29. Z. Tadmor, *J. Appl. Polymer Sci.* **18** (1974) 1753.
30. W. Rose, *Nature* **191** (1961) 242.
31. I. T. Barrie, *Plast. Polym.* **38** (1970) 47.
32. C. G. Gogos, quoted in ref. 29.
33. H. Keskkula and J. W. Norton, *J. Appl. Polymer Sci.* **2** (1959) 289.
34. R. A. Horsley, D. J. Lee and P. B. Wright, SCI Monograph No. 5 (1959) 63.
35. E. M. Hagerman, *Plast. Engng* **29**(10) (1973) 67.
36. R. L. Ballman, R. L. Kruse and W. P. Taggart, *Poly. Engng Sci.* **10** (1970) 154.
37. T. F. Reed, H. E. Bair and R. G. Vadimsky, *ACS Poly. Prepr.* **14**(2) (1973) 1074.

CHAPTER 12

ELECTROPLATING

Electroplating of plastics is a well-established commercial process, especially in the automobile industry, where it is used to manufacture radiator grilles, hub caps, lamp housings, door handles, body trim and other items that were previously made in steel or die-cast zinc-based alloy. Other applications include domestic appliances, plumbing, jewellery and furniture parts. The market in electroplated plastics is dominated by ABS, with some competition from rubber-toughened PPO. A number of homopolymers, including nylons, polypropylene, polysulphone and acetal resin, are also plated.[1]

Electroplated plastics offer advantages over plated metals on the one hand, and unplated plastics on the other. Plastics have replaced metals mainly because of lower production costs, even though the basic price of the polymer is usually higher. The operations involved in moulding and plating ABS are simpler and cheaper than the equivalent operations with metal parts. Weight savings are another advantage, which has been exploited especially in the American automobile industry, which is following a trend towards smaller and lighter cars. In addition, injection moulding offers greater flexibility in design, which is further extended by the ability to plate components in selected areas only. Finally, the problem of corrosion is reduced, although not completely eliminated: whilst the polymer is resistant, copper coatings in particular are not.[2]

Apart from the purely decorative aspect, electroplating confers a number of advantages upon the plastics component. Abrasion resistance is greatly improved—a considerable gain in applications such as radiator grilles. The polymer is shielded from ultraviolet light, thus avoiding ageing problems in outdoor applications. Furthermore, the plated metal layer, although thin, substantially increases the rigidity and strength of the moulding, especially when the shape is broad and flat. There are, of course, problems. Plastics will not withstand high temperatures, and are prone to warping even at moderate temperatures. ABS is preferred to polypropylene for electroplated items largely because it has better dimensional stability. One of the disadvantages of plating ABS is that it tends to embrittle the polymer. However, the most common problem with electroplated plastics is peeling of the metallic layer.

12.1 OUTLINE OF ELECTROPLATING PROCESS

Electroplating processes vary in detail, but are broadly similar in outline. There are three basic stages: etching, chemical deposition of metal and electro-deposition of metal. The following scheme illustrates the manner in which these processes can be carried out:

Cleaning: Degrease in mild alkaline cleaner for 3 min at 65 °C. Wash in water. Dip into dilute sulphuric acid at room temperature to neutralise any remaining alkali.[3]

Etching: Immerse in oxidising acid etch for 5 min at 70 °C. Rinse. Composition of etch: H_2SO_4, H_2O, CrO_3.

Sensitisation: Immerse for 1–2 min in $0.05M$ $SnCl_2/0.5M$ HCl solution at room temperature. Rinse. Stannous ions are adsorbed onto the etched surface at this stage.

Activation: Immerse for 1–2 min in $0.002M$ $PdCl_2/0.1M$ HCl solution at room temperature. Rinse. The palladous ions are reduced to palladium metal by the adsorbed stannous ions, which become oxidised to stannic.

Electroless   Immerse for 10–15 min at room temperature in the
metal deposition: following solution:

  10 g/litre copper sulphate;
  10 g/litre sodium hydroxide;
  10 ml/litre formaldehyde solution;
  50 g/litre sodium potassium tartrate.

The tartrate forms a complex with the cupric ion. Cupric ions are reduced by the formaldehyde to metallic copper, which nucleates on the palladium and forms a continuous thin layer on the ABS.

Electroplating: Coat with 5–25 μm copper, 5–25 μm nickel, and
                           0·25 μm chromium, using standard plating baths.
                           The continuous conducting copper layer formed
                           during the previous stage acts as the initial electrode.

The process outlined above was used during the early development of
electroplating as a commercial process for ABS polymers, and has since
been modified in detail, although the basic steps remain the same. For peel
strength measurements, a 50 μm thick layer of copper is plated onto the
polymer in order to ensure sufficient mechanical strength in the peeled strip.

The critical stage in the process is etching, which is discussed in detail
below. Radioactive tracer studies show that significant amounts of
chromium, of the order of 1–2 $\mu g/cm^2$, remain on the surface after etching,
possibly in the form of a complex with polar groups on the surface.[4] This
quantity is equivalent to a layer about 2 nm thick. Similar quantities of tin
are adsorbed during sensitisation. Finally, in the activation stage, small
particles of metallic palladium about 4 nm in diameter are deposited on the
surface.[5] The tin and palladium provide catalytic nuclei for the chemical
deposition of copper, which forms a continuous layer approximately 50 nm
thick. This continuous copper layer is the electrode in the electroplating
process proper.

12.1.1 Etching

Etching of ABS polymers by chromium trioxide was discussed in Chapter 3
as a method for preparing specimens for microscopy. Chromium trioxide is
a strong oxidising agent, which attacks the rubber particles selectively,
leaving etch pits in the surface of the moulding. The styrene–acrylonitrile
copolymer matrix is little affected by CrO_3 alone, but is attacked by a
chromic–sulphuric acid mixture. The attack on SAN is slower than that on
the rubber, with the result that interconnecting channels are formed,
reaching into the surface layer of the moulding, as shown in Fig. 12.1.
Control of etching is essential. Times, temperatures and compositions of
etches must be chosen to give the optimum structure in the etched surface.
Figure 12.1a illustrates under-etching, in which the pattern of channels is
barely developed, whilst Fig. 12.1c shows the effect of over-etching, which
results in an excessive removal of material from the surface layer, and
consequently a weakening of the structure. The preferred etch pattern is
shown in Fig. 12.1b: the acid has penetrated sufficiently to provide a good
mechanical key for the metallic layer.

Figure 12.2 is a section through an etched ABS moulding, at the stage
immediately before electroplating.[6] The electron micrograph shows how

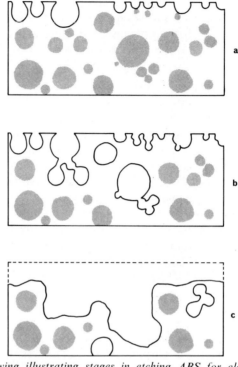

Fig. 12.1   *Drawing illustrating stages in etching ABS for electroplating:* (a)
underetched; (b) *optimum etching;* (c) *overetched.*

the etch has penetrated the surface, attacking the rubber particles, and
forming narrow interconnecting channels through the matrix. Some of the
channels are obviously out of the plane of the section. Chemically deposited
copper can be seen lining the newly etched surfaces, ready to receive the
electrodeposited copper at the next stage of the process. The result will be a
mechanically interlocked network of plastic and metal approximately 1 μm
thick, which provides the basis for the adhesion between the two materials.

The depth of etching depends not only on the temperature and duration
of etching, and the composition of the etch, but also on the structure of the
polymer. The important factors are:

(a)   volume fraction of rubber;
(b)   chemical composition of matrix;
(c)   degree of orientation in the moulded surface;
(d)   rubber particle size.

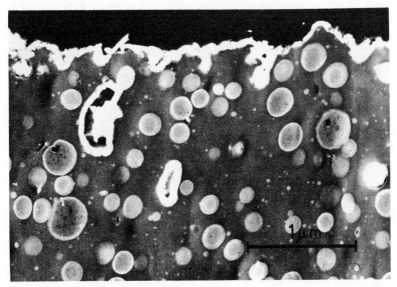

Fig. 12.2 *Osmium-stained section through the surface of an etched ABS moulding after electroless deposition of copper. Copper forms a layer about 50 nm thick on the surface and fills etch cavities to a depth of 1·5 μm below surface (from K. Kato, ref. 6, reproduced with permission).*

Since the attack on the rubber is rapid, the rate of etching increases with rubber content. However, if the rubber content is too high, the surface may become over-etched. In an ordinary ABS polymer, the rate-controlling factor is the attack on the matrix. After oxidising a rubber particle, the acid slowly etches the surrounding matrix, enlarging the etch cavity until it reaches a neighbouring rubber particle, and thus penetrates further into the interior. This does not happen in HIPS, because polystyrene is resistant to the etch. The activity of the chromic–sulphuric acid mixture towards ABS increases with acrylonitrile level: whilst polystyrene is almost inert, SAN copolymer of the 'atropic' composition is etched satisfactorily within 5 min at 70 C. Blends of HIPS with poly(2,6-dimethyl-1,4-phenylene oxide) provide an even more interesting example of the effects of matrix reactivity. HIPS is unsuitable for electroplating by existing processes, because only the rubber particles in the surface are etched by the standard reagents, but a blend of HIPS with PPO, which is attacked by the etch, provides an excellent basis for electroplating.

Molecular orientation is a third factor affecting etching behaviour.

Molecular orientation parallel to the surface of a moulding reduces the rate of etching, but is not usually a major consideration in practice, as orientation is undesirable for other reasons, and components to be electroplated are therefore moulded under conditions specifically designed to reduce orientation.[7]

12.2   PEEL STRENGTH

The standard method for measuring the resistance of the metallic layer to peeling is the Jacquet test, which is carried out on plated plaques using a tensile testing machine.[8] Two parallel grooves 2·5 cm apart are cut through the plated layer into the surface of the polymer, a third cut is made to connect these grooves, and tongue of metal is prised from the plaque. The force required to peel the metal strip from the polymer is then measured, using a roller system to ensure that the surface of the plaque remains at $90°$ to the applied force.

Peel strengths in the region of 20 N/cm are generally regarded as satisfactory. Components plated to this standard give good performance in service. A peel strength of 20 N/cm width of strip corresponds to the expenditure of 200 kJ/m^2 of energy in detaching the metal coating from the polymeric base. Failure occurs not at the metal–polymer interface, as might be expected, but within the polymer surface. This point can be demonstrated in a number of ways. With the aid of a diamond knife and the exercise of considerable care, it is possible to section through a partly peeled plaque near the tip of the tear, and to observe in the electron microscope a thin layer of polymer adhering to the underside of the peeled copper strip, as illustrated in Fig. 12.3.

A simpler method is to dissolve the peeled copper strip in dilute nitric acid. If the bond between the metal and polymer is a good one, removal of the copper leaves behind a continuous film of polymer about 1 μm thick.[4] If, on the other hand, the bond is a poor one, dissolving away the metal leaves either a ragged, broken polymer film or none at all. Weak failures of this type are observed when the polymer is inadequately etched: a progressive deterioration in the quality of the adherent polymer film is seen, for example, when either the time or the temperature of etching is reduced from the recommended value.

These experiments show that the function of the etch is to provide a good bond for the copper coating, so that peeling occurs by cohesive failure within the polymer rather than by loss of adhesion at the polymer–metal interface. The bond appears to derive largely from mechanical interlocking between the two materials, but it is possible that there is some chemical bonding.

The peeled surface of an ABS electroplated plaque usually shows evidence of stress-whitening. This whitening, which is not observed when peel strengths are low, is evidence of plastic yielding by multiple craze

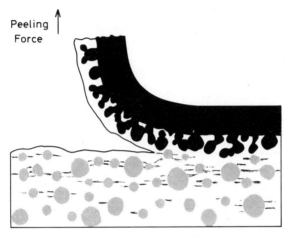

Fig. 12.3   Drawing illustrating peeling of the metal coating from an electroplated ABS plaque showing polymeric layer adhering to underside of metal.

formation during the peeling process. There is an obvious connection between peel strength and fracture toughness. Both measure the work done in yielding and fracture at the tip of a propagating crack, and have relatively high values in the case of ABS because the added rubber increases the resistance of the material to crack propagation. The peeling test differs from an ordinary fracture experiment in a number of ways: the presence of the metal layer inhibits yielding on one side of the propagating tear, the crack tip radius is different, and the stressing system is different. Nevertheless, the similarities are more striking than the differences. In particular, the $\mathscr{G}_{IC}$ values obtained in fracture mechanics tests, which are in the region of $50 \, \text{kJ/m}^2$ for ABS polymers,[9] are of the same order as the '$\mathscr{G}_{IC}$ (peeling)' values measured in the Jacquet test.

During peeling, the metal layer is deformed plastically in flexure, an effect which contributes both directly and indirectly to the measured peel strength. At the adhesion edge, the underside of the strip is in tension, and the upper side is in compression. The strip straightens by a combination of elastic and plastic deformation as it is pulled from the surface, and the stresses are thereby reversed. On removal of the tension, the strip curls elastically, with the original upper side concave. The effects of this deformation in the metallic layer are discussed briefly by Atkinson *et al.*[4] A more detailed analysis of the problem has been published by Duke.[10] The peel strength increases with the thickness of the metal layer not simply because more energy is absorbed in plastic deformation of the metal, but also because the increased radius of curvature of the metal strip at the tip of the tear leads to an increase in the amount of crazing in the polymer. There is some evidence that the thickness of the detached polymer film increases with the thickness of the metal layer, as a result of the change in curvature. In effect, the crack becomes blunted, and the energy absorption therefore increases.

It is clear that after making due allowance for energy absorbed plastically by the copper, the peel strength essentially measures the fracture resistance of the surface layer of the ABS moulding. In the peel test, the tensile direction is normal to the surface, and therefore normal to the orientation

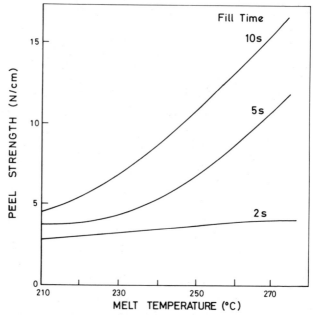

Fig. 12.4   *Curves showing the effects of mould filling time and melt temperature on the peel strength of electroplated ABS injection mouldings. Mould temperature 40 C (from P. A. M. Ellis, ref. 11, reproduced with permission).*

direction. It follows that maximum peel strength is achieved when the surface layer is isotropic, and that injection moulding conditions should be chosen to approach this condition as closely as possible.

The relationships between peel strength and injection moulding conditions have been investigated by Ebneth and Moll[7] and by Ellis.[11] The two most important factors are melt temperature and injection rate. Mould temperature and injection pressure also affect adhesion, but to a lesser extent. Figure 12.4 illustrates the effects of melt temperature and injection rate in ABS:[11] increasing the temperature and reducing the flow rate results in a lowering of surface orientation, and therefore a higher peel strength.

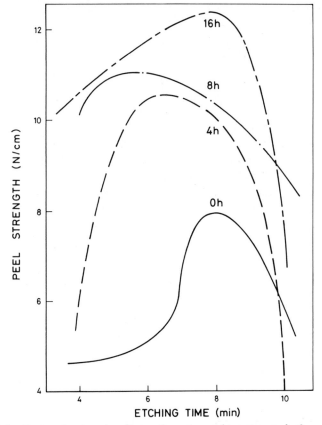

Fig. 12.5 Curves showing the effects of varying etching time and of annealing at 80 °C for 0–16 h on the peel strength of ABS injection mouldings (from H. Ebneth and R. A. Moll, ref. 7, reproduced with permission).

Wall thickness is another important factor: molecular orientation tends to be higher in thin-walled components, so that peel strengths are lower.

Factors affecting peel strength must be balanced against other considerations, especially processing costs and surface quality. Electroplating emphasises moulding defects such as sink marks and weld lines, and it is necessary to choose moulding conditions that avoid such defects. Very high melt temperatures and low injection rates tend to cause sink marks, and in order to eliminate this problem the recommended conditions for moulding ABS are a melt temperature of 240–250 °C, a mould temperature of 70–80 °C and moderate injection rates. Where shape and other considerations make it impossible to mould the article without high surface orientation, the peel strength can be increased significantly by annealing the ABS moulding below T_g before plating. Figure 12.5 illustrates the way in which etching and annealing times affect the bonding between the metal coating and the ABS moulding.

12.3 THERMAL CYCLING

Thermal cycling is a primary cause of failure in electroplated plastics, especially in automobile parts, which are frequently subjected to major fluctuations in temperature within relatively short periods. If the adhesion between metal and polymer is inadequate, thermal stresses can lead to blistering. Alternatively, the polymer might crack without separating from the electroplated layer, causing a crease in the metal, or the metal itself might crack.[3]

The basic reason for the failures is the difference in linear coefficient of thermal expansion between metals and polymers. Typical values are $1.7 \times 10^{-5} °K^{-1}$ for copper and $9.0 \times 10^{-5} °K^{-1}$ for ABS, approximately a factor of five. Differential expansion of the two layers inevitably sets up stresses in electroplated components.

Various thermal cycling tests have been devised to assess performance of plated articles. The following is a typical cycle for ABS:[3]

(a) place plated component in oven at 93 °C for 1 h;
(b) allow to cool in air at 20 °C for 30 min;
(c) place in cold box at −40 °C for 1 h;
(d) warm in air at 20 °C for 30 min.

The cycle is repeated a predetermined number of times, usually between 3 and 10, and the component is then inspected visually. The upper temperature of 93 °C (200 °F) is chosen as being just below the T_g of ABS, and $-40\,^{\circ}\mathrm{C}$ $(-40\,^{\circ}\mathrm{F})$ represents the lowest temperature likely to be encountered by automobiles in inhabited regions of the globe.

The properties tested by thermal cycling are very similar to those characterised by the Jacquet peel strength measurement. The advantage of the thermal cycling test is that it can be applied to any plated article, regardless of shape, whereas the Jacquet test is restricted to flat mouldings. Thermal cycling is therefore suitable for quality control testing, whereas the Jacquet test is more useful for research and development work.

The most important factor determining resistance to failure in thermal cycling is the adhesion between metal and polymer, which is also measured by the peel test. A peel strength of 8 N/cm appears to be sufficient to ensure adequate performance in thermal cycling under ordinary conditions.[3] Poorly etched mouldings, which have very low peel strengths, will sometimes blister during the plating operation.

The properties of the metallic layer are also important. Chromium is very brittle, and nickel is also prone to fracture at low strains. A typical value of elongation at break for bright nickel is $0.5\,\%$, compared with $40\,\%$ for bright copper. For this reason, copper is often preferred as an undercoat in contact with the plastics layer even when a nickel finish is required. The alternative is to deposit 'semi-bright' nickel, which is a slightly porous and more ductile form of the metal, as the undercoat. The tendency to failure in thermal cycling appears to be greatly reduced when the metal in contact with the ABS is fairly ductile, and can accommodate the thermal stresses and strains by yielding to a limited extent.

More fundamental solutions to the problem of thermal cycling failures have been proposed. In addition to improvements in plating procedure and component design, modified polymers having lower coefficients of expansion have been developed. Weston showed that considerable improvements in resistance to thermal cycling could be achieved by adding inorganic titanate fibres to ABS.[12] The fibres were single crystal whiskers with diameters below 1 μm, and aspect (length/diameter) ratios of about 40, which were able to reduce the linear expansion coefficient of ABS by a factor of two when dispersed randomly in the polymer at a volume loading of $6\,\%$. The modification in properties was due to the low coefficient of expansion and the high Young's modulus of the titanate crystal.

REFERENCES

1. C. O. Port, *Mod. Plast. Int.* **4**(5) (1974) 70.
2. G. D. R. Jarrett, *Trans. J. Plast. Inst.* **35** (1967) 561.

3. T. E. Such and C. Baldwin, *Trans. J. Plast. Inst.* **35** (1967) 553.
4. E. D. Atkinson, P. R. Brooks, T. D. Lewis, R. R. Smith and K. White, *Trans. J. Plast. Inst.* **35** (1967) 549.
5. A. Rantell, *Poly. Paint Colour J.* **163** (1973) 831.
6. K. Kato, *Polymer* **8** (1967) 33.
7. H. Ebneth and R. A. Moll, *Trans. J. Plast. Inst.* **35** (1967) 543.
8. P. A. Jacquet, *Trans. Electrochem. Soc.* **66** (1934) 393.
9. E. Plati and J. G. Williams, *Poly. Engng Sci.* **15** (1975) 470.
10. A. J. Duke, *J. Appl. Polymer Sci.* **18** (1974) 3019.
11. P. A. M. Ellis, *Trans. J. Plast. Inst.* **35** (1967) 537.
12. N. E. Weston, *SPE J.* **28**(12) (1972) 37.

SI UNITS CONVERSION TABLE

Quantity	To convert from	to	multiply by
Compliance	m^2/N	in^2/lbf	6.89×10^3
Energy	J	ft-lbf	0.738
Fracture Energy	J/m^2	erg/cm^2	10^3
Impact Strength	J/cm	ft-lbf/in	1.87
Length	nm	Ångstroms	10
	nm	mm	10^{-6}
	μm	Ångstroms	10^4
	μm	mm	10^{-3}
Peel Strength	N/cm	lbf/in	0.571
Specific Heat	J/g/deg K	cal/g/deg C	0.239
Stress, Pressure	N/m^2	lbf/in^2	1.45×10^{-4}
	MN/m^2	psi	1.45×10^2
	MN/m^2	atmospheres	10.198
	GN/m^2	psi	1.45×10^5
Stress Intensity Factor	$MN/m^{3/2}$	$psi\ (in)^{1/2}$	910.06
Time	ks	seconds	10^3
	Ms	seconds	10^6
Viscosity	$N.s/m^2$	poise	10
Weight	Tonnes	lbm (avoirdupois)	$2\,204.6$
		long tons	0.984

AUTHOR INDEX

347

SUBJECT INDEX